Nuclear Magnetic Resonance Imaging in Medicine and Biology

NMR imaging in Chemistry, Biology & Medicine

Nuclear Magnetic Resonance Imaging in Medicine and Biology

PETER G. MORRIS

Department of Biochemistry
University of Cambridge

CLARENDON PRESS · OXFORD · 1986

Oxford University Press, Walton Street, Oxford OX2 6DP

Oxford New York Toronto
Delhi Bombay Calcutta Madras Karachi
Kuala Lumpur Singapore Hong Kong Tokyo
Nairobi Dar es Salaam Cape Town
Melbourne Auckland

and associated companies in
Beirut Berlin Ibadan Nicosia

Oxford is a trade mark of Oxford University Press

Published in the United States
by Oxford University Press, New York

British Library Cataloguing in Publication Data

Nuclear magnetic resonance imaging in medicine and biology.
1. Nuclear magnetic resonance—Diagnostic use 2. Imaging systems in medicine
I. Title
616.07'575 RC78.7.N83
ISBN 0-19-855155-X

Library of Congress Cataloging in Publication Data

Morris, P. G.
Nuclear magnetic resonance.
Includes bibliographies and index.
1. Magnetic resonance imaging. I. Title.
II. Title: Imaging in medicine and biology.
RC78.7.N83M68 1985 616.07'5 85-331
ISBN 0-19-855155-X

Filmset and printed in Northern Ireland by
The Universities Press (Belfast) Ltd.

Preface

Since the first publications on the subject in 1973, the field of nuclear magnetic resonance (n.m.r) imaging has grown rapidly to the point that clinical imaging systems have become commercially available and have been installed in upwards of two hundred locations worldwide. Many different techniques have evolved during this rapid growth and it is the aim of this book to give the reader a sound physical understanding of the theory and practice of n.m.r. imaging. A decision was taken to write the book from first principles and the author hopes that the n.m.r. specialists among his readers will forgive the basic discussion of n.m.r. theory given in Chapter 2. This includes an introduction to nuclear magnetism, Fourier transform n.m.r., and to spin–spin and spin–lattice relaxation times which are central to an understanding of tissue contrast. High resolution techniques are also briefly discussed here and are illustrated with ^{1}H, ^{13}C, ^{19}F, and ^{31}P *in vivo* spectra recorded from a wide range of biomedical systems. Clinical applications of high resolution n.m.r. are also described. Chapter 3 deals with point and line imaging methods. Although these are now largely of historical interest only, this chapter does serve to introduce the important concepts of t.m.r. and of selective excitation. Chapter 4 discusses, at greater length, those two- and three-dimensional techniques currently finding application in commercial systems, namely projection reconstruction and Fourier imaging. The highly efficient technique of echo-planar imaging which enables images to be acquired in times from as short as 10 ms and the implications of this for real-time cardiac studies are also discussed. The different aspects of n.m.r. imaging instrumentation are discussed in Chapter 5. This includes a major section on magnet design; the magnet is arguably the single most important item, certainly it is the most expensive. Other sections treat magnetic field gradient generation, r.f. transmitter and receiver electronics, signal processing, and image display. Estimates of S/N and imaging time are also given. Tissue contrast and the pulse sequences designed to elicit it are described in Chapter 6 where extensive examples of clinical applications, a discussion of contrast enhancing agents, flow effects, and an appraisal of potential hazards are also given.

The book will appeal primarily to readers with a background in the physical sciences with a desire for a detailed understanding of n.m.r. imaging techniques; hospital physicists for example. With the omission of the basic theory section (Chapter 2) it should also be of interest to n.m.r.

spectroscopists working in the biomedical field. The value of the book to the clinician is largely limited to the discussion of applications which appears in Chapter 6. Although only one chapter this does represent a significant fraction of the text and could be read with little reference to the rest of the book. For the reader seeking a deeper understanding of the physical principles involved however, extensive cross-referencing is provided.

In a field which is developing so rapidly, with improvements in technique and in image quality appearing almost on a month-by-month basis it is virtually impossible to deliver a state-of-the-art textbook. Nevertheless every effort has been made to update the text at its many stages of revision and references have been included up to the second half of 1984. Many images have been generously provided from many different sources, both commercial and academic. They represent the subject at different stages of development and a critical intercomparison would be quite out of place; this is certainly not the author's intention. The reader should be especially careful on this score and should not base any judgement of a particular commercial system on the images shown in this book.

Many friends and colleagues have consented to read through some or all of the text which has benefited greatly from their comments. The omissions and errors which have escaped their efforts must however remain the responsibility of the author. For the sake of future editions the author invites his readers to point out any errors which come to their attention. Comments and criticisms will also be gratefully received by him. Especial thanks are due to the following: Drs J. Feeney and T. Frenkiel, MRC Biomedical NMR Centre; Professor D. Wilkie and Dr M. J. Dawson, Department of Physiology, University College London; Drs J. R. Griffiths and A. N. Stevens, Department of Biochemistry, St George's Hospital Medical School; Professor H. S. Bachelard and Dr D. W. G. Cox, Department of Biochemistry, St. Thomas's Hospital Medical School; Dr A. Jasinski, Institute of Nuclear Physics, Krakow; Dr G. Bydder, Department of Radiology, Hammersmith Hospital; Dr I. Young, Picker International, Wembley; Professor R. J. P. Williams, Inorganic Chemistry Laboratory, University of Oxford; and Dr C. Kemp, Institute of Ophthalmology, London. Thanks are also due to the many authors, publishers and commercial concerns who have provided images or who have consented to allow reproduction of their original figures. I am very grateful to Mr M. Tatham and the staff of the Photographic Section at NIMR and to Mr J. W. Brock who was responsible for much of the original artwork. I am also indebted to Mrs I. Godstone for typing the first draft of the book and to Mrs M. Fickling for coping with the many revisions in her customary efficient manner. Lastly, I would like to express my very special thanks to my wife Liz and daughter Claire for surviving without me for so long.

Mill Hill, London P. G. M.
September 1984

To Liz and Claire

Contents

1. General introduction

1.1. Imaging techniques

Everyone is to some extent familiar with the emission, absorption, and reflection of beams of energy, whether electromagnetic or acoustic in origin. Often one can unwittingly experience all three phenomena first thing in the morning when the bedroom curtains are drawn and a first glance is taken in the bathroom mirror. The three principal medical imaging modalities, radioisotope, X-ray, and ultrasound, are based on these simple principles. Thus, for radioisotope (gamma camera) imaging a radioactively-labelled compound is introduced into the patient and its distribution is monitored through the detection of the emitted radiation (using an array of lead collimators, crystal detectors, and their attendant photomultipliers). In the case of conventional X-ray systems the X-ray beam is progressively absorbed or "attenuated' as it traverses the subject. The distribution of intensity across the emergent beam, which can be recorded on X-ray sensitive film, reflects the nature of the object through which it passes. In ultrasonography one relies on the reflection of sound waves from discontinuities much in the same way that 'echo sounding' is used by the Navy for the detection of submarines or by trawlermen for the location of fish shoals.

Although the nature of the radiation may be unfamiliar or the details of the scattering processes obscure, the principles governing image formation are easily understood. This is true of many of the other imaging modalities and for the more sophisticated computerized tomographic (CT) versions of those already mentioned. Unfortunately, it is not the case for nuclear magnetic resonance (n.m.r.). Certainly one does irradiate the subject with electromagnetic radiation, but the absorption is a resonance phenomenon occurring at a single frequency or discrete set of frequencies rather like the emission or absorption of light in atomic spectroscopy: for example, the yellow light emitted by the relaxing electron in an excited sodium atom consists of two frequency components at 589.593 and 588.996 nm—the sodium D lines. In both types of spectroscopy the resonance effect derives from the same quantum mechanical origin, namely the existence of discrete sets of energy levels between which transitions can be observed. In the n.m.r. case, however, we are not dealing with electronic transitions but with nuclear ones. The particular property of the nucleus with which we are concerned is its angular momentum or spin. Since nuclei are positively charged, this spin has associated with it a current loop and hence a magnetic moment, a phenomenon known as nuclear paramagnetism. The different energy levels arise from the various possible orientations (two in the case of a spin $\frac{1}{2}$ nucleus such as the proton) of the nuclear magnetic moment in an applied magnetic field. (This lifting up of the degeneracy of energy levels by a magnetic field is known as the Zeeman effect.) The magnitude of the energy splitting, or equivalently (through the Planck relationship $E = h\nu$, where E is the energy, ν the frequency and h Planck's constant) the frequency of absorption, is directly proportional to the magnetic field strength. Even for the highest magnetic fields (14 T) currently used in n.m.r., the energy-splitting is small and the frequencies occupy the r.f. region of the electromagnetic spectrum. This is both a strength and a weakness: a strength because no ionizing radiation is involved, a weakness because the sensitivity is very low when compared with other spectroscopic techniques.

Provided the electrical conductivity of the sample is not too high (n.m.r. is useless for the examination of bulk metals and can play no role analogous to that of X- or γ-rays in metallurgical radiography), the r.f. will penetrate the sample virtually unattenuated. Fortunately, this is the case for the human body up to moderately high frequencies, and n.m.r. therefore provides an alternative 'window' for medical diagnostic investigations.

1.2. Development of n.m.r.

N.m.r. was seen for the first time in 1945 by two independent groups: one at Stanford under the leadership of Bloch (Bloch, Hansen, and Packard 1946), the other at MIT under Purcell (Purcell, Torrey, and Pound 1946). Both observations were made within a few days of each other and were reported in the same issue of *Physical Review*. Bloch and Purcell later shared the 1952 Nobel Prize for physics in recognition of their pioneering achievements.

The key discovery which led to the single most important application of n.m.r. to date, in the field of analytical spectroscopy, was that nuclei in chemically-distinct sites resonate at slightly different frequencies. The first observations of these chemical shifts were made in 1949 by Proctor and Yu, and independently by Dickinson. Their results, by a strange quirk of fate, were again published in the same issue of *Physical Review*. The first commercial n.m.r. spectrometer appeared in 1953. Proton n.m.r. has since become virtually indispensable in the organic chemistry laboratory. Following the development of Fourier transform techniques in the 1960s, ^{13}C n.m.r. now threatens to supersede infra-red spectroscopy as the method of choice for molecular fingerprinting. High-resolution n.m.r., as this branch of n.m.r. spectroscopy has come to be known, requires an extremely uniform magnetic field (certainly one part in 10^7, often one part in 10^9 or 10^{10}) if the small chemical-shift effects are not to be masked by magnetic inhomogeneity broadening.

High-resolution n.m.r. methods although not extensively applied, can be of direct use in the clinical situation; for example, in the analysis of bulk fluids such as urine or blood, or of tissue biopsy specimens. Ideally, however, one would like to make these measurements *in vivo*; the distribution of important metabolites could be mapped, fluxes through biochemical pathways determined, and responses to intervention or treatment monitored. To quote a standard biochemical text: 'The development of techniques for measuring the activity of enzymes within a living cell should be the aim of every clinical biochemist'. Such is the stuff that biochemists' dreams are made of! The technique of topical magnetic resonance (t.m.r.; Greek *Topos*, a place), recently developed by Oxford Research Systems, has taken an important first step in this direction, allowing metabolite concentrations to be measured at a single defined location.

Whilst it is, in principle, possible to retain such information in a full-scale imaging study, the static field homogeneity requirement is difficult to meet and the additional complexity required of the imaging sequence, coupled with the inherently low sensitivity of the method as a whole, has limited research in this area. The strategy in imaging studies has therefore been to measure the total intensity of the nuclear resonance as a function of position or, in other words, the number or concentration of spins within a given spatial region. The lack of sensitivity has resulted in attention being almost exclusively focused on the proton both because of its abundance in biological systems and also by reason of its high inherent sensitivity (second only to tritium). Of course, the human body contains a tremendous number of protons in all sorts of different chemical guises. However, unless special line-narrowing techniques are used, the only ones which contribute to the n.m.r. response are those undergoing rapid motion. In practice, this means largely water protons with lesser contributions from the protons of mobile lipids. Whilst the human body might seem a solid enough entity, from the n.m.r. viewpoint it has decidedly liquid-like characteristics!

1.3. N.m.r. imaging methods

Thus far we have given no indication as to how n.m.r. signals from different spatial regions can be separated. This important conceptual step, which paved the way for the development of n.m.r. imaging systems, was taken in 1973 by two independent groups. Lauterbur at the State University of New York at Stonybrook, who was the first to publish an n.m.r. image, and Mansfield and Grannell who were involved in n.m.r. 'diffraction' studies at Nottingham University. Both groups realized that, since the resonance frequency is proportional to the strength of the applied magnetic field, a magnetic field gradient would give rise to a range of resonant frequencies which reflected the spatial distribution of the contributing spins.

In contrast to the situation with X-ray or γ-ray systems, n.m.r. preserves phase as well as intensity information. A spatial co-ordinate can thus be encoded as a frequency or as a cumulated phase. Since a separate procedure is required for each direction along which resolution is desired, a great diversity of different schemes for n.m.r. imaging have evolved. Fortunately, the majority of these are now only of historical interest, with attention having been limited to a small number of the most promising alternatives; namely projection reconstruction, Fourier imaging in its various forms, and echo-planar imaging.

N.m.r. as an imaging technique is unusual in another important respect in that the physical property which is actually imaged is to some extent a matter for choice. One option is to select spin density, in which case the signal depends solely on the number of 'n.m.r. visible' spins in a given region. However, the spin system has associated with it other important parameters, including two (principal) relaxation times, which can be combined with the spin density information or used as the basis for an image in their own right. Equally one can arrange to incorporate local diffusion effects or bulk flow; in short, any microscopic or macroscopic property of the spin system which is measurable by n.m.r. can be imaged. Determining which property will be most useful for a particular application will require a considerable research effort. However, in the case of medical imaging, clear pointers towards suitable diagnostic protocols are already beginning to emerge from the first clinical trials.

1.4. Spatial resolution and imaging times

An important question, to be asked of any imaging modality, is the spatial resolution which it can achieve. Since, in a conventional imaging system, this resolution is roughly equal to the wavelength of irradiation used (hence in microscopy the progression from white light through ultraviolet to electron microscopes), one might suppose that, because of the radiofrequencies used for n.m.r., the resolution would be tens or even hundreds of metres; hardly a desirable characteristic for a potential imaging method. In fact, the resolution in n.m.r. imaging is determined by the strength of the applied field gradient: there is no theoretical lower limit to what can be achieved, just a practical one of decreasing sensitivity; as the number of image points is increased so the amount of material contributing signal to each one is reduced. It is possible to develop a pseudo wave description of n.m.r. imaging in which the resolution *is* given by the (pseudo)wavelength which in turn is proportional to the field gradient strength. At the time of writing the spatial resolution offered by commercial n.m.r. imaging systems is typically 1–3 mm in slices of 5–12 mm thickness.

Related to the question of resolution is the one of imaging time. There are two times which are of importance; first, the minimum time required to acquire sufficient data to reconstruct an image, and secondly the time necessary to achieve a given signal-to-noise ratio (S/N). The first or minimum performance time is an important consideration in those applications involving motion—for example, whole-body scanning—and requires considerable ingenuity to minimize. Thus far, only one of the proposed methods, echo-planar imaging, allows a complete image to be reconstructed from a single n.m.r. record. The minimum performance time is impressive ($\sim$10 ms) and has enabled remarkable n.m.r. film sequences of cardiac motion to be recorded *in vivo.* When it comes to achieving a high S/N per unit time, a number of methods, including echo-planar, projection reconstruction and Fourier imaging, approach the theoretical optimum. However, S/N is not the whole story—an image has to display contrast if it is to be diagnostically useful. A number of pulse sequences have been designed to elicit different forms of tissue contrast and these often add considerably to the imaging time. Currently, a 256×256 image of acceptable contrast and S/N would take in the region of 30 s–10 min to obtain. Simultaneous multislice studies can usually be performed with little extra time-penalty. However, isotropic three-dimensional (volume) images can take several tens of minutes.

1.5. Development of n.m.r. imaging

Since the over-riding requirement of high-resolution n.m.r. is for good field homogeneity, conventional n.m.r. systems have been based on the use of small sample volumes, 5- or 10-mm sample tubes being typical. The development of n.m.r. medical imaging instrumentation has therefore been primarily one of scaling up the initial demonstrative experiments over some two orders of magnitude. This has been no mean achievement, all the more so when one considers the rapidity of the progress. Following the first publications in 1973, n.m.r. cross-sections through fingers demonstrated human anatomic detail for the first time in 1976. Intermediate-sized systems (1–10 cm) appeared shortly afterwards, allowing scans of human limbs and small laboratory animals. The ultimate goal of whole-body imaging was first achieved by Damadian using his f.o.n.a.r. technique in 1977 (Damadian, Goldsmith, and Minkoff 1977), a mere four years after the subject's inception. He was followed shortly afterwards by Mansfield and colleagues using a line-scanning method (Mansfield, Pykett, Morris, and Coupland 1978), and by many other academic and commercial concerns since.

Industry has played its part in this development, with companies such as Oxford Instruments responding quickly and effectively to the new demand for wide-bore magnets. However, with few exceptions (notably a team then at EMI, now with Picker International), the pioneering n.m.r. work was carried out in academic institutions. During the late 1970s many of the major manufacturers of medical scanning equipment, who had hitherto only maintained a watching brief, became involved in the development of n.m.r. imaging systems, either directly or through the support of academic research.

The improvement in image quality over the last few years which has resulted from this increased financial investment has been quite remarkable. Figure 1.1 illustrates the point; the fine anatomical detail of the

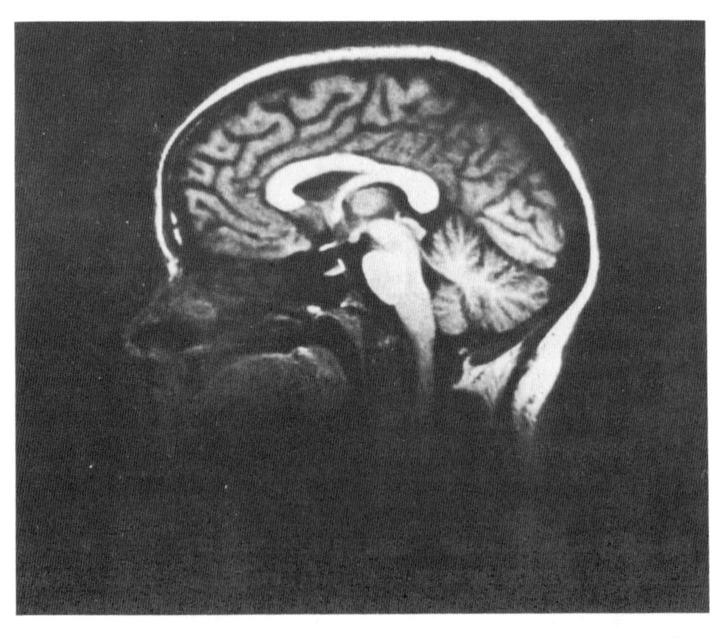

Fig. 1.1. Midline sagittal image of the human head recorded on a 1.5 T proton n.m.r. scanner using an inversion recovery method (TI = 700 ms, TR = 1.5 s). The slice thickness was 5 mm. Note the excellent grey-white matter contrast in the cerebellum. (Reproduced with permission from the General Electric Company, May 1984.)

human head is beautifully resolved in this midline sagittal section. Note particularly the clear discrimination between white and grey matter.

In 1983, preliminary clinical trials with prototype instruments showed sufficient promise to enable a number of manufacturers to seek and obtain FDA approval, enabling their imaging systems to be made commercially available in the USA.

1.6. Commercial exploitation

Clinical n.m.r. imaging systems are currently available in a range of options (differing mainly in the frequency of operation) from several companies. Table 1.1 lists the principal manufacturers at the time of writing. Many of the early systems employed resistive magnets operating at field strengths in the range 0.04–0.2 T (corresponding to a proton frequency range of 1.7–8.5 MHz).

Such systems are still available commercially and are capable of yielding excellent image quality. The recent trend, however, has been towards higher operating fields (0.5–2.0 T). Currently, 90 per cent of magnets supplied by Oxford Instruments, the leading magnet manufacturer, are of the superconducting variety. Over 200 have been delivered at the time of writing. Such magnets are expensive, accounting typically for 30 per cent of the total system cost. They also entail substantial installation and

Table 1.1
Principal manufacturers of n.m.r. imaging systems.

Asahi, Kyoto, Japan
Bruker Medizintechnik GmbH, Rheinstetten, W. Germany
CGR Koch and Sterzel, Essen, W. Germany
Diasonics Inc., San Francisco, California, U.S.A.
Elscint Ltd., Haifa, Israel
Fonar Corp., Melville, New York, U.S.A.
General Electric Co., Milwaukee, Wisconsin, U.S.A.
Hitachi Medical Corp., Tokyo, Japan
Instrumentarium Oy, Helsinki, Finland
M & D Technology, Aberdeen, Scotland
Oxford Research Systems,[1] Abingdon, England
Philips Medical Systems, Eindhoven, The Netherlands
Picker International Inc., Highland Heights, Ohio, U.S.A.
Sanyo Electrical Medical Co. Ltd., Osaka, Japan
Shimadzu, Tokyo, Japan
Siemens Aktiengesellschaft, Erlangen, W. Germany
Technicare, Solon, Ohio, U.S.A.
Toshiba Medical Systems, Tokyo, Japan

[1] ORS offer systems for *in vivo* spectroscopy with an imaging option.

running expenses. N.m.r. imaging systems as a whole are therefore expensive, costing perhaps 30 per cent more on average than an X-ray CT scanner. In round terms the price of a 0.15 T resistive n.m.r. system might be £700 000 whereas a superconducting system could cost £1M–£1.5M depending on the choice of field strength (1984 prices).

Although n.m.r. scanning times are significantly longer than those for X-ray CT, multislicing can often be performed so that patient throughput is generally similar for the two modalities. The cost per scan is likely to be rather higher for n.m.r. However, given its outstanding clinical potential (it is already the method of choice for a number of applications) the future of n.m.r. imaging seems assured. The sales figures for 1983/84 certainly reflect this optimism.

1.7. Scope of text

Chapter 2 gives an introduction to basic n.m.r. theory, from first principles. An attempt has been made to back up otherwise bald statements of principle with a physical explanation at a level and depth which reflects the importance of the concept to an overall understanding of n.m.r. imaging. Thus the treatment of the Fourier transform may seem simplistic and laboured to those familiar with its use, but this is deemed to be justified in view of its central importance. However, chemical shifts and scalar couplings which, in a standard n.m.r. treatment would occupy a sizeable portion of the text, here receive only scant attention.

Whereas the first four sections of Chapter 2 present a coherent (it is hoped) introduction to the n.m.r. experiment, the next two are concerned with special properties of spin systems. In particular, the relaxation times are crucially important to an understanding of the origins of tissue contrast and conventional n.m.r. techniques used for their measurement provide the basis for n.m.r. imaging schemes designed to elicit greater contrast; hence their discussion here.

In the remaining section of Chapter 2 a number of examples are given of the application of high-resolution n.m.r. methods to living systems. Whereas these examples cannot be described as imaging studies, such methods do have important clinical application, and many imaging systems currently offer a spectroscopic facility. Additionally, full n.m.r. imaging techniques which preserve chemical shift information are becoming available (some preliminary examples are shown in § 6.7). It was therefore felt that a demonstration of the power of high-resolution methods was not wholly out of place.

The description of the many imaging techniques has been divided between Chapters 3 and 4, Chapter 3 dealing with the point and line methods, and Chapter 4 with planar and full three-dimensional ones. The accent is placed on spatial encoding rather than on generation of specific types of tissue contrast, or on flow or chemical-shift studies. The modifications needed to encompass such measurements are common to many of the techniques and are treated generally in Chapter 6. Specific details are given, however, for some of the more important methods, Fourier imaging included.

The coverage of techniques does not aim to be totally comprehensive, with the accent being placed on the most promising of those currently in use; these are to be found in Chapter 4. Chapter 3 does, however, include a description of t.m.r., not strictly an imaging technique in the sense that in t.m.r. attention is restricted to a single volume element, but nevertheless of major interest since it permits the concentrations and reaction rates of important metabolites to be measured *in vivo.* Other important concepts such as selective irradiation and steady-state free precession are also included there.

Instrumentation is the subject of Chapter 5. A modular approach has been adopted in which each major element (magnet, field gradients, transmitter, receiver, control, and display systems) has been treated separately. Some discussion of the influence of operation frequency on S/N is given and an estimate is made of the imaging times which can be achieved with an optimum imaging system.

The discussion of imaging applications is reserved for Chapter 6. The first section deals with the origins of tissue contrast and the second with general aspects of n.m.r. imaging; for example, pulse sequences designed to generate specific types of tissue contrast are described and the dependence of the image on pulse parameters discussed. Also included in this section is a discussion of n.m.r. contrast agents. There is enormous potential and interest in this area and preliminary clinical trials of promising agents have already been undertaken. N.m.r. imaging is especially sensitive to flow phenomena allowing imaging of the vascular system and, in some cases, measurement of flow velocity in principal vessels. Such experiments are discussed in § 6.3 and clinical examples are given in §§ 6.5 and 6.6. Small scale n.m.r. imaging is dealt with in § 6.4, which traces the development of the subject and describes applications to small animal and other studies. The clinical potential of n.m.r. imaging is illustrated in §§ 6.5 and 6.6 for head and whole-body studies respectively, whilst in § 6.7 recent developments in multinuclear and chemical-shift imaging are discussed. In the final section, potential hazards of n.m.r. are identified and recommendations for safe operating procedures given.

Illustrative results have kindly been provided by a number of concerns, both academic and commercial. They span a considerable time period in a technology which has seen rapid and major improvements over the last few years. For the most part, they have been obtained using prototype machines, and their quality is often not representative of that to be expected from state of the art production instruments. Critical inter-comparisons would therefore be premature and have been avoided in this monograph.

Similar reasoning applies to the comparison of n.m.r. with other imaging modalities; a proper and conclusive evaluation will only be possible following publication of the extensive clinical trials now under-way in a number of institutions. Nevertheless, the opportunity has not been wasted to point out particular advantages (and some disadvantages) where they are already apparent.

1.8. Advice to the reader

This monograph is introductory and treats the subject from first principles. However, the accent is placed on physical descriptions of imaging methods and techniques and it will therefore appeal primarily to readers interested in developing a thorough understanding of how n.m.r. imaging works. It is envisaged that these will include hospital physicists and n.m.r. spectroscopists working in the biomedical area. The adventurous clinician, provided that he does not allow himself to become too embroiled in what for his purposes is unnecessary detail, should also find much of interest, particularly in Chapter 6.

N.m.r. imaging continues to evolve at a remarkable pace. Whereas basic principles may remain unaltered, techniques and image quality are improving on a month by month basis. At such a stage in a subject's development one can take very little 'as read'. The only safe advice is to follow the literature. It is to be hoped that this monograph will take the reader to the stage where he finds this possible. For scientists with a broad interest in biomedical applications of n.m.r. imaging and spectroscopy, the Royal Society of Chemistry include a chapter on 'N.m.r. of living systems' in their annual 'Specialist Periodical Reports on N.M.R.' (Morris 1984).

References

Bloch, F., Hansen, W. W., and Packard, M. E. (1946). *Phys. Rev.* **69,** 127.

Damadian, R., Goldsmith, M., and Minkoff, L. (1977). *Physiol. Chem. Phys.* **9,** 97.

Dickinson, W. C. (1950). *Phys. Rev.* **77,** 736.

Lauterbur, P. C. (1973). *Nature, Lond.* **242,** 190.

Mansfield, P., and Grannell, P. K. (1973). *J. Phys.* C **6,** L422.

——— Pykett, I. L., Morris, P. G., and Coupland, R. E. (1978). *Br. J. Radiol.* **51,** 921.

Morris, P. G. (1984). 'N.m.r. of Living Systems', in Royal Society of Chemistry *Specialist Periodical Reports on N.M.R.* **13,** 348.

Proctor, W. G., and Yu, F. C. (1950). *Phys. Rev.* **77,** 717.

Purcell, E. M., Torrey, H. C., and Pound, R. V. (1946). *Phys. Rev.* **69,** 37.

2. Introduction to n.m.r.

2.1. Nuclear spin and magnetic moments

Nuclear magnetic resonance (n.m.r.) spectroscopy relies on the paramagnetic properties of the nucleons (neutrons and protons) which make up atomic nuclei. Just as electrons possess intrinsic spin and associated magnetic moment, so too do the nucleons. In fact, the spin angular momentum and magnetic moment vectors, denoted **J** and **μ** respectively, are colinear and proportional. The scalar constant of proportionality γ is known as the magnetogyric ratio. Thus

$$\boldsymbol{\mu} = \gamma \mathbf{J}. \tag{2.1}$$

In order to understand which parameters affect γ we can consider as a simple model of the proton a spinning ball of mass m with radius a and uniformly distributed charge e. As the ball rotates, each element of mass traces out a circular path and contributes to the spin angular momentum. Similarly, each element of charge traces out a circular path or current loop which contributes to the magnetic moment.

Consider the ring of material of radius r illustrated in Fig. 2.1. The elemental spin angular momentum δJ arising from rotation of this ring is

$$\delta J = \frac{4\pi^2 \rho_m}{t_r} r^4 \sin^3\theta \, \mathrm{d}r \, \mathrm{d}\theta, \tag{2.2}$$

where t_r is the period of rotation and ρ_m the (mass) density given by

$$\rho_m = 3m/4\pi a^3. \tag{2.3}$$

The total spin angular momentum J is the sum or integral of all the elemental contributions δJ. Thus

$$J = \frac{4\pi^2 \rho_m}{t_r} \int_0^{\pi} \int_0^{a} r^4 \sin^3\theta \, \mathrm{d}r \, \mathrm{d}\theta. \tag{2.4}$$

The element of magnetic moment $\delta\mu$ contributed by the same ring of material is given by $\delta\mu = i\,\delta S$ where i and δS are the ring current and area respectively. Thus

$$\delta\mu = \frac{2\pi^2 \rho_e}{t_r} r^4 \sin^3\theta \, \mathrm{d}r \, \mathrm{d}\theta, \tag{2.5}$$

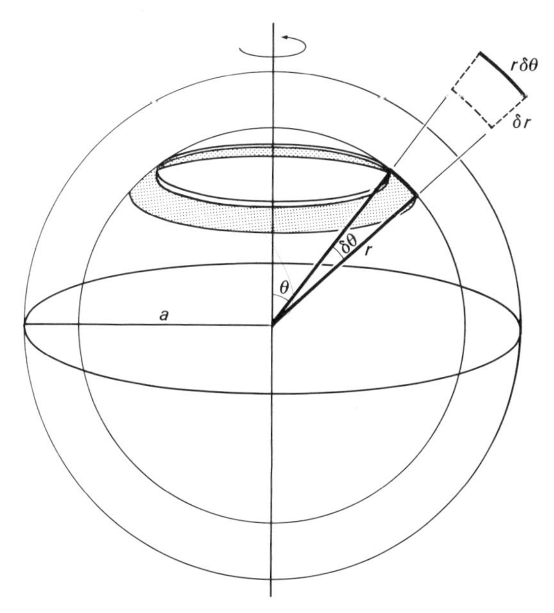

Fig. 2.1. The charged spinning ball model of the proton. Inset shows the cross-section of the ring of radius r located within the proton sphere of radius a.

where ρ_e is the charge density given by

$$\rho_e = 3e/4\pi a^3. \tag{2.6}$$

The total magnetic moment $\delta\mu$ is therefore

$$\mu = \frac{2\pi^2\rho_e}{t_r}\int_0^{\pi}\int_0^{a} r^4 \sin^3\theta \, dr \, d\theta. \tag{2.7}$$

From eqns (2.3), (2.4), (2.6), and (2.7) we see that for this classical model of the proton:

$$\gamma = \mu/J = e/2m. \tag{2.8}$$

Thus γ depends only on the charge-to-mass ratio. The $\frac{1}{2}$ factor should not be taken too seriously, however, as our simple model cannot be expected to be accurately representative of a true proton.

When looked at in the light of this model, the neutron poses something of a problem, since, although it is without charge, it nevertheless has a magnetic moment. To some extent this can be reconciled with the classical model if the neutron is considered as a proton orbited by a negatively-charged pion. This picture also accounts for the negative sign of the neutron's magnetogyric ratio. When one comes to consider nuclei other than the proton, a quantum mechanical treatment is necessary in order to understand the manner in which the spins of the elementary particles combine. It turns out that, just as in the electronic case, the total angular momentum is quantized in units of $\hbar/2$, where $2\pi\hbar$ is Planck's constant. Spin quantum numbers which are integral, half-integral, or zero are possible. In the latter eventuality, the nucleus would have no angular momentum and hence no magnetic moment: it would therefore not be directly observable by n.m.r. Such is the case for the common isotopes of carbon ($^{12}_{6}C$) and oxygen ($^{16}_{8}O$), for example (see § 2.3.3). The magnetogyric ratio is still given by eqn (2.8) with the addition of a correction factor g_N, known as the nuclear g factor. It can be calculated from the details of the nuclear structure, and is generally a small number. However, unlike the electronic case (Landé g factor) it can be either positive or negative corresponding to magnetic moments which are parallel or antiparallel to the angular momentum. (Some of the earliest n.m.r. experiments were designed to probe nuclear structure through determination of nuclear g factors.)

It can be seen from eqn (2.8) that the magnetic moment of a proton should be smaller than that of an electron by a factor equal to the ratio of their respective masses, namely 1836. (In fact, the experimentally-determined factor is 658.) Nuclear paramagnetism is thus two to three orders of magnitude weaker than its electronic counterpart and is the reason for electron spin resonance frequencies lying in the microwave (GHz) region of the electromagnetic spectrum, whereas n.m.r. frequencies normally occupy the radiofrequency (MHz) band.

Another important consequence is that electron–nuclear dipole interactions are much stronger than nuclear–nuclear interactions. This means that trace amounts of paramagnetic impurities can dominate the relaxation behaviour (see § 2.5) of n.m.r. samples. In fact, high-resolution n.m.r. spectroscopists generally go to great lengths in order to exclude such impurities, for example by using the 'freeze pump thaw' method to remove dissolved oxygen, and chelating agents, such as ethylene diamine tetraacetic acid, to remove paramagnetic cations. Electron paramagnetism does, however, have a more positive aspect from the n.m.r. viewpoint. For example, the use of rare earth ions such as praseodymium, dysprosium, or europium as chemical-shift reagents and, in the context of n.m.r. imaging, the use of free radicals and paramagnetic centres such as Mn^{2+}, Fe^{3+} and Gd^{3+} as contrast agents (see § 6.3.5.).

2.2. Behaviour in static magnetic fields

2.2.1. *Uniform magnetic fields*

Everyone knows that if a bar magnet (for example, a compass needle) is placed in a uniform magnetic field, such as the earth's field when considered over a restricted spatial region, then a couple acts on the magnet to align it with the field direction (magnetic N–S in the case of our compass), this being the position of minimum energy or stable equilibrium. In the same way, when a nuclear magnetic moment is placed in a magnetic field $\mathbf{B}_0$, it experiences a couple tending to align it with the field direction. However, because of the presence of spin angular momentum the motion is not rotational but precessional, very much like that of a gyroscope. The analogy can be further pursued by deriving the precessional frequencies for the two cases.

Taking the more familiar example of a gyroscope first (Fig. 2.2(a), (b)): two forces act on the gyroscope, a gravitational force of magnitude mg

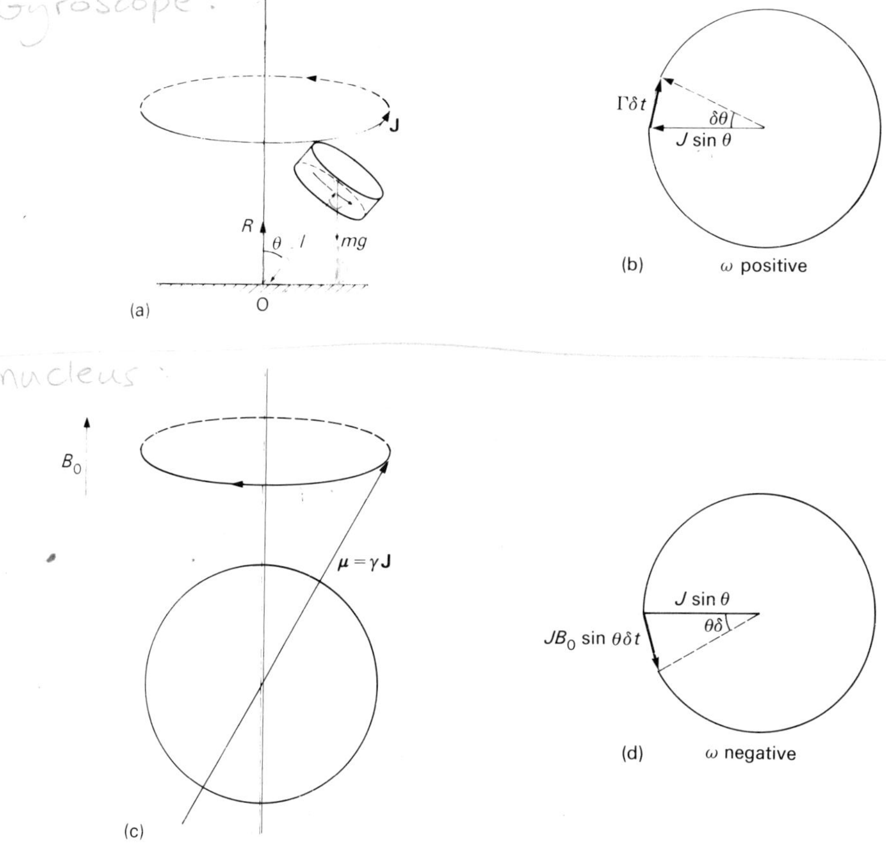

Fig. 2.2. Gyroscopic and nuclear precession. (a) Gyroscopic precession in the earth's gravitational field. (b) Motion of the tip of the angular momentum vector J; ω is positive, corresponding to a clockwise rotation for an observer looking out from the centre of the co-ordinate system O. (c) Precession of a nucleus in a uniform magnetic field B_0. (d) Sense of rotation now anti-clockwise, ω negative.

through the centre of gravity, and an equal and opposite reaction R through the point of contact of the spindle with the table. Here m and g are the mass of the gyroscope and the acceleration due to gravity respectively. Together these two forces constitute a couple Γ of magnitude $mgl \sin \theta$ (see Fig. 2.2(a)) acting normal to the plane of the paper. In a time δt this results in an additional component $\Gamma\, \delta t$ of angular momentum in the horizontal plane. This is in effect a clockwise rotation of the horizontal component of J through an angle $\delta\phi = \Gamma\, \delta t/J \sin \theta$ (see Fig. 2.2(b)). Thus, the angular precessional frequency ω is in this case given by

$$\omega = \frac{\delta\phi}{\delta t} = \Gamma/J \sin \theta = mgl/J. \tag{2.9}$$

In the case of the nuclear spin, similar reasoning applies (see Figs 2.2(c), (d)). The couple Γ is now given by $\Gamma = |\boldsymbol{\mu} \times \mathbf{B}| = \gamma B \sin \theta$ and the angular precession or Larmor frequency is therefore

$$\omega_0 = -\gamma B. \tag{2.10}$$

Note that since the couple now acts to align the spin with the magnetic field and is in the opposite sense to the gravitational couple acting on the gyroscope, the precession is anticlockwise (denoted by the negative sign in eqn (2.10)) when γ is positive. The frequency of precession is known as the Larmor frequency and the equation as the Larmor equation. It is the fundamental relationship of n.m.r. and states that the precessional frequency is directly proportional to the magnetic field.

In conventional n.m.r. one takes advantage of the fact that B refers to the field *at* the nucleus and includes any perturbation due to the electronic environment. B is proportional to the applied field B_0 and can be written as $B = B_0(1-\sigma)$ (see § 2.6.1). Thus, provided the static magnetic field can be made sufficiently uniform, and modern superconducting magnets are capable of homogeneities in the region of one part in 10^9 over sample volumes of a few cm^3, these small interactions can be observed and in fact form the basis of n.m.r. analytical spectroscopy. Some of these interactions will be discussed in §§ 2.5 and 2.6. For the present, however, such interactions are neglected so that $B = B_0$.

N.m.r. imaging makes use of eqn (2.10) in a rather different manner through the use of non-uniform magnetic fields.

2.2.2. *Magnetic field gradients*

If the static magnetic field is made to vary in space then, because of the Larmor relationship, the angular frequency of precession ω_0 will reflect this same spatial dependence. In other words, the frequency at which a particular spin precesses, which, as will be shown in the next section, is also the frequency of electromagnetic radiation which it absorbs, is determined by the position of the spin within the object to be imaged. Consider the simplest type of spatial dependence; a uniform magnetic field gradient $G = \partial B_z/\partial x = \text{constant}$. The net magnetic field which is taken to lie in the z-direction is then:

$$B_z = B_0 + Gx, \tag{2.11}$$

as illustrated in Fig. 2.3(a). Consider now what happens when an object is placed into this magnetic field. For simplicity the object is chosen to be of constant cross-section normal to z, as illustrated in Fig. 2.3(b). The spin density will thus only be a function of x, y and can be written as $n(x, y)$

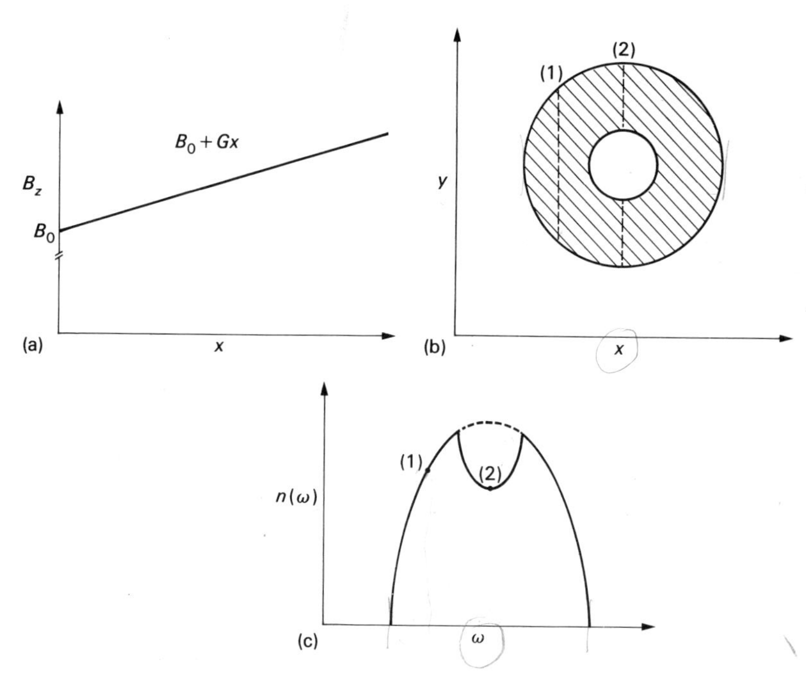

Fig. 2.3. Application of a linear field gradient to a two-dimensional sample. (a) Magnetic field variation due to gradient. The break in the vertical axis indicates that the field variation is normally small compared with B_0. (b) The two-dimensional sample: the spin density ρ is unity within the shaded region and zero everywhere outside. (c) The number of spins which absorb at frequency ω; points 1 and 2 are the line integrals of the corresponding regions shown in (b). The profile is a half-ellipse with a smaller semi-elliptic intrusion corresponding to the central occlusion of the annulus. In the absence of this occlusion (i.e. for a disc-shaped sample) the profile is a pure half-ellipse (broken line).

spins per unit volume. The angular frequency of precession will be the same for *all* spins lying in a plane normal to the x-direction and will be given by:

$$\omega_0(x) = \gamma(B_0 + Gx). \tag{2.12}$$

If the number of spins $n(\omega)$ which precess/absorb at a particular angular frequency $\omega_0(x)$ is measured, it will therefore tell us how many spins lie in the corresponding plane, namely:

$$n(\omega) = n(x) = \int n(x, y)\,dy. \tag{2.13}$$

This is illustrated in Fig. 2.3 where the numbered lines in the object of (b) give rise to the correspondingly numbered points in the spectrum (c). This spectrum is known as a (one-dimensional) profile or projection of the object and can in itself be quite informative. For example, Fig. 2.4 shows a one-dimensional projection through a line of five water-filled capillaries obtained by proton n.m.r. in a field gradient G of magnitude 1.28 mT/m. Each tube of constant cross-section, results in a semi-elliptic profile whose minor axis gives the internal diameter of the capillary in frequency units $\Delta\omega$. We can convert this to a real width Δx knowing γ

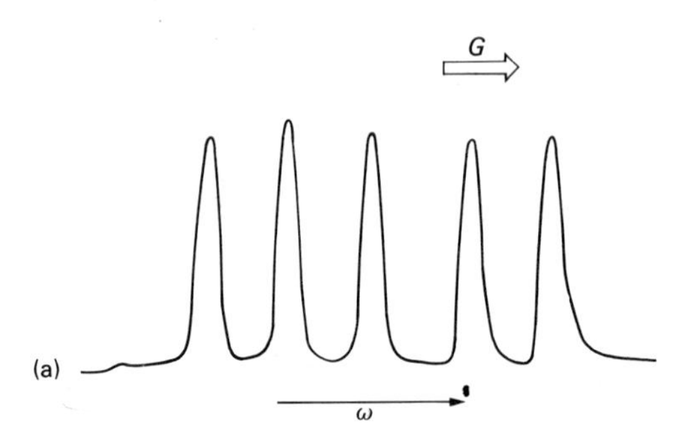

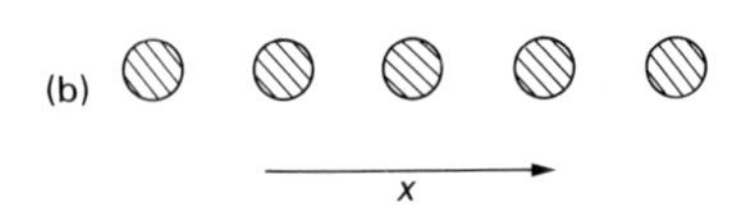

Fig. 2.4. (a) The one-dimensional projection obtained by recording the n.m.r. spectrum in the presence of a linear field gradient G from the row of five water-filled capillaries shown in (b).

and G. Thus

$$\Delta\omega = \gamma\, \Delta B = \gamma G\, \Delta x, \tag{2.14a}$$

so that

$$\Delta x = \Delta\omega/\gamma G. \tag{2.14b}$$

In the example shown it is found to be 0.105 cm. The physical separation of the tubes can be determined in like manner by measuring the frequency separation of the maxima and, in this case, it is found to have an average value of 0.25 cm. Finally, it is worth noting that the area of each profile is proportional to the total number of spins contained within each capillary; in this case all are nominally the same.

A rather simple object was chosen for the above illustration. It had a high degree of symmetry so the spin density was a particularly straightforward function:

$$n(x, y, z) = 1 \quad \text{within a capillary,}$$
$$n(x, y, z) = 0 \quad \text{everywhere else,}$$

and it was aligned so that the line of five tubes corresponded with the field gradient direction. In this simple situation a single projection yielded all the information we could wish to learn about the object. However, in the more general case of an object with low symmetry and a distribution of spin densities this is no longer true, at least for static field gradients. It is clearly then necessary to obtain more information and one way in which this can be achieved is to look at the object 'from different directions' by collecting projections taken in different orientations of the field gradient. For example, if the set of five capillaries were to be observed in a field gradient orthogonal to that shown in Fig. 2.4, i.e. $G = \partial B_z/\partial y$, a single elliptic profile would be obtained with the same minor axis as before but with five times the amplitude. Intermediate gradient orientations would show the individual profiles moving together, overlapping and finally coalescing to the single central line. The general problem of how these different projections are combined or reconstructed to yield an image of the object is the subject of § 4.1.

Another approach to the study of general objects is to simplify the problem by restricting observation to a limited region of the sample. In the most extreme case this would correspond to looking at a single volume element of the final image. This can be achieved by introducing time varying gradients and filtering the n.m.r. signal as in the case of the sensitive point method (§ 3.2.3) or by using profiling coils to shape the magnetic field so that all parts of the sample experience a strong non-linear field gradient apart from those occupying a small central region or resonance aperture. This is the basis of the f.o.n.a.r. method in which an image is built up by moving the object through the resonance aperture (§ 3.2.1). A similar principle is employed by the topical magnetic resonance (t.m.r.) technique for obtaining high-resolution spectra from localized regions within a sample (§ 3.2.2).

The detailed discussion of f.o.n.a.r., t.m.r. and other techniques is reserved for Chapters 3 and 4. It is, however, worth mentioning at this stage that it is possible to form an image of an object from a single projection by making use of time-dependent field gradients. The method used to achieve this is known as echo-planar imaging. It is the fastest method so far developed and is capable of producing n.m.r. images in a few tens of milliseconds. We describe it in detail in § 4.3.

2.3. Radiofrequency absorption

2.3.1. *Quantum description*

Thus far a classical approach has been adopted for the description of n.m.r. Whereas such a description is perfectly adequate for the large numbers of non- or weakly-interacting spins with which we are normally concerned, the quantum nature of individual nuclear spins should nevertheless be recognized. As mentioned above, one consequence of this is that the nuclear spin is itself quantized.

Let $\hat{J}$ be the quantum mechanical operator corresponding to the angular momentum. A dimensionless spin operator $\hat{I} = \hat{J}/\hbar$ can then be defined with a corresponding eigenvalue equation:

$$\hat{I}\,|\psi\rangle = [I(I+1)]^{\frac{1}{2}}\,|\psi\rangle. \tag{2.15}$$

Here $|\psi\rangle$ are the nuclear spin wavefunctions, and the spin quantum number I can only have integral or half-integral values. The application of an operator to a wavefunction corresponds, classically, to a measurement of the corresponding physical quantity. Thus, restated, eqn (2.15) says that if the nuclear spin angular momentum of a nucleus in state $|\psi\rangle$ is measured it will be found to have a value of $[I(I+1)]^{\frac{1}{2}}$. It is possible to excite transitions between nuclear states corresponding to different I values and this is made use of in nuclear orientation and Mossbauer spectroscopy, for example. The energies involved are large, typically lying in the gamma-ray region of the electromagnetic spectrum. We, however, will be exclusively concerned with the ground state.

The z component of the spin angular momentum I_z can be measured by applying the operator $\hat{I}_z$ to the wavefunction. Thus:

$$\hat{I}_z\,|\psi\rangle = m\,|\psi\rangle. \tag{2.16}$$

The appropriate quantum number in this case is m which can have

$(2I+1)$ different values: $-I, -(I-1)\ldots(I-1), I$. For example, a proton has $(2 \cdot \frac{1}{2}+1)=2$ possible I_z states, namely $m=+\frac{1}{2}$ and $m=-\frac{1}{2}$. ^{23}Na, on the other hand, has four possible states corresponding to $m=\frac{3}{2}, -\frac{1}{2}, \frac{1}{2}, \frac{3}{2}$. In general, the state $|\psi\rangle$ of a particular nucleus will not be a pure eigenfunction of $\hat{I}_z$ but will be some linear combination of them. Thus for a proton:

$$|\psi\rangle = a\,|\tfrac{1}{2}\rangle + b\,|-\tfrac{1}{2}\rangle, \tag{2.17}$$

where $|\frac{1}{2}\rangle$ and $|-\frac{1}{2}\rangle$ are the states corresponding to $m=\frac{1}{2}$ and $m=-\frac{1}{2}$ respectively, and a, b are (complex) constants. What happens when the I_z measuring operator $\hat{I}_z$ is applied? The result is

$$\hat{I}_z\,|\psi\rangle = a/2\,|\tfrac{1}{2}\rangle - b/2\,|-\tfrac{1}{2}\rangle, \tag{2.18}$$

which is to be interpreted in the sense that a value for I_z of $\frac{1}{2}$ will be obtained with a probability of $|a|^2$ and a value of $-\frac{1}{2}$ with a probability of $|b|^2$. Note that since the proton must be in one or other of the two states

$$|a|^2 + |b|^2 = 1. \tag{2.19}$$

Note also that, although the average or expectation value of I_z which we would obtain from repeated measurements starting with the spin system in the same initial state would be given by

$$\langle\psi|\,I_z\,|\psi\rangle = [|a|^2 - |b|^2]\cdot\tfrac{1}{2}, \tag{2.20}$$

each individual measurement would tell us only that the spin was either in state $|+\frac{1}{2}\rangle$ or $|-\frac{1}{2}\rangle$. Further, having found the spin to be in state $|+\frac{1}{2}\rangle$ say, a second measurement on the same nucleus would merely confirm that it was in that state. Thus the very act of measurement alters the spin wavefunction from $|\psi\rangle = a\,|+\frac{1}{2}\rangle + b\,|-\frac{1}{2}\rangle$ to $|\psi\rangle = |+\frac{1}{2}\rangle$. It can be seen therefore that it is often not possible to measure properties of small-scale objects without affecting them. In particular, if a number of measurements are to be made, the results may well depend on the order in which they are carried out. In the case of $\hat{I}$ and $\hat{I}_z$ this is not so; the operators are said to commute. However, for $\hat{I}_z$ and $\hat{I}_x$ say, it would be true. These difficulties, arising from the quantum nature of small particles, disappear when we come to deal with macroscopic objects which typically contain Avogadro's number of spins. However, for the moment, we continue to deal with a single particle and consider what happens when a spin $\frac{1}{2}$ nucleus such as a proton is placed in a magnetic field.

The classical energy of interaction for a magnetic moment in a magnetic field of strength B_0 along the z-direction is

$$E = -\boldsymbol{\mu}\cdot\mathbf{B}_0 = -\gamma J_z B_0. \tag{2.21}$$

The operator corresponding to the energy is the Hamiltonian $\hat{H}$ and the equivalent quantum mechanical expression to eqn (2.21) is:

$$\hat{H} = -\gamma\hbar\hat{I}_z B_0. \tag{2.22}$$

Since $\hat{I}_z$ is the only operator appearing on the right hand side of eqn (2.22), $\hat{H}$ will have the same eigenfunctions as $\hat{I}_z$ and the eigenvalues of $\hat{H}$ will be related to those of $\hat{I}_z$ through the constant of proportionality $-\gamma\hbar B_0$. Thus the energy eigenvalues are

$$E_{+\frac{1}{2}} = -\gamma\hbar B_0/2, \tag{2.23a}$$

$$E_{-\frac{1}{2}} = \ \gamma\hbar B_0/2, \tag{2.23b}$$

for the respective situations where I_z is parallel ($|+\frac{1}{2}\rangle$; $E_{+\frac{1}{2}}$) or antiparallel ($|-\frac{1}{2}\rangle$; $E_{-\frac{1}{2}}$) to the magnetic field direction. The difference in energy between the two states $\Delta E = \gamma\hbar B_0$ is known as the Zeeman splitting and is illustrated in Fig. 2.5.

In order to excite a transition from one Zeeman state to another it is necessary to supply an energy quantum

$$\Delta E = \hbar\omega = \gamma\hbar B_0, \tag{2.24}$$

where ω is the angular frequency of the applied irradiation. Thus $\omega = \gamma B_0$, and we note that the absorption frequency is identical with the frequency of precession (viz $\omega = \omega_0$). It seems reasonable, at least intuitively, that the spin should interact with an electromagnetic field rotating at its precessional frequency.

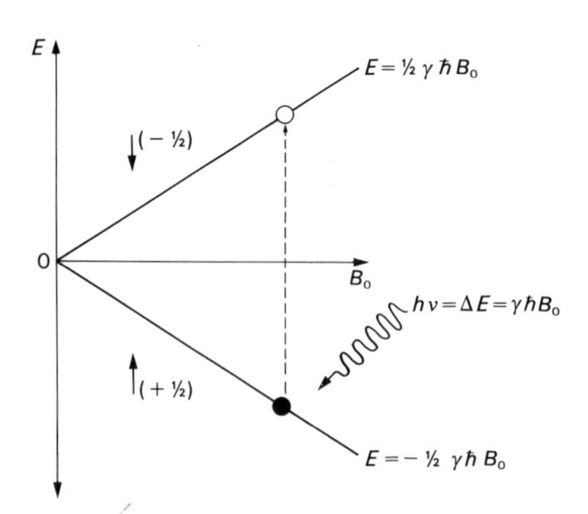

Fig. 2.5. The energy-level diagram for a spin $\frac{1}{2}$ system. The Zeeman energy-splitting increases linearly with applied magnetic field. Transitions between the two Zeeman states can be induced by electromagnetic energy of frequency ν given by $\nu = (\gamma/2\pi)B_0$.

2.3.2. *Distribution of spin states*

An electromagnetic field applied at the resonance frequency to a proton-containing sample will be just as likely to stimulate a transition from $|-\frac{1}{2}\rangle$ to $|+\frac{1}{2}\rangle$ states as vice versa. Thus, in order for a net absorption (or emission) of energy to occur, there must be a difference in population of the two energy levels.

If it is assumed that the spin system is in thermal equilibrium with the lattice (the mechanism by which this comes about will be discussed in § 2.5.2) the spins will be distributed amongst the Zeeman energy levels according to the Boltzmann distribution. If there are $n_\uparrow$ and $n_\downarrow$ spins with I_z parallel and antiparallel to $\boldsymbol{B}_0$ respectively then

$$n_\downarrow/n_\uparrow = \exp(-\Delta E/kT)$$
$$= \exp(-\gamma\hbar B_0/kT), \tag{2.25}$$

where k is Boltzmann's constant and T is the absolute temperature. The total number of spins is given by

$$n = n_\uparrow + n_\downarrow, \tag{2.26}$$

and the difference in population is

$$n_\uparrow - n_\downarrow = n\frac{\{1-\exp(-\gamma\hbar B_0/kT)\}}{\{1+\exp(-\gamma\hbar B_0/kT)\}} \tag{2.27a}$$
$$= n\tanh(\gamma\hbar B_0/2kT). \tag{2.27b}$$

At body temperature (310 K), in a magnetic field of 1 T (10 000 Gauss), the fractional excess of protons $(n_\uparrow - n_\downarrow)/n$ in the low-energy state is only 3.295×10^{-6}. This arises because the Zeeman energy $\gamma\hbar B_0$ is small compared with the quanta of thermal energy which are of magnitude kT. In fact, for this case, the Zeeman energy corresponds to an equivalent temperature (E/k) of a mere 2.04×10^{-3} K (~2 mK). In the earth's magnetic field which in London is 20 μT in the horizontal plane and 40 μT in the vertical, the proton resonance frequency is about 2 kHz ($\gamma/2\pi$ for the proton is 42.573 MHz/T) and the corresponding temperature is 10^{-7} K (0.1 μK). Even in the strongest magnetic fields currently used for n.m.r. (14.093 T corresponding to a resonance frequency of 600 MHz) the equivalent temperature is still only 0.0287 K.

These very small population differences are the fundamental reason for the poor sensitivity of the n.m.r. technique compared with other branches of spectroscopy. It is therefore worth examining eqn (2.27*b*) more closely to see how matters might be improved. Since, as has been shown, $\gamma\hbar B_0 \ll kT$, the tanh function can be approximated by its argument so that:

$$(n_\uparrow - n_\downarrow)/n \simeq \gamma\hbar B_0/2kT. \tag{2.28}$$

The bulk nuclear magnetization M_0 is then given by:

$$M_0 \simeq (n_\uparrow - n_\downarrow)\gamma\hbar m$$
$$\simeq \gamma^2\hbar^2 B_0 n/4kT, \qquad (2.29)$$

and the nuclear susceptibility χ by

$$\chi = M_0/B_0 \simeq \gamma^2\hbar^2 n/4kT. \qquad (2.30)$$

The general results for a nucleus of spin I are:

$$M_0 \simeq \gamma^2\hbar^2 B_0 nI(I+1)/3kT, \qquad (2.31)$$

and

$$\chi \simeq \gamma^2\hbar^2 nI(I+1)/3kT. \qquad (2.32)$$

It may be noted that the magnetic susceptibility is proportional to $1/T$ and therefore obeys Curie's Law (at least for non-interacting spins) so that one way of improving the sensitivity is to reduce the temperature of the sample. Since this reduction needs to be substantial to achieve any reasonable improvement this is hardly a practical method for most biological samples. An alternative is to use larger magnetic fields. In fact, the last ten years has seen a steady increase in state-of-the-art spectrometer frequencies from 60 to 600 MHz for ^{1}H, both for reasons of improved sensitivity and also for the increased chemical shift dispersion (see § 2.6.1) which is invaluable for high-resolution solution studies. However, when, as in the case of n.m.r. imaging, one has to deal with macroscopic objects which have finite conductivity, 'skin-depth' problems arise (see § 5.7). The strategy then is to use the highest frequency (and hence field) consistent with complete r.f. penetration of the sample. The optimization of operating conditions is discussed further in Chapter 5 where we shall see that for human whole-body studies, the appropriate frequency range currently extends up to about 85 MHz.

2.3.3. N.m.r. sensitivity of nuclei

For nuclei other than the simple proton or deuteron the magnitude of the nuclear spin, quantized in units of $\hbar/2$, is governed by three simple rules, namely:

(i) if the mass number A is odd, the spin is half-integral
(ii) if A is even but the atomic number Z is odd, the spin is integral
(iii) if both A and Z are even the spin is zero.

This means that most elements have at least one naturally-occurring isotope which possesses a non-zero nuclear spin and is therefore observable by n.m.r. Table 2.1 lists nuclear properties for those n.m.r. observable isotopes which are potentially of biomedical interest. Note that the common isotopes of carbon (^{12}C) oxygen (^{16}O) and sulphur (^{32}S) are ruled

Table 2.1

N.m.r. sensitivities of nuclei of biomedical interest.

Nucleus	*Spin*	$\gamma/2\pi$ (MHz/T)	*Natural abundance* (%)	*Rel. sensitivity at constant field*	*Rel. sensitivity at constant frequency*	*Typical human physiological concentration of the element*[1]	*Rel. imaging sensitivity at constant frequency*[2]
^{1}H	$\frac{1}{2}$	42·57	99·98	1	1	100 M	1
^{2}H	1	6·54	0·015	$2{\cdot}4\times10^{-6}$	$6{\cdot}4\times10^{-5}$	100 M	6×10^{-5}
						(10 mM)	– (4×10^{-5})
^{3}H	$\frac{1}{2}$	45·41	0	—	—	(10 mM)	– (1×10^{-4})
^{13}C	$\frac{1}{2}$	10·71	1·108	$2{\cdot}5\times10^{-4}$	$2{\cdot}8\times10^{-3}$	10 mM	3×10^{-7} (3×10^{-5})
^{14}N	1	3·08	99·63	$1{\cdot}9\times10^{-3}$	$1{\cdot}9\times10^{-1}$	10 mM	2×10^{-5}
^{15}N	$\frac{1}{2}$	−4·31	0·37	$6{\cdot}8\times10^{-6}$	$3{\cdot}7\times10^{-4}$	10 mM	4×10^{-8} (1×10^{-5})
^{17}O	$\frac{5}{2}$	−5·77	0·037	$1{\cdot}9\times10^{-5}$	$5{\cdot}9\times10^{-4}$	50 M	3×10^{-4}
^{19}F	$\frac{1}{2}$	40·05	100	$8{\cdot}5\times10^{-1}$	$9{\cdot}4\times10^{-1}$	(10 mM)	9×10^{-5}
^{23}Na	$\frac{3}{2}$	11·26	100	$1{\cdot}3\times10^{-1}$	1·3	80 mM	1×10^{-3}
^{25}Mg	$\frac{5}{2}$	−2·61	10·13	$5{\cdot}5\times10^{-4}$	$7{\cdot}2\times10^{-2}$	1 mM	7×10^{-7}
^{31}P	$\frac{1}{2}$	17·23	100	$8{\cdot}3\times10^{-2}$	$4{\cdot}0\times10^{-1}$	10 mM	4×10^{-5}
^{33}S	$\frac{3}{2}$	3·27	0·76	$3{\cdot}3\times10^{-5}$	$2{\cdot}9\times10^{-3}$	0·3 mM	9×10^{-9}
^{35}Cl	$\frac{3}{2}$	4·17	75·53	$6{\cdot}3\times10^{-3}$	$3{\cdot}7\times10^{-1}$	20 mM	8×10^{-5}
^{39}K	$\frac{3}{2}$	1·99	93·1	1×10^{-3}	$2{\cdot}2\times10^{-1}$	45 mM	1×10^{-4}
^{43}Ca	$\frac{7}{2}$	−2·86	0·145	$1{\cdot}8\times10^{-5}$	$2{\cdot}0\times10^{-3}$	0·5 mM	1×10^{-8}

[1] Figures in brackets refer to artificially introduced material.
[2] Figures in brackets correspond to 100 per cent enriched isotope.

out. Fortunately, however, these elements do have other isotopes with nuclear spin, namely ^{13}C, ^{17}O, and ^{33}S.

Clearly, the signal-to-noise (S/N) ratio of the final n.m.r. image will not depend exclusively on the magnitude of the nuclear magnetization but will also reflect the efficiency of the detection system, the required spatial resolution, and other factors. We defer the detailed calculations of S/N until Chapter 5. However, it is instructive to compare at this stage the relative sensitivities of different n.m.r. nuclei, viz the S/N available from a given number of atoms of a particular element relative to that available in the same time from an equal number of protons.

Many conventional n.m.r. spectrometers operate with a superconducting magnet which is run up to field on delivery and thereafter operates in the persistent mode. Thus to observe different nuclei one changes the frequency and it is appropriate to compare sensitivities at constant field in which case they are proportional to $\gamma^{11/4}I(I+1)\times$percentage abundance of the n.m.r. isotope (Hoult and Richards 1976). However, for n.m.r. imaging one ideally selects the highest frequency which the object size and electrical characteristics permit and in this case the dependence is $\gamma I(I+1)\times$percentage abundance of the n.m.r. isotope. Table 2.1 shows the constant frequency and field results together with an n.m.r. imaging sensitivity obtained by multiplying the constant frequency value by typical physiological concentrations for the element in question. This last column is included as a rough guide only. Note that it is the material present in free solution which is important: the presence of an element in large quantity does not guarantee its n.m.r. visibility. For example, most of the naturally-occurring ^{19}F in the body is present as fluorapatite in teeth. Similarly, much of the calcium is bound to proteins rendering it solid-like and invisible as far as conventional imaging studies are concerned.

In the case of sodium one can consider a 'sodium image' in which the mobile sodium spin density is mapped over a body or head cross-section. However, this may be inappropriate for nuclei such as ^{13}C, ^{15}N, and ^{31}P, where one is likely to be concerned with imaging a specific metabolite rather than the total distribution of the element. For example, a period of ischaemia can lead to a complete loss of phosphocreatine with a corresponding increase in inorganic phosphate. The total phosphorus content, however, remains virtually constant. For this reason the 'typical' physiological concentration for these nuclei has been taken as 10 mM in Table 2.1. The imaging sensitivity scales linearly with the number of contributing spins and can thus be easily adjusted for any other desired concentration of metabolite. It is possible to introduce an isotopically-enriched or tagged substrate (^{19}F is particularly promising in this respect). Figures for this case (again at the 10 mM level) are also shown in Table 2.1.

As a final cautionary note it should be mentioned that those nuclei with spins greater than one-half will have nuclear quadrupole moments. This can lead to splitting of the resonance in motionally restricted environments, e.g. deuterated membrane lipids, or line-broadening due to quadrupolar relaxation effects. The latter effect can be severe and has led, for example, to the widespread use of ^{15}N (spin $\frac{1}{2}$) rather than the more abundant ^{14}N (spin 1) for nitrogen n.m.r. studies.

Looking at the figures in the final column of Table 2.1, it might seem a hopeless task to try and image anything other than proton concentration. We shall see, however, that the dependence of imaging sensitivity on spatial resolution is extremely strong, third-order in fact, so that ^{23}Na and ^{31}P images, for example, may be achieved with a resolution of about 1 cm rather than the 1 mm which can be achieved with protons. See § 6.7 for examples.

2.4. Detection of n.m.r.

2.4.1. Continuous wave and Fourier transform methods

We now turn to a preliminary discussion of the detection and processing of n.m.r. signals, a subject which often causes the non-specialist considerable difficulties. These generally arise from the somewhat unfamiliar methods used to overcome the low sensitivity and also to preserve phase information which is essential to many imaging techniques and which is often unavailable in other branches of spectroscopy. We begin by discussing the conceptually simpler continuous wave (c.w.) n.m.r. experiments before turning to the pulsed Fourier transform (F.t.) methods which have largely superseded them.

In a standard spectroscopic experiment, the sample is subjected to electromagnetic irradiation, the frequency of which is scanned through the region of interest. In optical spectroscopy, for example, this frequency sweep can be achieved with a 'white' source and a rotating prism. By measuring the amount of radiation passing through the sample and comparing it with the unattenuated level, or by using a double beam apparatus, the sample absorption can be determined.

Similar experiments can be performed by n.m.r. However, rather than vary the operating frequency of the spectrometer, it is more convenient to sweep the magnetic field, bringing the n.m.r. distinguishable nuclei successively into resonance at fixed frequency. Methods of irradiation and detection are also somewhat different. As the frequency range occupies the r.f. band, wavelengths are typically tens of metres. Transmitter and detector antennae of these dimensions are clearly not a practical proposition for most biomedical samples. The solution is not to use a radiated

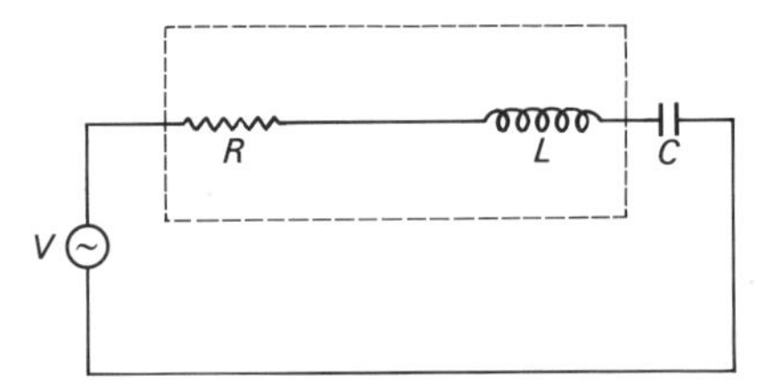

Fig. 2.6. A resonant circuit consisting of resistance R, inductance L and capacitance C in series with a voltage source V.

wave at all but to place the n.m.r. sample at the centre of a r.f. coil (inductor) which forms part of a tuned circuit resonant at the Larmor frequency.

Figure 2.6 shows such a circuit consisting of an inductor and capacitor connected in series with a voltage source. The resistance R, assumed small, is a measure of the lossiness of the reactive components and arises primarily from the resistance of the coil windings rather than from the capacitor which, for this application, can be chosen to have a very high quality factor Q (see below). At the resonance frequency $\omega_0/2\pi$, the voltage appearing across the inductance L is QV, where V is the r.f. generator voltage and

$$Q = \omega_0 L/R. \tag{2.33}$$

For an *unloaded* circuit, Q will typically lie in the range 100–500. It is a measure of the ratio of energy stored in the inductor to energy dissipated in the circuit per r.f. cycle. When the inductor is loaded by insertion of a sample, the observed Q value will drop. This arises from the additional absorption of r.f. energy by the sample, resulting in an increased effective resistance and hence a decrease in Q. The magnitude of this effect will depend on the dimensions of the sample, its conductivity, and the resonance frequency itself. In the case of biological studies such as human imaging the samples have a substantial ionic conductivity and a typical reduction in Q might be to 50 from an unloaded value of 200. As the frequency and conductivity increase so the r.f. tends to become restricted to regions near the surface of the sample—the so-called 'skin depth' effect. As well as increasing the effective series resistance and hence degrading the S/N in the final image, such effects can give rise to phase and intensity anomalies. This is further discussed in § 5.7.

When the n.m.r. sample is positioned within the coil and the static magnetic field is adjusted for resonance, additional r.f. energy is absorbed by the nuclear spin system, giving rise to a further drop in Q. This manifests itself as a proportional drop in voltage across the coil which could be monitored directly (Q meter detection). However, the change is rather small compared with the voltage normally present so this method is prone to drift and other problems. A popular method of overcoming this difficulty is to offset the unwanted component by making the tuned circuit part of an r.f. bridge network.

If the nuclear spins had no means of losing Zeeman energy, the net absorption of energy from the r.f. field would rapidly cease as the spin populations became equalized. The system would be said to be saturated. In fact, energy is lost to the lattice by a process known as spin–lattice relaxation (see § 2.5.1) and it is this which governs the rate of absorption of energy by the spin system.

An alternative method of detection for c.w. n.m.r. makes use of a crossed-coil system. The sample is again contained within a r.f. coil (the transmitter coil) fed with c.w. r.f. power. The magnetic field undergoes a slow field sweep and, as the resonance condition is approached, nuclear magnetization is nutated away from the static field direction. This component of magnetization, precessing at the Larmor frequency, induces an e.m.f. in a second (receiver) coil placed at right angles to the first in order to minimize direct pick-up of r.f. energy. The e.m.f. is then detected and amplified. This is the principle of nuclear induction.

Thus far we have discussed c.w. methods which were used exclusively in commercial instruments until the 1960s when the advantages of F.t. (Fourier transform) n.m.r. became more widely appreciated—so much so in fact, that nowadays all state-of-the-art n.m.r. instruments operate on this principle. Instead of measuring the absorption of the spin system as a function of frequency (or magnetic field), the F.t. method involves simultaneous irradiation of the sample with a wide band of frequencies achieved by applying a short intense r.f. pulse (the bandwidth being inversely proportional to the pulse duration). The signal induced in the receiver coil, via the nuclear induction process, is the sum of responses to all frequency compounds and can be 'unravelled' by a process known as Fourier transformation. Since an appreciation of the F.t. method is central to an understanding of many n.m.r. imaging techniques, it is discussed in some detail in the subsequent four sections.

It is mentioned here, however, that since transmitter and receiver functions are temporally separated in F.t. n.m.r., it is possible to make use of a single coil for both purposes. This has been the practice for high-resolution superconducting n.m.r. spectrometers. However, the conflicting requirements of receiver and transmitter systems for n.m.r. imaging have sometimes led to a retention of the two-coil system (see Chapter 5 for a more detailed discussion).

2.4.2. *Theoretical analysis*

2.4.2.1. The rotating frame concept. As has been shown in § 2.2, the nuclear magnetization vector precesses about the static magnetic field direction at the Larmor frequency (typically a number of MHz). It is now necessary to consider in detail how this magnetization evolves under the action of an applied radiofrequency field, and it is clearly a difficult conceptual problem in the laboratory frame. The situation is rather akin to a parent wishing to communicate with a son or daughter on a fair-ground roundabout. His best solution is to get onto the roundabout himself. Likewise, our best means of investigating the motion of the nuclear magnetization vectors is from the standpoint of a reference frame revolving about the static field direction at the Larmor frequency $\omega_0 = -\gamma \boldsymbol{B}_0$. This is known as the rotating frame, whose co-ordinates are designated by the primed quantities x', y', z': the corresponding laboratory frame co-ordinates being x, y, z.

The individual nuclear magnetic moments appear in this representation as stationary vectors distributed over the surface of a double cone as shown in Fig. 2.7. Since these vectors are randomly distributed in the $x'y'$-plane (i.e. have random phases) and there are more z'-components parallel to $\boldsymbol{B}_0$ than antiparallel to it, the net magnetization, $\boldsymbol{M}_0$ has a magnitude given by eqn (2.29) and a direction parallel to $\boldsymbol{B}_0$. It has been shown in § 2.3 that the nuclear magnetization should interact with the magnetic component of an electromagnetic field whose frequency is ω_0. For experimental convenience a linearly-polarized r.f. field is often used. This may be thought of as being generated by two counter-rotating components of magnitude $\boldsymbol{B}_1$, namely $\boldsymbol{B}_1[\cos \omega_0 t, \; -\sin \omega_0 t, 0]$ and $\boldsymbol{B}_1[\cos \omega_0 t, +\sin \omega_0 t, 0]$ as illustrated in Fig. 2.8(a). As might be intuitively expected, it is the component rotating in the same sense as the nuclear moments which is effective and the other component is henceforth ignored. The effective component of the resonant r.f. field $\boldsymbol{B}_1$ can

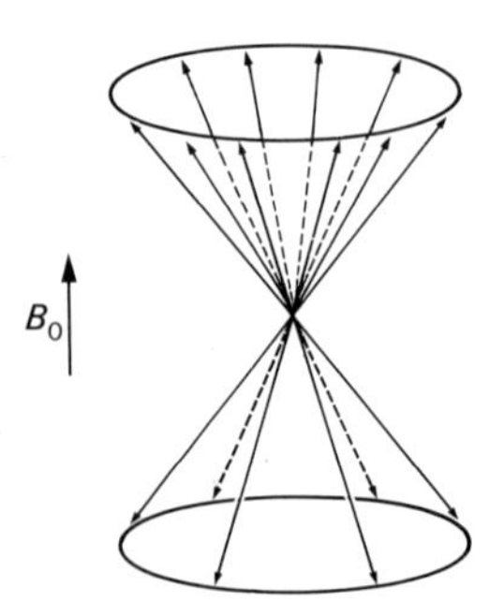

Fig. 2.7. The distribution of individual nuclear magnetic moments in the rotating frame for a spin $\frac{1}{2}$ nucleus. The semi-vertical angle of the cones is 54° 44′ to give the correct projection on the z'-axis. The excess of spins in the low-energy state is greatly exaggerated for clarity.

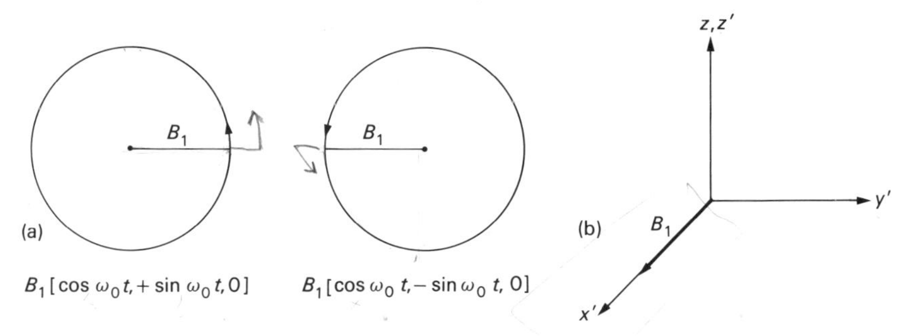

Fig. 2.8. The two counter-rotating field components into which a linearly-polarized field can be decomposed.

therefore be represented as a stationary vector of amplitude B_1 in the x' direction of the rotating frame (see Fig. 2.8(b)).

2.4.2.2. Radiofrequency pulses. The application of a pulse of r.f. irradiation to the spin system is now considered. The response will be a precession of the bulk magnetization in the $y'z'$-plane of the rotating frame, rather as the nuclear moments themselves precess about the static magnetic field in the laboratory frame.

The analysis can be developed more formally as follows: let the net magnetization in the rotating frame be $\mathbf{M} = M_{x'} \cdot \mathbf{i} + M_{y'} \cdot \mathbf{j} + M_{z'} \cdot \mathbf{k}$ where $\mathbf{i}$, $\mathbf{j}$, $\mathbf{k}$ are unit vectors along the x', y', z' axis of the rotating frame and $M_{x'}$, $M_{y'}$, $M_{z'}$ are the corresponding components of magnetization along these directions. Then:

$$\frac{\mathrm{d}\mathbf{M}}{\mathrm{d}t} = \left(\mathbf{i}\frac{\partial M_{x'}}{\partial t} + \mathbf{j}\frac{\partial M_{y'}}{\partial t} + \mathbf{k}\frac{\partial M_{z'}}{\partial t}\right) + \left(M_{x'}\frac{\partial \mathbf{i}}{\partial t} + M_{y'}\frac{\partial \mathbf{j}}{\partial t} + M_{z'}\frac{\partial \mathbf{k}}{\partial t}\right). \quad (2.34)$$

For a reference frame rotating at angular frequency $\boldsymbol{\omega}$,

$$\frac{\partial \mathbf{i}}{\partial t} = \boldsymbol{\omega} \times \mathbf{i}, \quad (2.35)$$

and similarly for $\mathbf{i}$ and $\mathbf{k}$, so that:

$$\frac{\mathrm{d}\mathbf{M}}{\mathrm{d}t} = \frac{\partial \mathbf{M}}{\partial t} + (\boldsymbol{\omega} \times \mathbf{M}). \quad (2.36)$$

The equation of gyroscopic precession (see § 2.2.1) can now be written as:

$$\frac{\mathrm{d}\mathbf{M}}{\mathrm{d}t} = \gamma \mathbf{M} \times (\mathbf{B}_0 + \mathbf{B}_1). \quad (2.37)$$

Thus:

$$\frac{\partial \mathbf{M}}{\partial t} + (\boldsymbol{\omega} \times \mathbf{M}) = \gamma(\mathbf{M} \times (\mathbf{B}_0 + \mathbf{B}_1)); \tag{2.38}$$

so, if the resonance condition $\boldsymbol{\omega} = \gamma \mathbf{B}_0$ applies, then

$$\frac{\partial \mathbf{M}}{\partial t} = \gamma(\mathbf{M} \times \mathbf{B}_1), \tag{2.39}$$

and following the analysis of § 2.2.1 the frequency of precession of the bulk magnetization about B_1 is $\omega_1 = \gamma B_1$. If the pulse is applied for a time t_w then the magnetization will be

$$\mathbf{M} = M_0\{0, \sin \beta, \cos \beta\}, \tag{2.40}$$

where $\beta = \gamma B_1 t_w$. When $\beta = \pi/2$, $M = M_0\{0, 1, 0\}$, the magnetization lies in the y'-direction and one speaks of a $\pi/2$ or 90° pulse. When $\beta = \pi$, $M = M_0\{0, 0, -1\}$, the magnetization lies along $-z'$ and one speaks of a π or 180° pulse.

If the r.f. is not applied at the resonant frequency then the corresponding equation to (2.39) is

$$\frac{\partial \mathbf{M}}{\partial t} = \gamma(\mathbf{M} \times (\mathbf{B}_1 + \Delta \mathbf{B}_0)), \tag{2.41}$$

where $\Delta \mathbf{B}_0 = \Delta B_0 \mathbf{k}$ is the field offset from resonance. The magnetization in this case precesses about the vector $\mathbf{B}_1 + \Delta \mathbf{B}_0$.

In many applications of n.m.r. one is concerned with short intense r.f. pulses; thus for high-resolution solution work the time for a 90° pulse would generally be a few tens of microseconds, whereas in the case of instruments designed for solid-state studies, where it is necessary to excite a much wider bandwidth, a figure of 0.2–2 μs would be typical. These times are sufficiently short that relaxation effects (see § 2.5) can normally be neglected during the pulse itself, and, provided the r.f. is applied close enough to resonance such that $B_1 \gg \Delta B_0$, the simple expression of eqn (2.39) pertains. In subsequent discussions, unless otherwise stated, we shall have these conditions in mind when discussing r.f. pulses. However, it should be mentioned that long low-level r.f. pulses are of importance when *selective* excitation of a particular resonance or sample region is called for. These are known as selective or soft pulses. They are discussed in some detail in connection with those imaging techniques which make use of them in § 3.3.2.

2.4.2.3. Signal detection. During the r.f. pulse the nuclear spin system is unobservable, since the necessarily sensitive receiver is either saturated or else, for preference, is gated off. The receiver recovers in a (hopefully)

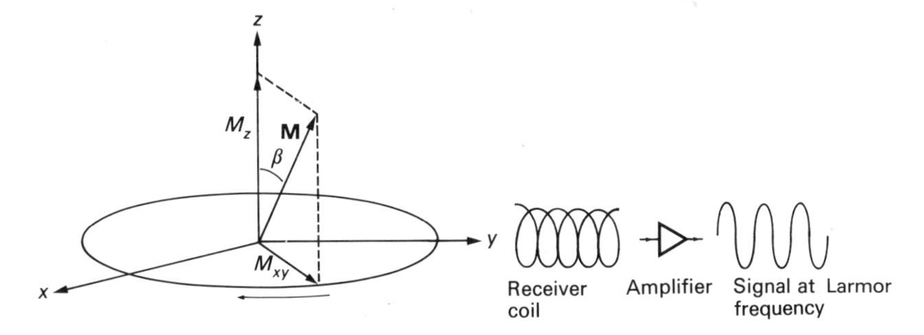

Fig. 2.9. Precession of bulk magnetization in the *laboratory* frame following application of a r.f. pulse. The horizontal component M_{xy} rotating at frequency $\omega_0/2\pi$ induces an e.m.f. at this frequency in the receiver coil.

short period known as the 'dead' time following the end of the pulse. The component of magnetization in the xy-plane M_{xy} can then be observed via the e.m.f. it induces at the Larmor frequency ω_0 in the receiver coil (see discussion of § 2.4.1 and Fig. 2.9). This r.f. e.m.f. can be amplified and converted to a low (audio) frequency by means of a phase sensitive detector (see Chapter 5) prior to digitization and computer processing. The phase sensitive detector (p.s.d.) is a device for measuring the component of magnetization in the $x'y'$-plane in the direction corresponding to the reference phase. Thus, if the signal is phase sensitively detected at the Larmor frequency with a reference phase corresponding to the y'-direction, the result would be a steady d.c. voltage as illustrated in Fig. 2.10(a). If the phase of the reference was shifted 90° to the x'-direction then no signal would be detected since the input signal does

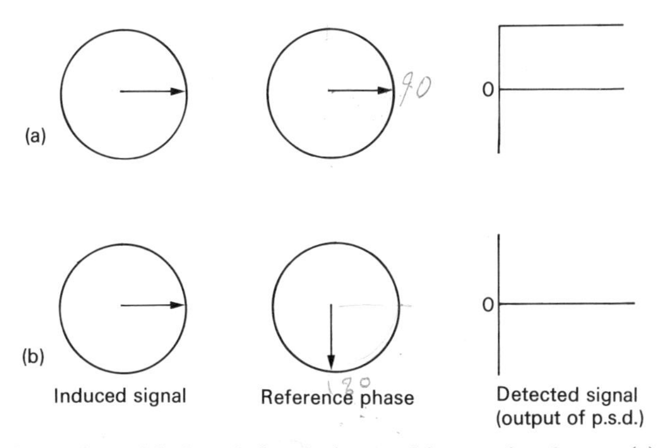

Fig. 2.10. Detection of induced signal observed in rotating frame. (a) Reference phase equal to signal phase. (b) Reference phase and signal phase in quadrature.

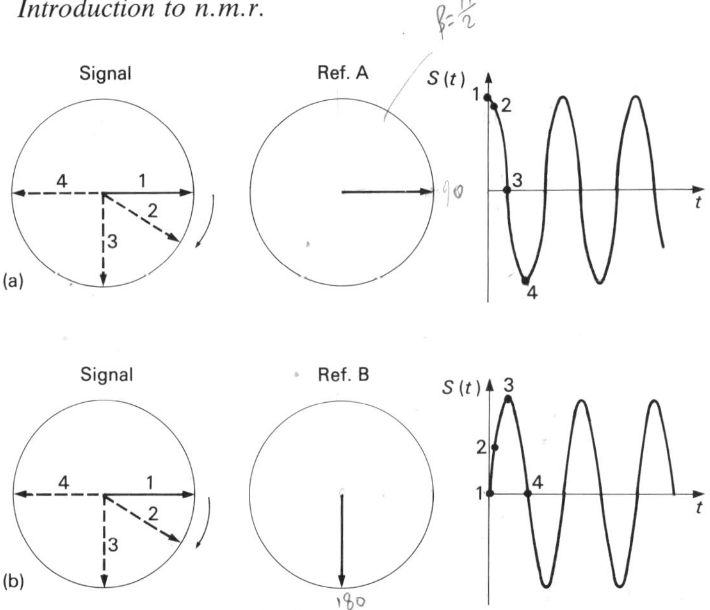

Fig. 2.11. Quadrature off-resonance nuclear induction signals. Reference phases in (a) and (b) are in quadrature.

not have a component in this direction (see Fig. 2.10(b)). In practice, one normally employs two p.s.d.s with reference phases in quadrature (90° apart). This method is known as quadrature detection and is desirable since it allows a $\sqrt{2}$ improvement in S/N to be achieved (see § 5.4.3). Suppose now that a second set of spins exist which are resonant at a slightly different frequency $\omega \neq \omega_0$. The same spectrometer frequency (the frequency of the r.f. pulse) is maintained and the same reference signals are fed to the p.s.d.s. Then, provided $|\Delta\omega| = |\omega - \omega_0| \ll \omega_1$, as discussed in § 2.4.2.2, the effect of the r.f. pulse is again a simple rotation through an angle β in the $y'z'$-plane. However, the subsequent evolution of the magnetization will be different. The effective field ΔB_0 will be $\Delta\omega/\gamma$ and the spins precess at a frequency of $\Delta\omega$ about the z' direction (eqn (2.41) with $\mathbf{B}_1 = 0$). The nuclear signal following phase sensitive detection is as illustrated in Fig. 2.11. Note that quadrature detection allows the direction of rotation to be ascertained so that it is possible to distinguish between spins which are above ($\Delta\omega$ positive) and below ($\Delta\omega$ negative) the spectrometer frequency.

2.4.2.4. Signal analysis. The concept of Fourier transformation is now introduced. We have seen that a set of spins which absorb at a single frequency (a spin isochromat) gives rise to a simple harmonic (sine/cosine) signal whose frequency is given by the (small) difference $\Delta\omega$ between the Larmor precessional frequency and the spectrometer reference frequency. Thus for the 'on resonance condition' ($\Delta\omega = 0$), the signal is a constant d.c. voltage and a frequency analysis would indicate a single zero frequency component (see Fig. 2.12(a)). For $\Delta\omega = \omega_a$ the signal would be a simple cosine function $\cos \omega_a t$ with the corresponding frequency picture consisting of a single component or delta function at ω_a (Fig. 2.12(b)). Similarly, a set of spins with frequency offset $\Delta\omega = \omega_b$ would give rise to a single delta-function at ω_b (Fig. 2.12(c)). Now suppose that half of the spins have a frequency offset ω_a and the other half an offset ω_b. The signal $S(t) = \frac{1}{2}\{\cos \omega_a t + \cos \omega_b t\}$ and the corresponding frequency distribution are illustrated in Fig. 2.12(d). Given the time function $S(t)$, it is non-trivial to deduce the frequency distribution, though, with a little practice, it can still be done by inspection. Thus the main (carrier) frequency could be measured as $(\omega_a + \omega_b)/2$ and the beat frequency as $(\omega_a - \omega_b)/2$. The time function $S(t)$ would therefore be recognized as $\cos(\omega_a + \omega_b)t/2 \cdot \cos(\omega_a - \omega_b)t/2$ which, using the well-known cosine identity:

$$\tfrac{1}{2}\{\cos \mathrm{A} + \cos \mathrm{B}\} = \cos\left\{\frac{\mathrm{A}+\mathrm{B}}{2}\right\}\cos\left\{\frac{\mathrm{A}-\mathrm{B}}{2}\right\}, \tag{2.42}$$

could be resolved into the individual frequency components ω_a and ω_b.

Of course, the process could be continued by adding third, fourth, and subsequent isochromats. The situation would then start to approach that of a real spin system where spectral lines have a finite width due to lifetime broadening, magnet imperfections etc. As the number of frequency components rises so the complexity of the time signal is increased as illustrated in Fig. 2.13. (Compare Figs 2.13(a), (c), with Fig. 2.12(d)). Notice, too, that as the separation of the frequency components is reduced, so the beat period increases in inverse proportion (see Figs 2.13(a),(b)). We return to this point below. Suppose now that we have recorded the time signal of Fig. 2.13(c)(i), how would we know that it originated from the corresponding frequency distribution of Fig. 2.13(c)(ii)? Let us, for the moment, simplify the problem by assuming that all the frequencies which contribute to the signal are discrete multiples of a certain base frequency ω_s. It will be seen later that this turns out to be a particularly appropriate assumption for n.m.r. spectrometers in which discrete sampling is employed. The time signal can then be written as:

$$S(t) = a_0 + \sum_{n=1}^{\infty} a_n \cos n\omega_s t, \tag{2.43}$$

where the a_n are the amplitudes of the discrete frequency components.

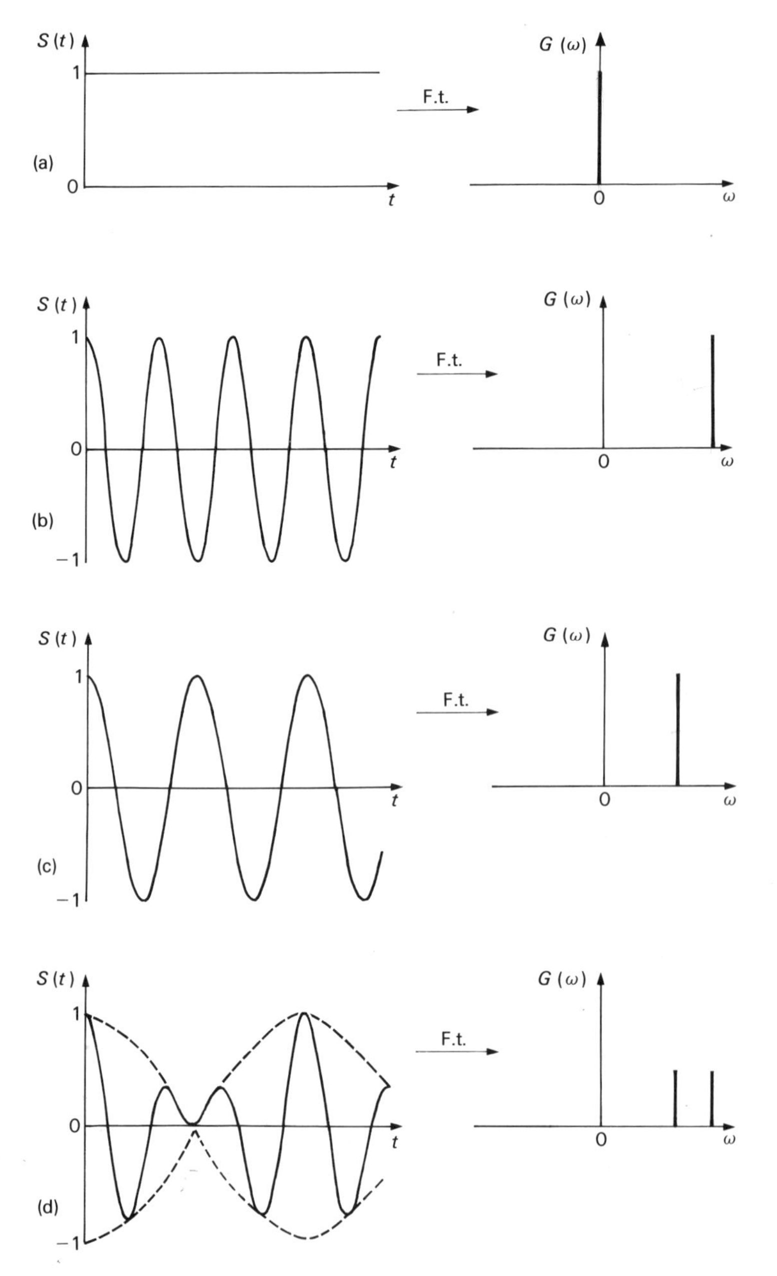

Fig. 2.12. (a) Constant signal and its F.t., a δ function at $\omega = 0$. (b) Cosine signal of frequency ω_a and its F.t., a δ function at $\omega = \omega_a$. (c) Cosine signal of frequency ω_b and its F.t., a δ function at $\omega = \omega_b$. (d) Signal and F.t. corresponding to half sum of (b) and (c).

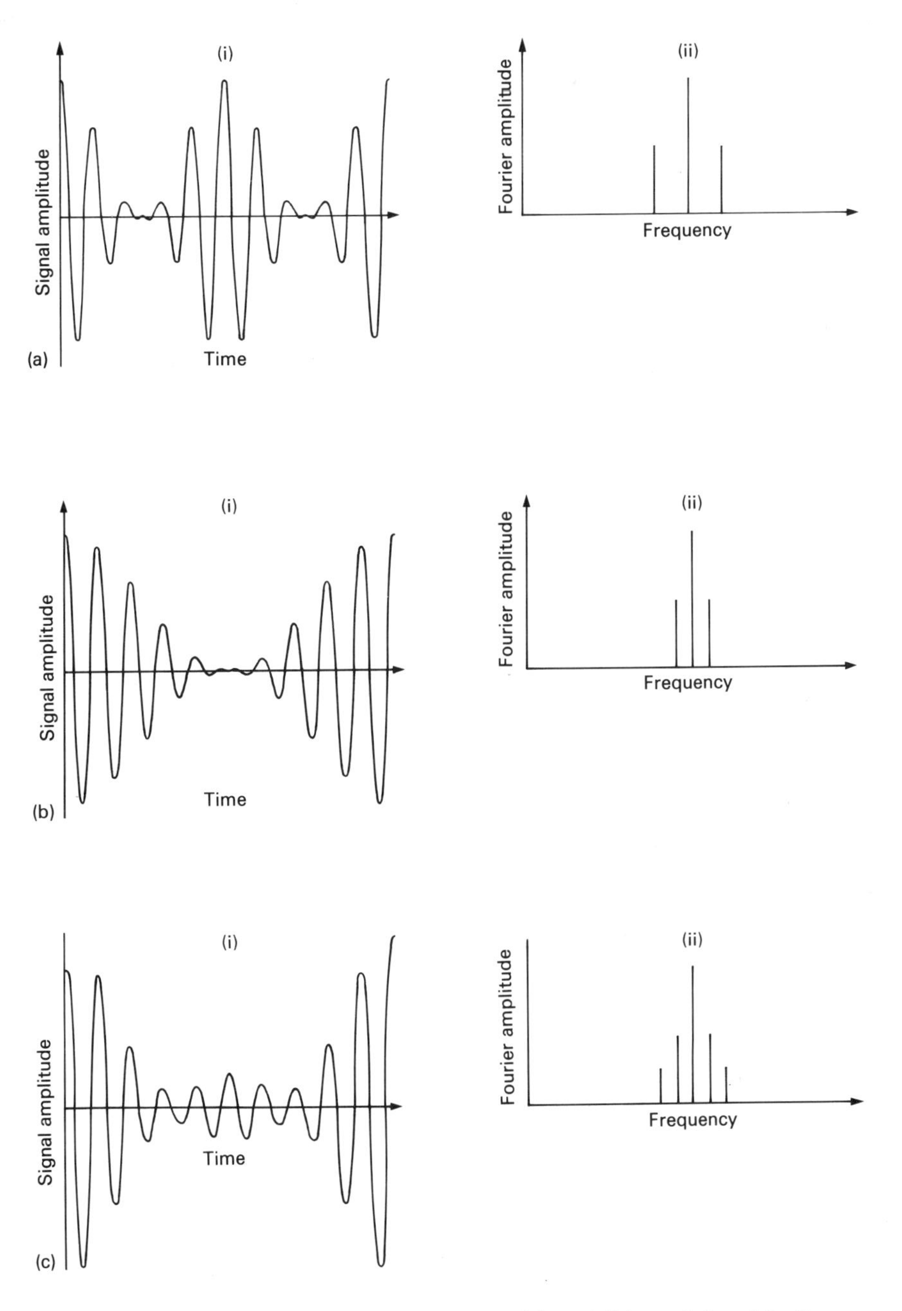

Fig. 2.13. Fourier transform pairs for three, (a) and (b), and five (c), discrete frequencies, approximating a continuous line profile.

The problem is to find this set of constants a_n. Use can be made of the orthogonality of the harmonic functions. Thus, if both sides of eqn (2.43) are multiplied by $\cos m\omega_s t$ and integrated over time between the limits $\pm\pi/\omega_s$ the result is:

$$\int_{-\pi/\omega_s}^{\pi/\omega_s} S(t)\cos m\omega_s t\,\mathrm{d}t = \int_{-\pi/\omega_s}^{\pi/\omega_s} a_0 \cos m\omega_s t\,\mathrm{d}t + \sum_{n=1}^{\infty} \int_{-\pi/\omega}^{\pi/\omega_s} a_n \cos n\omega_s t \cos m\omega_s t\,\mathrm{d}t. \quad (2.44)$$

The first term on the right vanishes for all $m \neq 0$ and the second for $m \neq n$ so that:

$$a_0 = \frac{\omega_s}{2\pi} \int_{-\pi/\omega_s}^{\pi/\omega_s} S(t)\,\mathrm{d}t,$$
$$a_{n\neq 0} = \frac{\omega_s}{\pi} \int_{-\pi/\omega_s}^{\pi/\omega_s} S(t)\cos n\omega_s t\,\mathrm{d}t. \quad (2.45)$$

The contribution of any spin isochromat at frequency $n\omega_s$ to the signal $S(t)$ is therefore obtained by multiplying by $\cos n\omega_s t$ and integrating. This is the process of (cosine) Fourier transformation. In a real system the situation is still more complicated, however, since the signal arises from a continuous distribution of frequency components. This can be accommodated by letting the interval ω_s between discrete frequency components become smaller and smaller until, in the limit $\omega_s \rightarrow 0$, the discrete sum of eqn (2.43) becomes an integral:

$$S(t) = \int_{-\infty}^{\infty} G(\omega)\cos \omega t\,\mathrm{d}\omega. \quad (2.46)$$

In this expression the lineshape function $G(\omega)$ replaces the set of discrete amplitudes a_n. It can be determined in like manner by inverse Fourier transformation:

$$G(\omega) = \frac{1}{2\pi} \int_{-\infty}^{\infty} S(t)\cos \omega t\,\mathrm{d}t. \quad (2.47)$$

Consider qualitatively the signal to be expected from the continuous distribution of frequencies centred on resonance as illustrated in Fig. 2.14(a). Following the application of a 90° pulse all the nuclear magnetization starts off aligned along the y'-direction of the rotating frame (Fig. 2.14(b)). The isochromat which is exactly on resonance ($\Delta\omega = 0$) remains in this direction, but the other components ($\Delta\omega \neq 0$) precess at their offset frequencies and start to spread out or dephase as illustrated in Figs 2.14(c), (d). Eventually, they become uniformly distributed around the whole of the $x'y'$-plane. The net magnetization and hence the induced

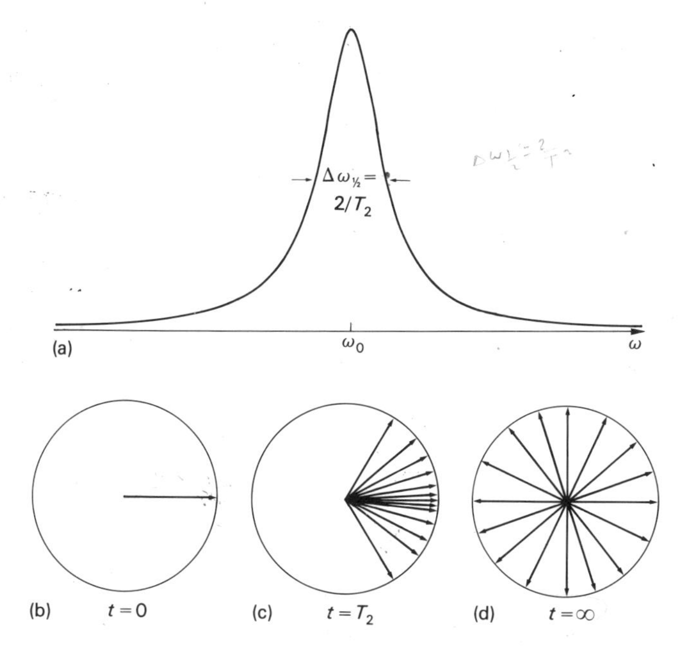

Fig. 2.14. Dephasing of magnetization with time (b), (c), (d) following application of 90° pulse to a Lorentzian line of width $\Delta\omega_{\frac{1}{2}} = 2/T_2$, (a). At $t = 0$ (b) all isochromats are in phase; after a time T_2 the isochromats corresponding to the half-height points (shown as extreme values) are two radians out of phase (c); at long times the isochromats are isotropically distributed and there is no net magnetization.

signal thus decays to zero. This phenomenon has given rise to the name free-induction decay or f.i.d. (free in the sense that it occurs in the absence of any driving r.f.). Note that in contrast to the discrete frequency case no rephasing of the magnetization occurs once it has decayed; no beats are therefore observed.

How long does this dephasing process take? This time is an important n.m.r. parameter known as the spin–spin relaxation time T_2; so-called because, in the absence of magnet imperfections, it arises from interactions between spins (see § 2.5.2). To get a simple estimate, suppose the linewidth (full width at half-height, f.w.h.h.) is $\Delta\omega_{\frac{1}{2}}$, then the isochromats corresponding to the half height points (see Fig. 2.14(a)) will be 2 radians out of phase and the intervening components distributed over the $x'y'$-plane, when $t = 2/\Delta\omega_{\frac{1}{2}}$. This is the spin–spin relaxation time. Note that $T_2 \propto 1/\Delta\omega_{\frac{1}{2}}$, i.e. the greater the spread in the frequency domain, the more rapidly the nuclear induction signal decays.

For a single spectral line in the absence of an applied field gradient the envelope of the f.i.d. is often exponential, i.e.

$$S(t) = a \exp(-t/T_2). \tag{2.48}$$

The spectral distribution $G(\omega)$ which gives rise to this is given by the Fourier transform of $S(t)$ (see eqn (2.47)). Thus, in the above case

$$G(\omega) = \frac{a \cdot \dfrac{1}{T_2}}{(\omega-\omega_0)^2 + \left(\dfrac{1}{T_2}\right)^2}, \tag{2.49}$$

a Lorentzian line of width $\Delta\omega_{\frac{1}{2}} = 2/T_2$ (see Fig. 2.14(a)).

For the purposes of n.m.r. imaging, a magnetic field gradient is generally applied to the sample in the manner of § 2.2.2. The lineshape will then give us the information sought about the structure of our object provided the gradient dominates the linewidth measured in its absence. This point is considered in a more quantitative fashion following a discussion of T_2 relaxation mechanisms (§ 2.5.2). However, assuming that this condition is fulfilled, it is of interest to enquire what f.i.d.s will be obtained from simple objects in the presence of a linear field gradient. As a first example, take a solid pipe of square cross-section with the field gradient applied in a direction normal to one of the faces (Fig. 2.15(a)(i)).

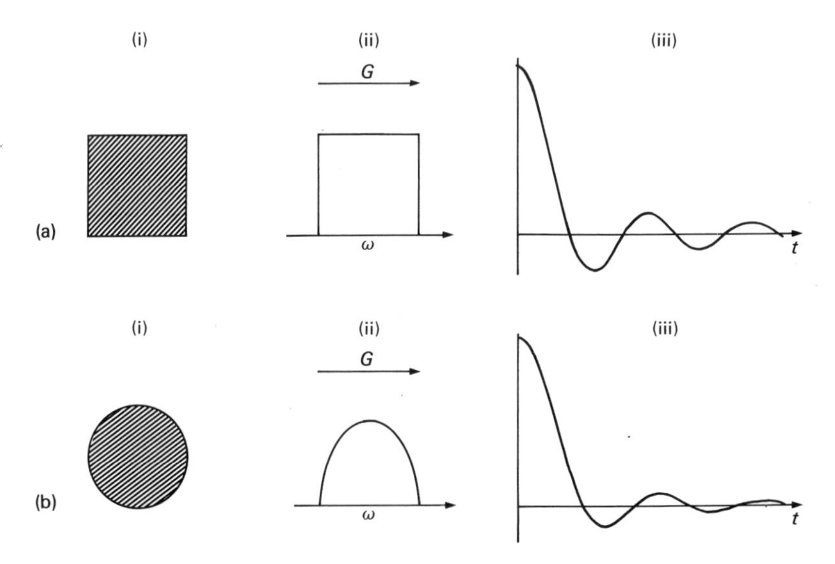

Fig. 2.15. Application of linear field gradient to simple objects: (a) square-sectioned pipe (i), 'top hat' projection profile (ii), and sinc function f.i.d. (iii); (b) circular-sectioned pipe (i), elliptic profile (ii), and Bessel function f.i.d. (iii).

The distribution of resonant frequencies will then be a simple 'top hat' function (Fig. 2.15(a)(ii)) and the f.i.d. will be given by its Fourier conjugate—a sinc function (Fig. 2.15(b)(iii)) with a central maximum whose width is again inversely related to that of the top hat. For a second example, consider the familiar rod of circular cross-section (Fig. 2.15(b)(i)) which, as we have seen in § 2.2.2, gives rise to semi-elliptic profiles (Fig. 2.15(b)(ii)). Their Fourier transforms are Bessel functions (Fig. 2.15(b)(iii). This fact was appreciated by Gabillard (1951) who used tubes of known diameter as a means for determining the strength of field gradients: experiments which in a sense were the converse of today's imaging studies.

2.5. Relaxation times

In this section a brief discussion of spin–spin and spin–lattice relaxation times is given. The reader interested in quantitative descriptions of the various relaxation mechanisms is referred to the excellent text by Abragam (1961).

2.5.1. Spin–lattice relaxation

Spin lattice relaxation is the process by which the spin population returns to its equilibrium Boltzmann distribution following the absorption of r.f. energy. In the case of a 90° pulse, this energy absorption leads to an equalization of the populations in a simple two-state system, a 180° pulse causes population inversion.

The main feature of the T_1 process is an exchange of spin (Zeeman) energy with the thermal motions of the molecules (the 'lattice') of which the nuclear spins are part. Qualitatively, it is possible to think of a spin as a magnetic dipole which 'sees' fluctuating electromagnetic fields arising from the motion of the other spin magnetic dipoles. The components at the Larmor frequency and for some mechanisms, in particular the important dipole–dipole one, at twice the Larmor frequency, can interact with the spins exactly as they do in the case of the applied r.f. pulse. For liquid systems such as water, with which we shall largely be concerned, typical T_1 relaxation times lie in the range 0.1–10 s. In solids, where the motion is considerably less, T_1 relaxation times of minutes or even hours are not uncommon.

Spin–lattice relaxation is generally a first-order process:

$$\frac{\mathrm{d}M_z}{\mathrm{d}t} = -\frac{(M_0 - M_z)}{T_1}, \tag{2.50}$$

and the magnetization recovery towards the equilibrium value M_0 is exponential. T_1 is dependent on two principal factors: the components of motion at ω_0 and $2\omega_0$, and the strength and nature of the magnetic interactions giving rise to relaxation. One of the most important mechanisms is the dipole–dipole interaction such as would occur between the two protons in a water molecule. For two like spins $\mathbf{I}_1, \mathbf{I}_2$, separated by a displacement vector $\mathbf{r}$, the interaction Hamiltonian can be written

$$H_{\mathrm{D}} = \frac{\gamma^2\hbar^2}{r^3}\left\{\mathbf{I}_1 \cdot \mathbf{I}_2 - \frac{3(\mathbf{I}_1 \cdot \mathbf{r})(\mathbf{I}_2 \cdot \mathbf{r})}{r^2}\right\}. \tag{2.51}$$

It can be split into six terms, A, B, C, D, E, and F, each of which connects different spin states in the energy-level diagram for a two-spin system (see Fig. 2.16). A change in the total spin energy and hence in spin populations can only be achieved via the C, D terms at ω_0 and the E, F terms at $2\omega_0$, the necessary energy, $\hbar\omega_0$ and $2\hbar\omega_0$ respectively, being exchanged with the motional energy of the lattice. Although terms A and the spin-flip term B are not involved with spin–lattice relaxation, they do contribute to spin–spin relaxation for which energy conserving, zero frequency processes are also important. For this simple case of two identical spins, the dipolar contribution to the relaxation rate can be written:

$$\frac{1}{T_1} = \tfrac{3}{2}\gamma^4\hbar^2 I(I+1)\{J_1(\omega_0) + J_2(2\omega_0)\}, \tag{2.52}$$

where $J_1(\omega)$ and $J_2(\omega)$ are spectral density functions (giving the components of motion at frequency ω). In the case of intramolecular dipole–dipole interaction (which usually predominates) only rotational motion need be considered and, if it is assumed isotropic, $J_1(\omega_0)$, $J_2(2\omega_0)$ can be

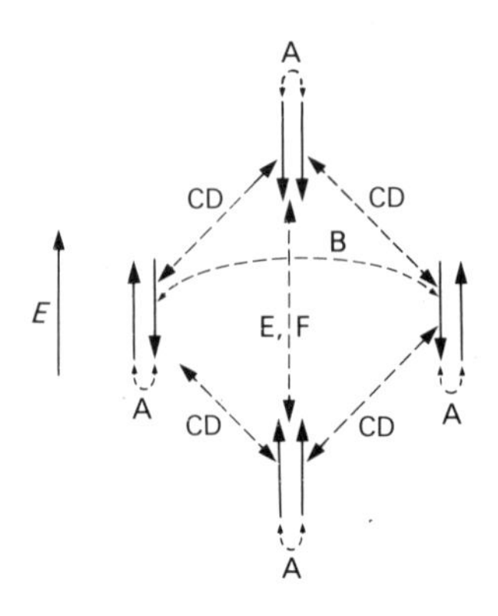

Fig. 2.16. The energy level diagram for a two-spin dipolar system. The lettering indicates which of the dipolar terms A, B, C, D, E, F are responsible for a particular transition.

simply evaluated giving

$$\frac{1}{T_1}=\frac{2\gamma^4\hbar^2 I(I+1)}{5r^6}\left\{\frac{\tau_c}{1+\omega_0^2\tau_c^2}+\frac{4\tau_c}{1+4\omega_0^2\tau_c^2}\right\}, \qquad (2.53)$$

where τ_c is the rotational correlation time. It is worth mentioning here, although a full discussion is deferred until Chapter 6, that in biological systems the situation is a good deal more complicated since water molecules can occur in a number of different exchanging or non-exchanging environments, for example, they may be bound to macromolecules which results in their having much longer correlation times.

In general, however, the T_1 relaxation remains exponential and is ultimately dependent on the dipole–dipole interaction.

For a discussion of other relaxation mechanisms see the texts by Abragam (1961), and Farrar and Becker (1971).

2.5.1.1. Measurement of T_1 relaxation times. T_1 relaxation times can be determined by conventional pulsed n.m.r. methods in a number of ways. Of these, the 180°–TI–90° or inversion recovery sequence is generally the method of choice since it is both simple and reliable. It is illustrated in Fig. 2.17, and consists of a 180° pulse to invert the magnetization followed by a variable time delay TI during which recovery takes place with a time constant, equal to the spin–lattice relaxation time T_1. The extent of the recovery is determined by applying a 90° inspection pulse to nutate the magnetization from the z'-axis into the $x'y'$-plane, where the signal is detected, digitized and stored. The initial amplitude of this f.i.d., which will be equal to the area under the spectrum obtained by Fourier transformation, is

$$M_{x'y'}(0)=M_{z'}(\tau)=M_0\{1-2\exp(-\mathrm{TI}/T_1)\}. \qquad (2.54)$$

If TI is varied over a suitable range, the spin–lattice relaxation time can be found by computer fitting the exponential recovery or, more simply (but less reliably), from the gradient of a logarithmic plot. It is necessary to wait for a period $>4T_1$ between repetitions of the inversion recovery sequence to ensure that the magnetization returns to its equilibrium value. This is not the case for the progressive saturation method in which 90° pulses are repeated at intervals TR which are short compared with T_1. After a few pulses an equilibrium condition is established with the z'-magnetization immediately preceding a pulse given by:

$$M_{z'}=M_0\{1-\exp(-\mathrm{TR}/T_1)\}. \qquad (2.55)$$

TR is the time available for recovery between pulses and, as it is reduced, the equilibrium magnetization falls: the spin system is 'progressively saturated'. The spin–lattice relaxation time can be extracted from a series

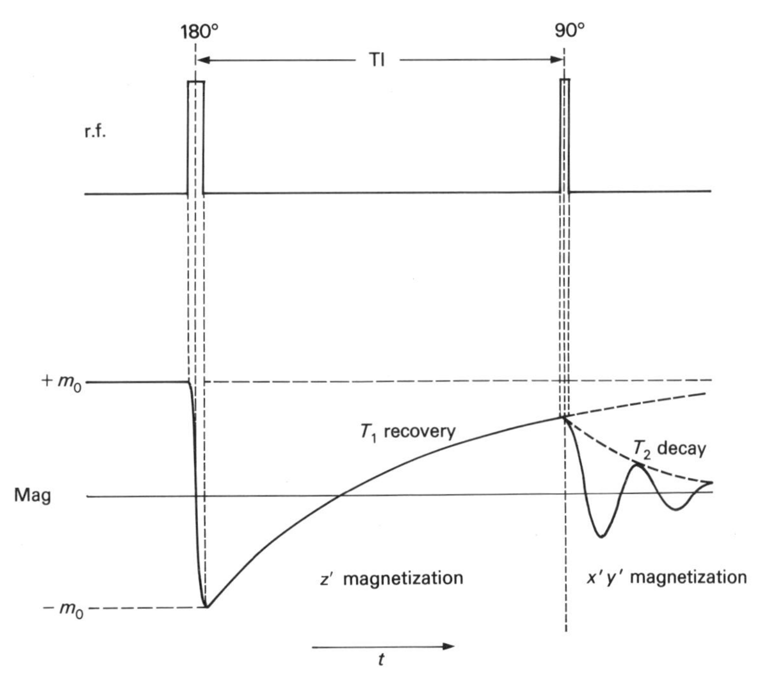

Fig. 2.17. The inversion recovery method for measuring T_1. The width of the r.f. pulses has been increased relative to the interpulse spacing TI for clarity.

of measurements at different repetition rates by similar procedures to those used with the inversion recovery method. Progressive saturation has the advantage of simplicity and speed. Among its disadvantages are: sensitivity to the precession of the 90° pulse setting and a limited range of T_1 values over which it can be applied; the interpulse interval cannot be reduced below either the acquisition time or the spin–spin relaxation time of the sample (see discussion of s.f.p. methods in § 3.2.3).

In many n.m.r. experiments sensitivity considerations force one to repeat f.i.d.s more regularly than a 4–5T_1 interval would allow. In such cases the spin system will be subject to partial saturation and, if there is a distribution of T_1s, there will be differential T_1 effects. This can account for a substantial amount of the tissue contrast seen in repeated f.i.d. or saturation recovery (SR)-type imaging experiments.

2.5.2. Spin–spin relaxation

Spin–spin relaxation is the process in which the net magnetization in the $x'y'$-plane of the rotating frame is destroyed. Similar mechanisms to those responsible for spin–lattice relaxation are involved. However, energy-conserving zero-frequency components along the z'-direction are also of great importance. These latter components vary from point to point in a random fashion causing corresponding variations in the total effective B_z field at different nuclear sites and lead to a dephasing of the spin isochromats. Clearly any inhomogeneity of the applied static magnetic field will have a similar effect. Spin–spin relaxation processes can therefore be divided into two categories: natural (including exchange processes) and those resulting from instrumental imperfections. Taking natural processes first, one of the principal contributors is again the dipole–dipole interaction. In solids this often gives rise to very large static field components which can be as much as several Gauss in the case of interacting protons. This corresponds to a spectral line with a width of some tens of kHz and a T_2 in the region of 100 μs. Note, however, that the decay is generally not exponential and often cannot even be characterized by a single relaxation time. However, T_2 is still a valuable concept and one can think of it, in rather nebulous fashion, as the 'average' time constant for decay.

The application of n.m.r. imaging to solids may therefore seem an impossible task since it is necessary to separate image regions with spectral lines which are naturally 20 kHz wide. The situation is not, however, hopeless since, unlike the liquid case which is governed by random thermal processes, the dephasing in solids is a quantum coherent phenomenon, being the evolution of the magnetization under the Hamiltonian H_D (see eqn (2.51)). It is possible to reverse the effects of this operator (or in other words make time run backwards!) by the proper application of a suitably chosen train of r.f. pulses (Haeberlen 1976). In view of the additional level of complexity involved, it is perhaps not surprising that solids have largely been ignored. Nevertheless, some of the early development of the subject by Mansfield and co-workers (Mansfield and Grannell 1973, 1975) was carried out with solids in mind. Wind and Yannoni (1979) have more recently demonstrated a point-imaging method whose selectivity is based on the strong frequency offset dependence of most line-narrowing sequences. It is also possible to make use of multiple quantum n.m.r. In this case the applied field gradient strength is effectively multiplied by the order of the transition observed, allowing the natural linewidth to be overcome.

In liquids the situation is fortunately quite different: the dipolar interaction is time-averaged to zero and the quantum coherent dephasing is no longer operative. Dipolar relaxation therefore arises from the effects of the randomly-fluctuating fields at frequencies 0, ω_0 and $2\omega_0$. For two like spins:

$$\frac{1}{T_2} = \gamma^4\hbar^2 I(I+1)\{\tfrac{3}{8}J_0(0) + \tfrac{15}{4}J_1(\omega_0) + \tfrac{3}{8}J_2(2\omega_0)\}, \tag{2.56}$$

which for the isotropic rotation, intramolecular interactions case reduces to:

$$\frac{1}{T_2}=\frac{\gamma^4\hbar^2 I(I+1)}{5r^6}\left\{3\tau_c+\frac{5\tau_c}{1+\omega_0^2\tau_c^2}+\frac{2\tau_c}{1+4\omega_0^2\tau_c^2}\right\}. \tag{2.57}$$

(Compare the analogous expressions (2.52) and (2.53) for the spin–lattice relaxation case.) In the extreme narrowing limit where $\omega_0\tau_c \gg 1$, corresponding to rapid motion such as would occur at room temperature for dilute solutions of small molecules, $T_1 = T_2$ and both are frequency independent. However, if there are low motional frequencies present, arising, for example, from binding to a slowly tumbling macromolecule or from exchange, either chemical or a physical exchange from one environment to another, then T_2 can be substantially shorter than T_1. In this brief discussion of spin–lattice relaxation time the difficulty of trying to identify the relaxation mechanisms which are of importance in biological systems has already been hinted at. In the case of water, with which this monograph is primarily concerned, the T_2 process is generally non-exponential. However, the decay can normally be resolved into two or more component exponentials which can be ascribed to water in different non-exchanging compartments. The major fractions have natural linewidths which are more or less frequency-independent and are typically in the range 10–50 Hz. When there are experimental imperfections present, there will be additional contributions to the linewidth. Thus the effective spin–spin relaxation time T_{2e} is:

$$\frac{1}{T_{2e}}=\frac{1}{T_{2n}}+\frac{1}{T_{2mi}}+\frac{1}{T_{2ms}}+\cdots, \tag{2.58}$$

where $1/T_{2n}$ is proportional to the rate of dephasing (or linewidth) brought about by natural means, $1/T_{2mi}$ is the additional line-broadening arising from the use of non-uniform static magnetic fields and $1/T_{2ms}$ is the contribution from field inhomogeneities arising from magnetic susceptibility effects.

Typically the magnetic field inhomogeneity in a magnet designed for whole-body imaging studies might be between 1 and 10 p.p.m. At an operating frequency of 21 MHz this amounts to between 21 and 210 Hz, i.e. of the same order of magnitude as the natural linewidth itself. Magnet shimming (the improvement of field homogeneity) although often a very time-consuming process is therefore a worthwhile exercise in optimizing an imaging system (see Chapter 5). We mention here that the decay due to magnetic inhomogeneity, unlike that arising from thermal motion, is a (classical) coherent process. This contribution to the linewidth can therefore be eliminated using a spin–echo type experiment as discussed in § 2.5.2.2.

Magnetic susceptibility effects arise from discontinuities in magnetic susceptibility which occur principally at structural boundaries within tissues. They are typically ~1 p.p.m. and, whereas they are of minimal importance in human imaging studies, they may well be crucial at the microscopic level where much larger static fields are required.

Once again, if it is assumed that the T_2-decay is, at least approximately, a first-order process, the decay of the $x'y'$-magnetization can be written as:

$$\frac{\mathrm{d}M_{x'}}{\mathrm{d}t} = -M_{x'}/T_2, \tag{2.59}$$

$$\frac{\mathrm{d}M_{y'}}{\mathrm{d}t} = -M_{y'}/T_2, \tag{2.60}$$

and the width of the spectral line (f.w.h.h.), $\Delta\nu_{\frac{1}{2}} = \Delta\omega_{\frac{1}{2}}/2\pi$, is given by

$$\Delta\nu_{\frac{1}{2}} = 1/\pi T_2. \tag{2.61}$$

Equations (2.59) and (2.60) are the Bloch equations governing the evolution of the $x'y'$ magnetization for liquid-like systems at resonance ($\Delta\mathbf{B}_0 = 0$) in the absence of an r.f. driving field. They can be extended to include off-resonance and r.f. effects through the addition of the term given by eqn (2.41). Spin–lattice relaxation can also be included to give a single vector equation (the full Bloch equation) describing the evolution in the rotating frame as:

$$\frac{\mathrm{d}\mathbf{M}}{\mathrm{d}t} = -\frac{M_{x'}}{T_2}\mathbf{i} - \frac{M_{y'}}{T_2}\mathbf{j} + \frac{(M_0 - M_{z'})}{T_1}\mathbf{k} + \gamma\mathbf{M}\times(\mathbf{B}_1 + \Delta\mathbf{B}_0). \tag{2.62}$$

This equation is of great importance since it allows one to predict the behaviour of the spin system under different r.f. pulse sequences and in the presence of applied field gradients, which enter via the field offset term $\Delta\mathbf{B}_0$. It forms the basis for a number of computer models of n.m.r. imaging procedures (see, for example, Bittoun and Taquin 1982) and also permits the simulation of imperfections in the static field, gradients, r.f. pulses, etc. An example of its use will be given in Chapter 3 in connection with selective pulse theory.

2.5.2.1. Choice of field gradients. Consider two adjacent volume elements (voxels) in the object to be imaged as illustrated in Fig. 2.18. The aim will be to apply a linear field gradient of sufficient magnitude to resolve the signals from these voxels. To a first approximation the material can be considered as concentrated at the centre of each element. Then, if the side of the element is δx and the field gradient G is applied along it, the requirement for resolution is that

$$G\,\delta x > 1/T_{2e}, \tag{2.63a}$$

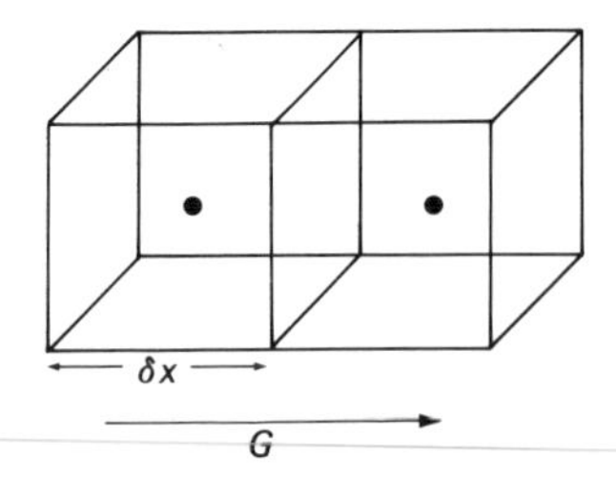

Fig. 2.18. The application of a field gradient G to two adjacent cubic volume elements of side δx.

or

$$G > 1/T_{2e}\,\delta x. \qquad (2.63b)$$

The finer the spatial resolution δx required, the greater the field gradient amplitude needed to achieve it. Equally, the greater the effective linewidth $1/T_{2e}$, the larger the gradient needed to overcome it. It will be shown in Chapter 5 that very large gradients are undesirable since they spread the total signal from the sample over a wide bandwidth and reduce the S/N per unit time.

Taking as a model of the human torso an elliptic section of major axis $L = 32$ cm and assuming that the full width is to be represented by 128 data points. (This number is generally a power of 2 to facilitate fast Fourier transformation and is typically in the range 32 (2^5) to 512 (2^9).) In this case $\delta x = 32/128$ cm $= 2.5$ mm. Taking as an example $1/T_{2e}$ as 40 Hz the requirement is that $G \gtrsim 1$ mT/m (0.1 G/cm).

2.5.2.2. Measurement of spin–spin relaxation times. In principle, it is possible to determine spin–spin relaxation times directly from the decay constant of the f.i.d. or alternatively from the linewidth of the corresponding spectral line, see eqn (2.61). However, there are uncertainties at the start of the f.i.d. arising both from the use of finite (non-delta function) r.f. pulses and also as a consequence of the receiver dead time (see § 4.1.4). These problems are particularly acute in the case of short T_2 values and lead to different design criteria for spectrometers intended for 'wide-line' (short T_2) work. For systems with naturally long T_2 values, in the sense $T_{2n} > T_{2mi}$ (see eqn (2.58)), the major problem arises from the additional line broadening due to the static magnetic field inhomogeneity. This results in an observed T_2, which is shorter than the natural one. Thus:

$$\frac{1}{T_2} = \frac{1}{T_{2n}} + \frac{1}{T_{2mi}}. \qquad (2.64)$$

As mentioned above, the decay due to magnetic inhomogeneity is a coherent process which can be eliminated using a spin-echo sequence. The simplest example of this type of experiment, for which only two r.f. pulses are required, was first discussed and demonstrated by Hahn in 1950 and is illustrated in Fig. 2.19. Following the initial 90° pulse applied along the x'-direction (Fig. 2.19(a)), the isochromats begin to dephase. This process is allowed to continue for a time TE/2 (Fig. 2.19(b)) at which point a 180° pulse is applied, again along the x'-direction (Fig. 2.19(c)), causing the distribution of magnetization to be 'reflected' about the $x'z'$-plane. The isochromats continue to move in their original sense and so rephase along the negative y'-axis after a further time TE/2 (Fig. 2.19(d)). The echo then seen (Fig. 2.19(e)) will be reduced in amplitude by an amount which depends solely on the natural T_2, i.e. on those processes which are not coherent and cannot therefore be refocused. As the time TE/2 between pulses is varied, the amplitude of the echo, which will be inverted if phase-sensitive detection is used, will vary as

$$M(\text{TE}) = M_0 \exp(-\text{TE}/T_{2n}). \tag{2.65}$$

The basic Hahn echo experiment can be extended as suggested by Carr

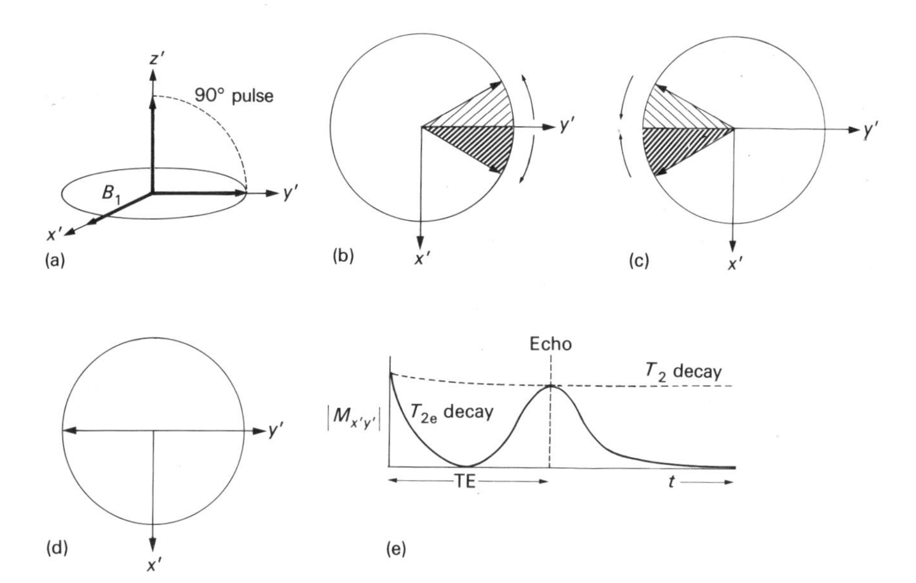

Fig. 2.19. The Hahn spin–echo experiment. (a) Application of a 90° pulse. (b) Dephasing of spins during period TE/2. (c) Position of spin isochromats following 180° pulse. (d) Negatively-refocused magnetization after further period TE/2. (e) Evolution of magnetization in $x'y'$-plane of rotating frame. Note the sign reversal following the 180° pulse has been removed for clarity.

and Purcell (1954), through the addition of further 180° pulses at intervals of TE giving a pulse sequence 90°–TE/2–180°–TE–180°–TE. . . . This produces a series of echoes of alternating sign at intervals of TE and allows T_2 to be determined in a single experiment. It also has the virtue of reducing errors due to the effects of diffusion. In a further modification of the method, Meiboom and Gill (1958) proposed the use of 180° pulses oriented along the y'-axis of the rotating frame, i.e. in phase quadrature with the preparatory 90° pulse. This has the additional merit of reducing the sensitivity of the experiment to pulse imperfections (mis-settings of the 180° pulse length). It is known as the Carr–Purcell–Meiboom–Gill (CPMG) sequence, and is the method of choice for the measurement of spin–spin relaxation time in high-resolution n.m.r. Both the simple spin-echo (SE) and the CPMG or multiple spin-echo (MSE) sequences find application in imaging studies where they are used to generate T_2-related contrast (see § 6.3.1).

2.6. Real spin systems

2.6.1. *Chemical shifts*

The electrons which constitute the nuclear environment give rise to small perturbations in the magnetic field at the spin site. This effect is known as the chemical shift and is generally described by the expression

$$B = (1-\sigma)B_0, \tag{2.66}$$

where σ is the shielding parameter.

It is this small quantity which forms the basis of n.m.r. analytical spectroscopy allowing identification and structural elucidation of chemical compounds. It is really a tensor quantity but, in the case of liquids or solutions with which we shall be exclusively concerned in this and the subsequent section, only the isotropic part, $\sigma = \frac{1}{3}(\sigma_{xx} + \sigma_{yy} + \sigma_{zz})$, is left unaveraged by thermal motion. (However, in asymmetric molecules, certain nuclei, notably ^{31}P, can have very large chemical shift anisotropies $\Delta\sigma = (\sigma_{\parallel} - \sigma_{\perp})$ which at high fields can provide an important mechanism for relaxation (see, for example, Farrar and Becker 1971).)

The chemical shift δ is normally expressed in parts per million (p.p.m.) as a relative difference in frequency from some reference compound—often tetramethylsilane (TMS) in the case of ^1H or ^{13}C spectroscopy. Thus:

$$\delta = \frac{\nu - \nu_{TMS}}{\nu_{TMS}} \cdot 10^6 \tag{2.67}$$

where ν_{TMS} and ν are, respectively, the resonance frequencies of TMS and the spectral line of interest. For ^{1}H the whole chemical shift range is some 10 p.p.m. Notice that the frequency shifts are proportional to the applied magnetic field strength $(\nu-\nu_{TMS})=(\gamma/2\pi)B_0(\sigma_{TMS}-\sigma)$. At 4 MHz this therefore corresponds to a range of only 40 Hz whereas at 500 MHz it increases to 5000 Hz. Figure 2.20 shows a 'high-resolution' ^{1}H spectrum obtained *in vivo* from a rat muscle at 200 MHz. At first sight there appears to be a single peak with a width of about 40 Hz arising from the tissue water. However, if the vertical gain on the display is increased (Fig. 2.20(b)), a second peak is apparent ~3 p.p.m. upfield from the main water resonance. (This corresponds to a shift of 3 p.p.m. to *lower* frequency—the inverse frequency scale being an unfortunate convention dating from the days of c.w. n.m.r. when the field was swept from low to high value.) This peak arises primarily from the —CH_2— backbone of the mobile fats (primarily triglycerides in fat droplets). If the gain is increased still further, we can begin to identify some of the metabolites present in solution, for example, lactic acid and creatine. What implications does this have for ^{1}H-imaging? Taking the case of a whole-body system operating at 21 MHz, it might be reasonable to expect the 0.5 T magnet to have a homogeneity of about 20 p.p.m. over the region of interest (see § 5.1 for typical magnet specifications). If we were to repeat our high-resolution muscle study in this magnet the separation of fat and water peaks would be about 60 Hz but both of the lines would be broadened by 420 Hz due

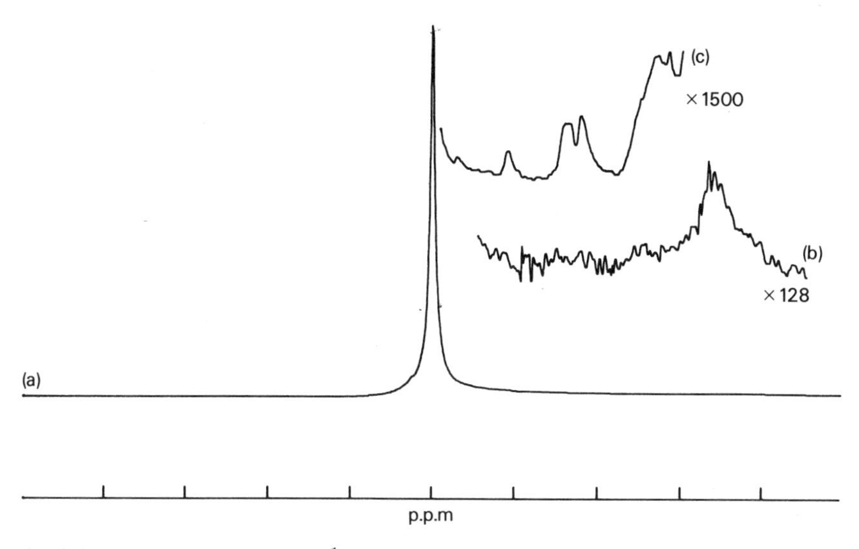

Fig. 2.20. The high-resolution ^{1}H n.m.r. spectrum of muscle. As the vertical gain is increased the signal from the metabolites becomes apparent.

to the static field inhomogeneity. They would not therefore be resolved and the fat would appear as a slight asymmetry in the water lineshape. In general, then, we cannot expect to separate the contributions from different chemically shifted protons. We therefore speak of imaging the mobile proton density with the understanding that 'mobile protons' include contributions from all signal sources. Chemical shift effects are nevertheless evident in some n.m.r. images. Head scans, for example, often display a ring artefact arising from the shifted fat signal. This is particularly apparent in high field Fourier imaging studies. The problem can be eliminated by using selective difference techniques or, less satisfactorily, by increasing the field gradient strength.

Of course, the metabolites are present in such small amounts relative to water ($\leqslant 10$ mM as opposed to 100 M) that they have little bearing on the interpretation of ^{1}H images. Mobile lipids, however, can exert an important influence. For example, they have T_1 values which differ from that of water so that in tissues in which they are a major constituent the T_1 recovery process will not be exponential. This is particularly apparent in mammary tissue where the fat content is very high. It can cause problems in interpretation of apparent T_1 changes since, unless care is taken to observe the whole recovery process, it is difficult to separate actual T_1 changes from changes in the relative amounts of fat and water.

With the important biological nuclei ^{13}C, ^{15}N, and ^{31}P the chemical shift dispersion is much greater: ~200, 500, and 50 p.p.m. respectively. Since the linewidths in tissue are reasonably small (~5–40 Hz) this suggests that it should be possible to image individual metabolites. However, as we have seen in § 2.3.3, their n.m.r. sensitivity is very low. Considerable progress has been made in this area nevertheless. See § 6.7 for further details.

2.6.2. *Scalar coupling*

Whereas the basic dipolar interaction is averaged to zero by thermal motion, there is a second-order effect known as scalar or J coupling which has a non-vanishing secular component. It takes the form

$$\hat{H}_J = hJ\mathbf{I}\cdot\mathbf{S} \tag{2.68}$$

where **I** and **S** are the spin operators and J the coupling constant. The interaction occurs between non-equivalent nuclei—either different spin species or the same spin species in different sites. It is a short-range interaction and occurs via the chemical bonds (unlike the dipole–dipole interaction which takes place 'through space').

Since both protons in the water molecule are equivalent there is no splitting of its n.m.r. resonance. ^{1}H-splittings (J couplings) in the range 0–20 Hz do, however, occur in many metabolites.

In the case of phosphorylated metabolites ^{31}P–^{31}P, scalar couplings of the order of 16 Hz can be seen; for example, in the –P–O–P– groups of ADP and ATP. ^{31}P–^{1}H couplings in naturally-occurring metabolites are smaller (<10 Hz) and are therefore not generally seen in *in vivo* n.m.r. studies.

The situation in which X–^{1}H couplings are most commonly observed is when X is ^{13}C. Thus a quaternary carbon atom appears as a singlet, a methine carbon (>CH—) as a doublet, a methylene carbon (CH_2—) as a triplet and a methyl carbon (CH_3—) as a quartet. Coupling constants and hence line splittings are typically 50–200 Hz. The splitting pattern in an undecoupled ^{13}C spectrum can be a very useful aid to assignment.

It is possible to derive some insight into the multiplicity of X–H^1 splittings from a simple qualitative picture. Take first the case of a single proton as illustrated in Fig. 2.21(a). The proton can be in one of two possible spin states (spin up (↑) or spin down (↓)). If it is in the up state it slightly decreases the field at the X nucleus and hence its Larmor frequency. Similarly, if the spin is down the reverse is true. The X resonance is thus split into two components of equal magnitude. Incidentally, the f.i.d. observed in such a case would look, in the absence of T_2 relaxation effects, like the beat pattern of Fig. 2.12(d); the carrier wave corresponding to the average frequency offset of the two components (i.e. the offset of the unsplit line) and the beat frequency to the coupling constant *J*. Next consider the case of two protons as illustrated in Fig. 2.21(b). The protons can either both be up, both down or one up, one down (with two possible permutations). The line is therefore split into a triplet with the central component unshifted and of twice the intensity of the outer two, which are shifted upfield in the ↓↓ case and downfield in the other. Similar reasoning can be applied for greater multiplicities, the intensity distribution over the multiplet being given by the binomial expansion for the corresponding number of spins. Although spin–spin splittings can give valuable structural information and are useful assign-

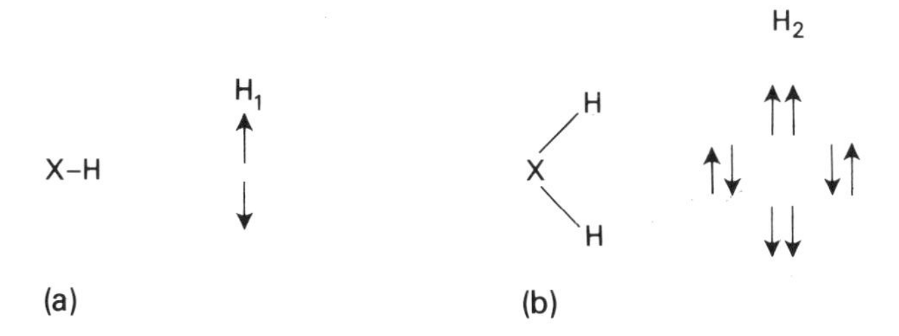

Fig. 2.21. Spin couplings. (a) Single proton coupled to an X nucleus; two possible orientations. (b) Pair of protons coupled to an X nucleus; four possible orientations, two of which are degenerate.

ment aids, they are undesirable from the S/N viewpoint (since the signal is distributed amongst a multiplicity of lines). In ^{13}C spectroscopy, therefore, it is common practice to irradiate the protons at their own Larmor frequency simultaneously with ^{13}C data acquisition. This has the effect of removing the splittings and is known as proton decoupling.

A number of other interactions such as the nuclear Overhauser effect (n.o.e.) and the quadrupolar interaction are also of importance in n.m.r. spectroscopy. However, we shall not encounter them further in this book and refer the interested reader to standard n.m.r. texts.

2.7. N.m.r. spectroscopic studies of living systems

We have indicated that there are considerable problems both in terms of sensitivity and in the requirement for high homogeneity over an extended region which make the imaging of specific metabolites extremely difficult. Thus, although multinuclear and chemical-shift imaging techniques are available, the subject remains very much at the development stage. Some preliminary results are discussed in § 6.7. It is, however, possible to perform more conventional *in vivo* n.m.r. studies by restricting observation to the small region of high homogeneity near the magnet centre. Standard high-resolution magnets may be used for studies at the cellular, perfused organ, or small animal level. Larger systems, including full-scale clinical instruments are available from Oxford Research Systems (see their Biospec range). Manufacturers of high field ($\geqslant 1.5$ T) ^{1}H imaging equipment generally also offer a spectroscopic option (see, for example, Bottomley, Hart, Edelstein, Schenck, Smith, Leue, Mueller, and Redington 1983).

Interest in high resolution n.m.r. studies of biological tissues began in 1973 with the study of red blood cells by Moon and Richards. It gained further momentum in 1974 with the demonstration by Hoult and coworkers at Oxford (Hoult, Busby, Gadian, Radda, Richards, and Seeley 1974) that ^{31}P spectra could be obtained from isolated skeletal muscle. Furthermore, and in contrast to the situation for protons, the spectra proved to consist of relatively few lines which were generally well-resolved and easily assigned. Although it would be out of place to enter into detailed discussion of how high-resolution ^{31}P and ^{13}C n.m.r. can be used to estimate the levels of important metabolites, measure the (unidirectional) fluxes through biochemical pathways (notably the glycolytic one), and correlate these with work output, oxygen supply and substrate type and level, a few illustrative examples should not be out of place. The subject has been treated much more extensively in a number of review articles (Iles, Stevens and Griffiths 1982; Gadian and Radda 1981) and in a recent monograph (Gadian 1982).

2.7.1. ^{31}P *n.m.r. studies*

Figure 2.22 gives an indication of the diversity of living systems to which high-resolution n.m.r. can be successfully applied (Mansour, Morris, Feeney, and Roberts 1982). It shows ^{31}P spectra obtained *in vivo* at 81 MHz from liver flukes (Fasciola hepatica) suspended in phosphate-free Krebs ringer within a standard 15 mm n.m.r. tube. Assignments are as indicated in the figure caption. Although the resonances of ATP, ADP, and the nicotinamide dinucleotides overlap to some extent, it is nevertheless possible to estimate the relative amounts of each. Thus peak J arises exclusively from the β-phosphate of ATP so that the area under this n.m.r. line will be proportional to [ATP]. Peaks F and G, arising from γATP and βADP respectively, are only partially resolved, but, if the total area is ascertained and the area under line J subtracted, a measure of [ADP] is obtained. Similarly, it is possible to estimate the concentration of the nicotinamide dinucleotides from the area of peak I seen just upfield of the αATP and ADP resonance (H). It is also possible to say, from the measured shifts, that the ATP is virtually completely magnesium-bound. (This method can be used to estimate the intracellular free Mg level; Gupta and Moore 1980). Peak C is due to internal inorganic phosphate. Apart from giving a measure of P_i, the position of this peak is very sensitive to pH (the pK lies in the physiological range, see Fig. 2.23) and therefore affords a measure of this quantity. The preferred reference, phosphocreatine (PCr), from which the shift of the P_i peak is measured, is absent in the liver fluke (itself an interesting observation). However, we can derive a qualitative estimate using as a reference the αATP/ADP resonance, which is fairly insensitive to pH over the normal physiological range. Thus the internal tissue pH is found to be 7.0. Another interesting feature of the P_i peak is its relatively large width (~40 Hz) indicating a range of environments with different pH (a spread of some 0.3–0.4 pH units centred about the mean value of 7.0). When glucose is added to the external medium it is rapidly metabolized. Note the rapid increase in sugar (hexose) phosphate (A) and decrease in organic phosphate in Fig. 2.22(b). On addition of serotonin (Fig. 2.22(c)) the sugar phosphate peak increases still further but the flukes are unable to maintain their ATP levels which are depleted as a result of the increased motility, readily observed and known to be induced by serotonin. This technique therefore allows one to study the hitherto poorly-understood metabolism of these important parasites and permits prospective treatments to be assessed.

Figure 2.24, in contrast, is a ^{31}P n.m.r. spectrum obtained from an isolated Langendorff-perfused ferret heart. Many of the same features are seen: phosphomonoesters (single peak), inorganic phosphate, ATP, and

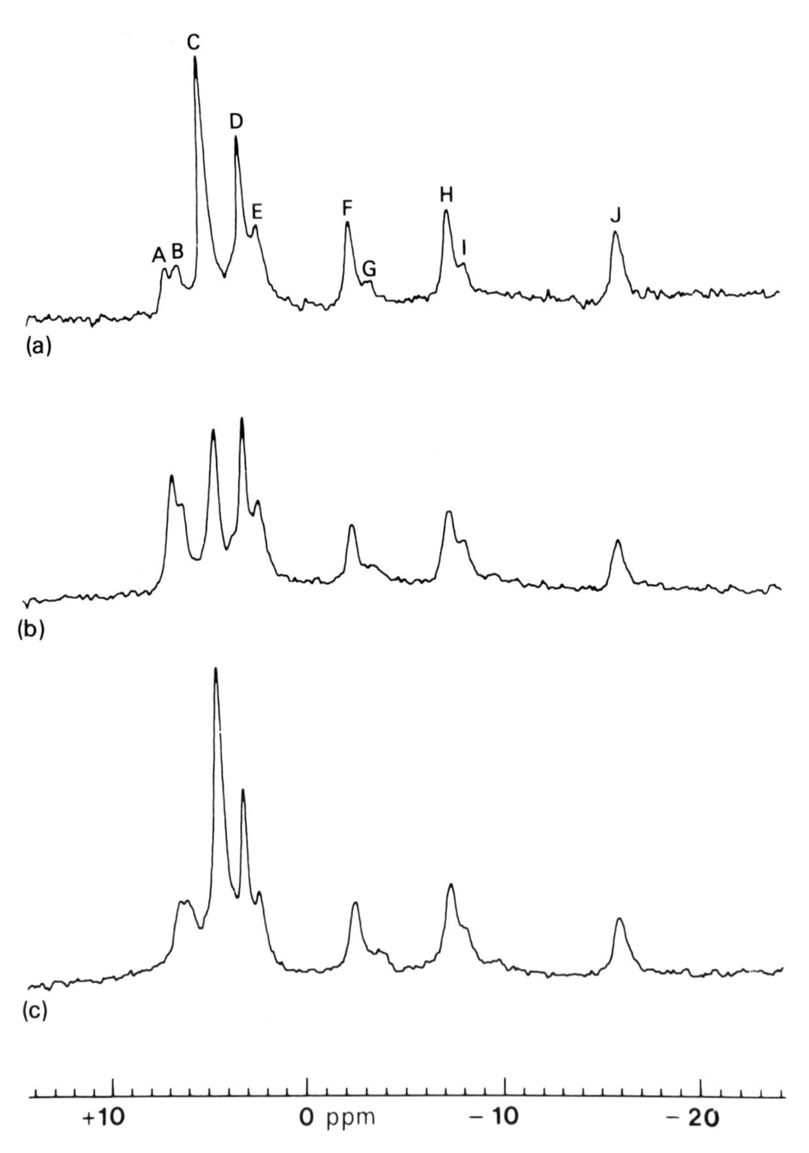

Fig. 2.22. The 81 MHz ^{31}P n.m.r. spectrum of fifteen liver flukes (a) in Krebs ringer solution; (b) in Krebs ringer containing 11 mM glucose recorded within 10 min of changing medium; (c) following addition of 1 mM serotonin. The chemical shifts are referenced to phosphocreatine and signal assignments are as follows: A, B sugar phosphates; C inorganic phosphate; D, E phosphodiesters; F γ-phosphate of ATP; G β-phosphate of ADP; H α-phosphate of ADP and ATP; I NAD/H, J β-phosphate of ATP. (From Mansour, Morris, Feeney, and Roberts 1982.)

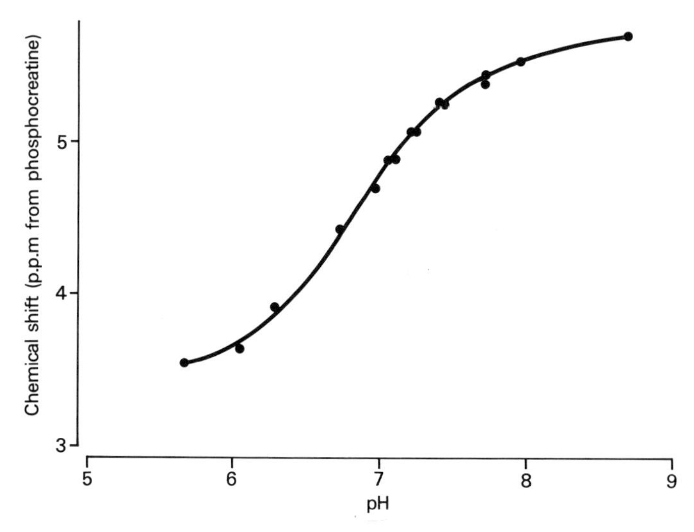

Fig. 2.23. Variation of the chemical shift of inorganic phosphate with solution pH at 37 °C. The medium contained 150 mM K^+, 5 mM Na^+, 1 mM Mg^{2+}. The chemical shift is expressed relative to phosphocreatine.

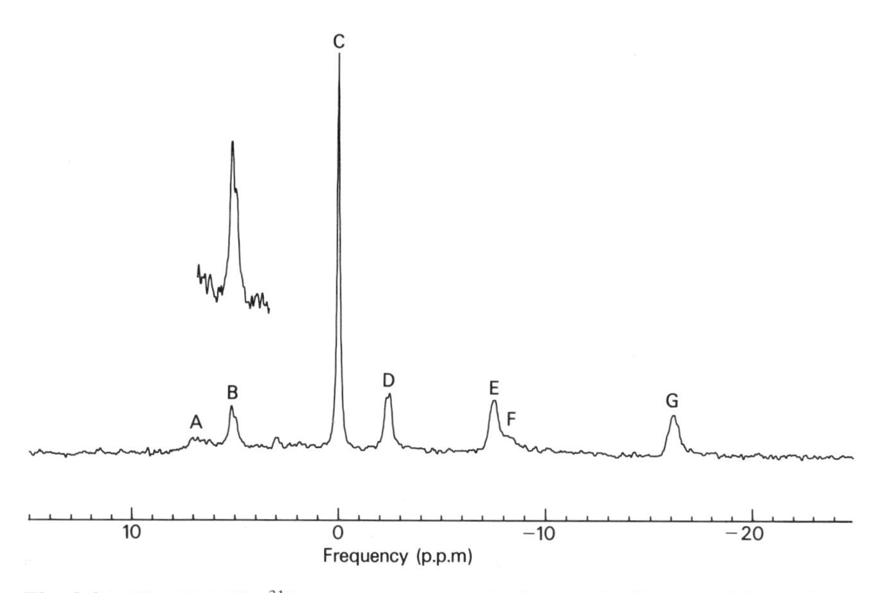

Fig. 2.24. The 81 MHz ^{31}P n.m.r. spectrum of a Langendorff perfused ferret heart (inset ×4). Signal assignments: A phosphomonoesters; B inorganic phosphate; C phosphocreatine; D, E, G γ-,α-,β-phosphates of ATP; F NAD/H.

nicotinamide dinucleotides, but there are important differences, notably the presence of phosphocreatine (PCr) at a concentration about 2.65 times that of ATP, and the virtual absence of any phosphodiesters (1.5 to 4.0 p.p.m. region of the spectrum). In addition, the areas of peaks D and G are the same within experimental error so in this case free ADP is not detected (ADP which was bound to sites on muscle fibres for example, would give a very broad signal which would be lost in the baseline noise. Its level can be assessed by n.m.r. analysis of perchloric acid extracts following freeze clamping of the organ. However, because the γ-phosphate of ATP is labile, this procedure is somewhat prone to error). The phosphocreatine acts as an energy store, transferring its high-energy phosphate group to ADP to give ATP which fuels the energy-consuming processes including, for example, muscular contraction. Thus:

$$\mathrm{PCr} + \mathrm{ADP} + \mathrm{H}^{+} \underset{}{\overset{\text{creatine kinase}}{\rightleftharpoons}} \mathrm{ATP} + \mathrm{Cr} \tag{2.69}$$

$$\mathrm{ATP} \rightarrow \mathrm{ADP} + \mathrm{P_i} + \Delta E. \tag{2.70}$$

There are reasons for believing that the reaction catalysed by the enzyme creatine kinase is in equilibrium (certainly this is the case in skeletal muscle). We can measure the concentrations of PCr and ATP, and if we assume they are in the same compartment as creatine kinase it is possible to calculate the concentration of free ADP. It turns out to be about 20 μM, substantially less than the 0.5 mM typically measured by the classical biochemical assays which determine total ADP. N.m.r. is the only method for estimating free ADP (or AMP) and it is the free ADP which is the controlling influence on the creatine kinase equilibrium, enabling the concentration of ATP to be maintained. The effectiveness with which this is done is beautifully illustrated in Fig. 2.25 which shows a ^{31}P n.m.r. time-course following introduction of 2 mM cyanide which blocks oxidative phosphorylation. Even when the PCr has been reduced to the level of background noise, no diminution of ATP is seen (see Orchard, Allen, and Morris 1985 for further details).

The inorganic phosphate peak in Fig. 2.24 is also of interest. It consists of two components: a sharp downfield spike corresponding to the phosphate in the buffer at pH 7.37 and a broader upfield line centred at pH 7.17 from the cardiac tissue itself. The amount of internal inorganic phosphate, is also very much less than that measured using conventional biochemical assays.

The beauty of such a system is that one can subject it to all manner of changes, for example in oxygen supply or substrate, and it is a valuable method for assessing the effects of drugs and other agents on cardiac metabolism. In our system the heart can be paced under computer control

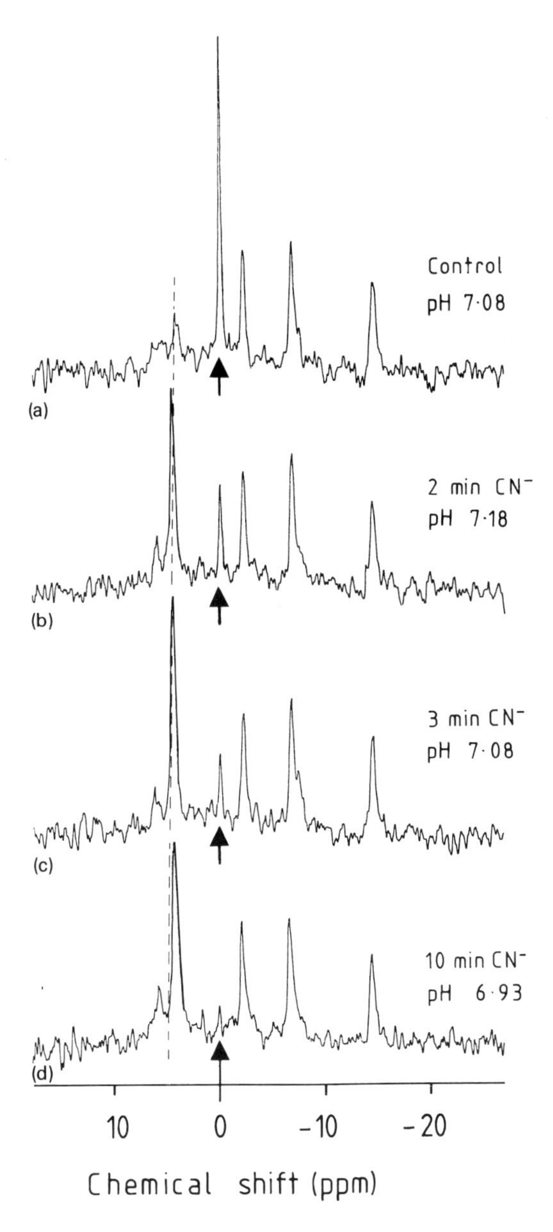

Fig. 2.25. 81 MHz ^{31}P n.m.r. spectrum of Langendorff perfused ferret heart. (a) Control spectrum; (b), (c), (d) time-course following addition of 2 mM NaCN. Note the slight alkalosis (shift to the left of the inorganic phosphate peak) in (b) prior to the more pronounced acidosis in (c), (d). See Fig. 2.24 for peak assignments. (From Orchard, Allen, and Morris 1985.)

allowing gated studies to be performed. Ventricular pressure is also monitored with the aid of a simple balloon catheter and transducer (Morris, Allen, and Orchard 1985).

It is possible to measure the unidirectional rate constants for eqn (2.69) by the methods of saturation or inversion transfer (see, for example, Forsén and Hoffman 1963; Campbell, Dobson, Ratcliffe and Williams 1978). These involve, respectively, destroying or inverting a single spectral line corresponding to either PCr or γATP (depending on whether the forward or reverse rate constant is required), and allowing a time for the spins to transfer from one chemical species to another, taking with them the 'memory' of the magnetic state in which they have been prepared, before acquiring the f.i.d. By varying the delay period it is possible to extract both the exchange and spin–lattice relaxation rates. For a discussion of rates determined in this way, see Ingwall (1982), Dawson (1983) and Dawson and Wilkie (1983).

Of course, neither of the above two experiments could be described as an imaging study since no element of spatial localization was involved and the entire sample therefore contributed to the signal.

The next logical development from a perfused system is to attempt to observe ^{31}P spectra from organs *in situ*. This clearly involves a degree of spatial selectivity and might justifiably be described as a point-imaging process. It can be carried out quite easily with a conventional magnet system by positioning the subject such that the organ of interest is located within the highly homogeneous central region. Then, if the organ is close to the periphery of the body—for example, a leg muscle—it can be studied with the aid of a surface coil (Ackerman, Grove, Wong, Gadian, and Radda 1980; see also § 3.2.2). If the organ of interest is more deeply embedded and there are intervening tissues which would contribute ^{31}P signals, minor surgery can be used to expose the organ which can then be surrounded by a conventional r.f. coil, if its shape permits (Ackerman, Bore, Gadian, Grove, and Radda 1980), or else a surface coil can again be used.

Figure 2.26(a) shows an example of an 81 MHz ^{31}P spectrum of a rat liver obtained by opening the abdomen and placing a two-turn, 1-cm-diam. surface coil directly onto the organ. A striking spectral feature is the underlying broad component (hump) which arises from the relatively immobile phospholipids (it is also observed in perfused brain slices (Cox, Morris, Feeney, and Bachelard 1983)). This phospholipid resonance can be removed, allowing the spectral features to be more easily discerned, by ^{31}P pre-irradiation for a period $\geqslant 1$ s as demonstrated Fig. 2.26(b). Phosphocreatine is absent and the adenosine nucleotide peaks are somewhat broader and their T_1 values shorter than is the case for other tissues, perhaps as a consequence of the higher concentrations of paramagnetic

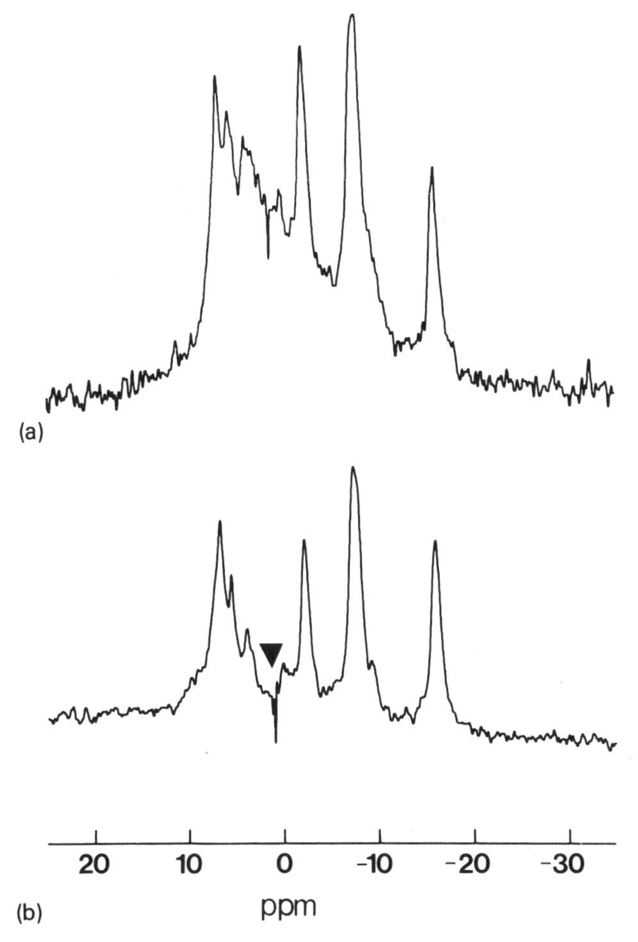

Fig. 2.26. 81 MHz ^{31}P n.m.r. spectrum of a rat liver obtained *in vivo* using a 1-cm-diameter surface coil. (a) Normal spectrum. (b) Spectrum obtained with 1.71 s ^{31}P pre-irradiation at the frequency indicated by the arrowhead. Assignments of principal peaks are from left to right; sugar phosphate; phosphodiesters, to left of arrow, and γ-ATP, β-ADP; α-ATP, ADP, NAD/H: β-ATP, to the right.

ions found in the liver. The time-course of metabolic events can also be followed, for example the production of fructose-1-phosphate following an infusion of fructose (Iles, Griffiths, Gadian, and Porteous 1980*a*; Iles, Griffiths, Stevens, Gadian, and Porteous 1980*b*). Various diseases of the liver can be investigated by this means, for example diabetes and glycogen storage disease.

Although all the regulatory mechanisms may remain intact when an organ is exposed, such a procedure can hardly be described as 'non-invasive'. It is, however, possible to add a further set of profiling coils to the main magnet in order to degrade the field away from the central region of high homogeneity in a controlled fashion. This technique was originally developed by Damadian and formed the basis of his f.o.n.a.r. imaging method (see § 3.2.1). It was later refined for high-resolution work by Oxford Research Systems under the name topical magnetic resonance or (t.m.r.) (see § 3.2.2). In these systems the position of the subject is again adjusted until the organ of interest coincides with the high-resolution resonance aperture. Signal emanating from this region yields a high-resolution spectrum, whereas that from the surrounding region contributes a broad low-resolution component which can be selectively removed using one of the many data enhancement routines available (see Gordon, Hanley, and Shaw 1982 for details). Additional spatial selectivity can be achieved provided the organ is favourably situated (i.e. near the surface) by the use of a surface coil in combination with (or instead of) the field profiling.

A 20 cm-bore horizontal superconducting t.m.r. system operating at 32 MHz for ^{31}P has been used to monitor the metabolic status of isolated human kidneys, allowing different storage media and conditions to be assessed and correlations to be made with renal performance following transplantation (Chan, French, Gadian, Morris, Radda, Bore, Ross, and Styles 1981; see also Bore, Sehr, Chan, Thulborn, Ross, and Radda 1981).

This magnet can also accommodate human limbs and has proved to be an excellent means for investigating muscular disorders. Figure 2.27 shows the TMR 32/200 system in the configuration used for such studies. The first clinical results were from a patient with a rare enzyme disorder known as McArdle's syndrome. This is a deficiency in muscle phosphorylase activity resulting in an inability to break down muscle glycogen to glucose (see Fig. 2.28). Such patients soon develop muscle fatigue on exercise. However, a 'second wind' is often possible through the use of blood borne glucose. Figure 2.29(a) shows a series of ^{31}P spectra obtained from the forearm of a normal subject undergoing a simple exercise regime (Ross, Radda, Gadian, Rocker, Esiri, and Falconer-Smith 1981). Under resting conditions the level of inorganic phosphate (see figure caption for peak assignments) is rather greater than for comparable animal studies but is nevertheless very much lower than measured by needle biopsy. Its shift relative to the PCr peak indicates a typical resting pH of 7.04. The muscle was subjected to ischaemic exercise with the aid of a sphygmomanometer cuff placed around the upper arm. Figure 2.29(a)(ii) shows the spectrum obtained during the first minute following the commencement of exercise. Since the production of ATP by the

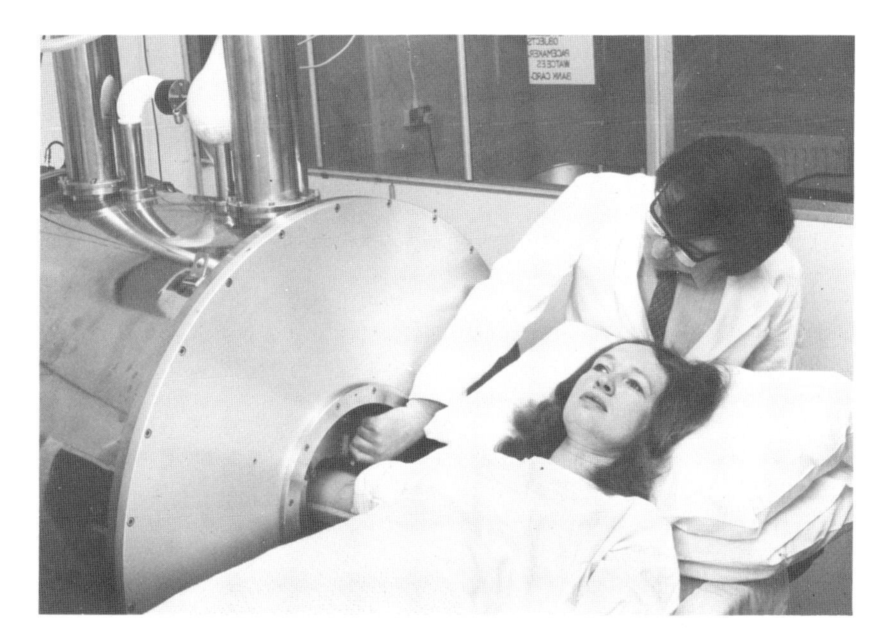

Fig. 2.27. A TMR 32/200 clinical n.m.r. spectrometer in the configuration used for studies of muscular disorders. (Oxford Research Systems, Abingdon, UK.)

Krebs cycle is blocked, the muscle is forced to derive it anaerobically via glycolysis and this leads to a build-up of lactic acid (see Fig. 2.28). The resulting acidosis is clearly demonstrated by the upfield shift of the P_i peak in (ii) to a position corresponding to a pH of 6.88. Note also the increased linewidth indicative of a spread of pH environments. After 1.5 minutes of exercise the pH falls to 6.43 and [PCr] to about one half of its resting value. On cessation of exercise and restoration of blood flow, the muscle returns to its resting state over a period of a few minutes. Figure 2.29(b) shows a similar series of spectra from the patient with McArdle's syndrome. The resting spectrum is similar, but, on commencement of exercise, the glycolytic pathway is unable to respond so there is no lactic acidosis evident in the ^{31}P n.m.r. spectrum; indeed, there is a slight alkaline shift which is probably due primarily to the proton absorption during PCr breakdown, This observation confirmed the diagnosis of McArdle's syndrome. Similar results have been seen in surface coil studies of phosphorylase kinase deficient I-strain mice (Stevens, Lutaya, Morris, Iles and Griffiths, 1983; see also the transient alkalosis due to PCr breakdown in Fig. 2.25).

The Oxford Group have also investigated examples of mitochondrial myopathy (arising from lesions in the electron transport chain) including

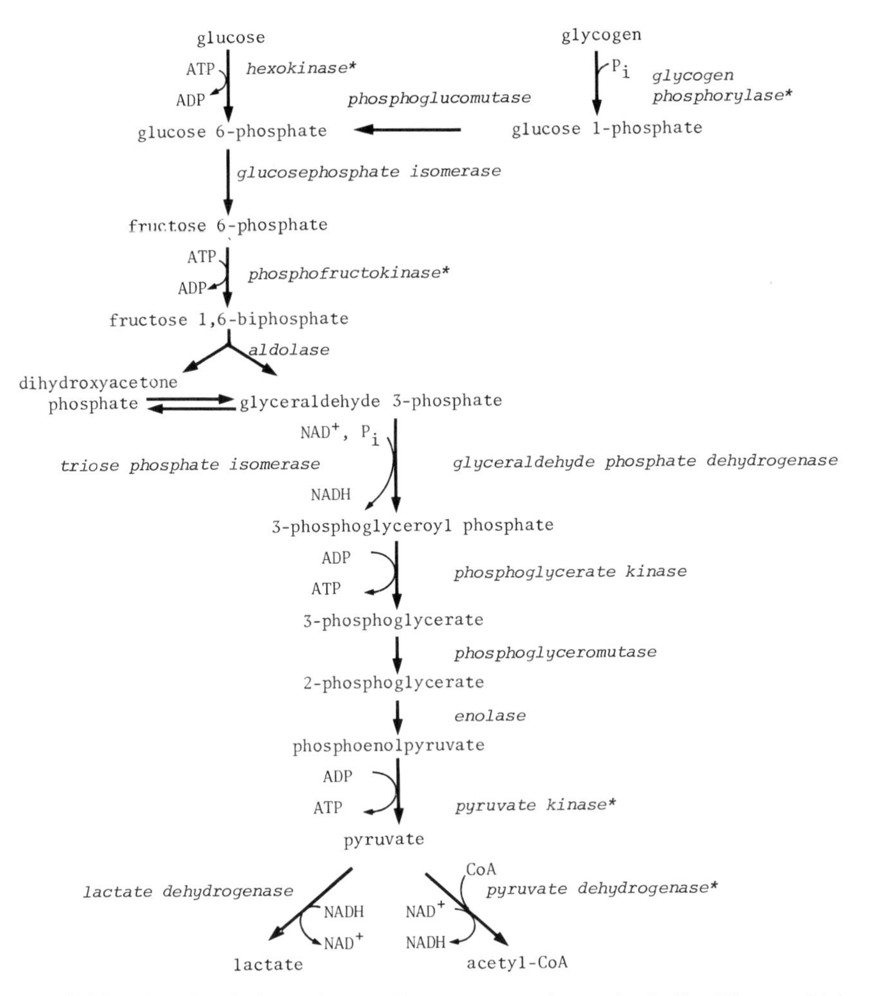

Fig. 2.28. The glycolytic pathway. Enzymes are shown in italic. Those which catalyse reversible reactions are indicated by an asterisk.

cases of NaDH-CoQ reductase deficiency (Gadian, Radda, Ross, Hockaday, Bore, Taylor and Styles, 1981; Radda, Bore, Gadian, Ross, Styles, Taylor, and Morgan-Hughes 1982).

Patients with phosphofructokinase deficiency have been studied (Edwards, Dawson, Wilkie, Gordon, and Shaw 1982; Chance, Eleff, Bank, Leigh, and Warnell 1982). Phosphofructokinase is the enzyme responsible for the conversion of fructose-6-phosphate to fructose 1,6-diphosphate (see Fig. 2.28) so that a deficiency leads to a blocking of glycolysis at this step. Mild aerobic exercise rapidly gives rise to a fiftyfold

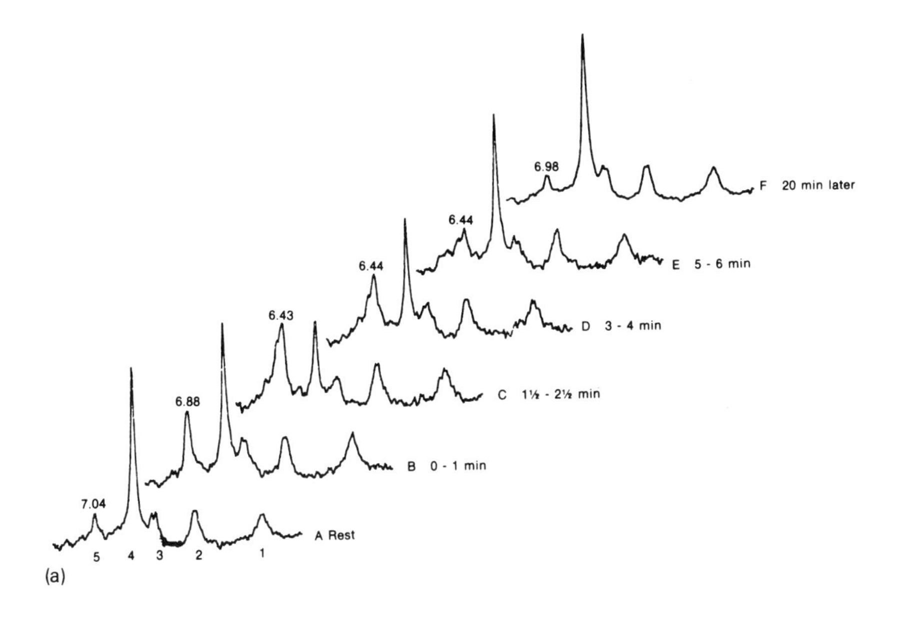

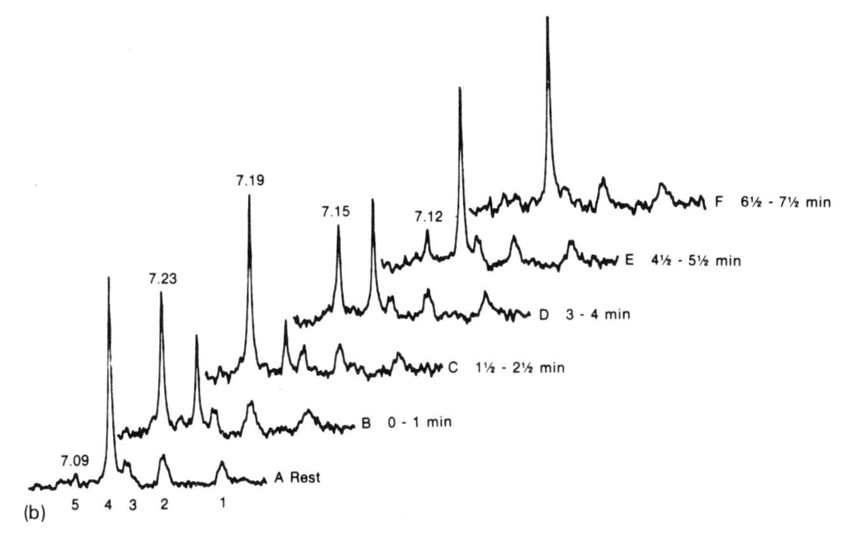

Fig. 2.29. (a) ^{31}P n.m.r. spectra obtained at 32.5 MHz from a normal human forearm. The first spectrum was recorded at rest, subsequent ones during and following ischaemic exercise which was maintained for $1\frac{1}{2}$ min. Arterial occlusion was continued to 3 min after which arterial flow was restored. (b) ^{31}P n.m.r. spectra from a patient with McArdle's syndrome. In this case exercise was maintained for $\frac{3}{4}$ min and arterial occlusion for 3 min. Measured pH values are given above each inorganic phosphate peak. (From Ross *et al.* 1981. Reprinted by permission of *The New England Journal of Medicine* **304,** 1338 (1981).)

increase in hexose phosphate levels, though in contrast to McArdle's syndrome there is virtually no change in high-energy phosphate (phosphocreatine and ATP) levels or in pH at the onset of muscular fatigue and pain. Patients with Duchenne dystrophy (the commonest form of muscular dystrophy) have been examined using both ^{1}H and ^{31}P n.m.r. (Newman, Bore, Chan, Gadian, Styles, Taylor, and Radda 1982). A number of abnormalities were observed: first, the ^{1}H spectrum showed a much higher proportion of fat relative to water, indicating the expected replacement of muscle tissue by fat; secondly, the concentrations of high-energy phosphates, estimated by division of the relevant ^{31}P peak areas by the ^{1}H water peak area, were much lower (5.5-fold reduction for PCr, 3.5-fold for ATP: however, this could be due to a higher water content in the connective tissues); thirdly, the proportion of PCr to ATP and of PCr to P_i was reduced, and fourthly, an extra peak, probably corresponding to glycerophosphorylcholine appeared in the phosphodiester region of the spectrum. In dystrophic chicken muscle, two such compounds have been identified, namely glycerophosphorylcholine and serine ethanolaminephosphodiester (Chalovich, Burt, Danon, Glonek, and Barany 1979; see also Fig. 2.22). It now seems less likely that phosphodiesters are specifically related to muscular dystrophy. Chalovich *et al.* (1979) failed to observe any in biopsies of dystrophic muscle and Burt, Pluskal, and Sreter (1982) have demonstrated that they are associated more with slow (type I) muscle fibres than with fast (type II) ones.

Another example of the clinical use of ^{31}P n.m.r. spectroscopy is in the study of paediatric brain. Cady, Costello, Dawson, Delpy, Hope, Reynolds, Tofts, and Wilkie (1983) have found that the ratio of phosphocreatine to inorganic phosphate levels is a sensitive indicator of cortical hypoxia: in the brain of normal babies this ratio has a value of 1.7 whereas in cases of severe birth asphyxia it generally lies in the range 0.2–1.0. It is restored as the clinical condition improves though in cases where it falls below 0.5 the prognosis is not good. Very high levels of a phosphomonoester, thought probably to be *o*-phosphorylethanolamine, are found in human neonatal brain (Chance, Younkin, Eleff, Warnell, and Delivoria-Pappadopoulos 1983). This substance lies on the pathway of lipid synthesis. Peripheral surface tumours have also been the subject of clinical ^{31}P surface coil investigations (Griffiths, Cady, Edwards, McCready, Wilkie, and Wiltshaw 1983). Animal tumour studies have indicated the possibilities for staging tumours and following their response to treatment (Ng, Evanochko, Hiramoto, Ghanta, Lilly, Lawson, Corbett, Durant, and Glickson 1982). Such methods are likely to become of increasing importance now that wide-bore high-resolution magnets are available for clinical use. Similarly, cardiovascular and other applications can now be explored and the full clinical potential of the method should soon become clear.

2.7.2. Nuclei other than ^{31}P

^{31}P n.m.r. spectra were the first to be studied using surface coil and t.m.r. type methods; there are however other n.m.r. nuclei worthy of attention. Thus, although ^{1}H n.m.r. spectra of tissues are dominated by the 100 M water resonance, it is nevertheless possible to use standard solvent suppression methods in order to observe the metabolites present at much lower concentrations (typically a factor of 10^5 down on water). For example, Shulman's group have used simple presaturation of the water signal to look at the *in vivo* rat brain using a surface-coil method at 360 MHz (Behar, Den Hollander, Stromski, Ogino, Shulman, Petroff, and Prichard 1983). Many resonances were identified including lactate, creatine, phosphocreatine, phosphoryl choline, and a number of amino acids.

One of the most informative nuclei to study is ^{13}C which, as we have seen, has a low n.m.r. sensitivity (low γ) and a low natural abundance (~1 per cent). Nevertheles, high-resolution natural abundance ^{13}C spectra can be obtained from biological tissues (Ugurbil, Brown, Den Hollander, Glynn, and Shulman 1978; Den Hollander, Brown, Ugurbil, and Shulman 1979; Cohen, Shulman, and McLoughlin 1979; Cohen, Ogawa, and Shulman 1979; Cohen, Glynn, and Shulman 1981*a*; Cohen, Rognstad, Shulman, and Katz 1981*b*; Alger, Sillerud, Behar, Gillies, Shulman, Gordon, Shaw, and Hanley 1981; Doyle, Chalovich, and Barany 1981; Edwards, Dawson, Wilkie, Gordon, and Shaw 1982; Stevens, Iles, Morris, and Griffiths 1982). An example is shown in Fig. 2.30. The comparatively narrow resonances must originate from molecules in rapid motion and correspond principally to the fats in triglyceride droplets and in the mobile components of membranes. The upper three spectra were obtained using proton decoupling (see § 2.6.2) but the last, a spectrum of a human arm, was obtained undecoupled due to the uncertainty concerning the extent of the r.f. heating effect with the relatively high powers required (see also comments of § 6.8.3). Four principal groups of signals can be discerned, corresponding to CH_3— (~14 p.p.m.), —CH_2— (~30 p.p.m.), —C=C— (~130 p.p.m.) and C=O (~170 p.p.m.). Note that in the undecoupled spectrum (Fig. 2.30(d)) the 1 : 2 : 1 triplet structure of the —CH_2— groups and the 1 : 1 doublet of the double-bonded carbons =C(H)— is apparent (see § 2.6.2). Smaller signals are also observed from choline —$N(CH_3)_3$, glycerol (two signals from

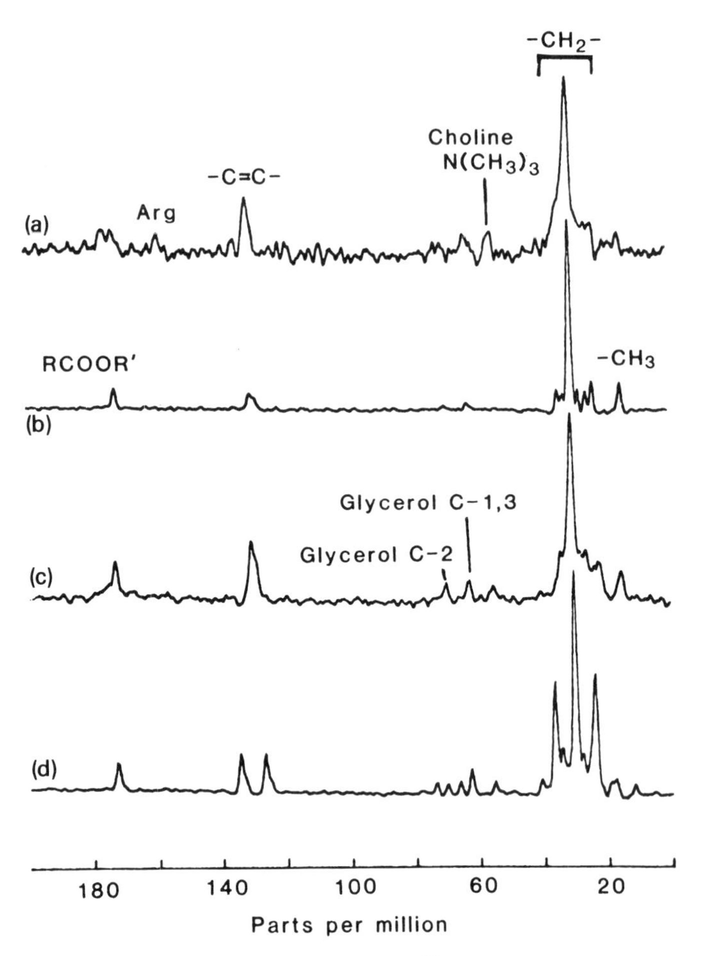

Fig. 2.30. *In vivo* 20.2 MHz natural abundance ^{31}C n.m.r. spectra of mammalian tissue obtained on a TMR 32 spectrometer using a 2.5-cm-diameter surface coil. (a), (b), (c) Proton decoupled spectra from head, abdomen, and hind leg respectively of an anaesthetized rat. (d) Undecoupled spectrum of human forearm. (From Alger *et al.* 1981. Copyright 1981 by the AAAS.)

carbons C1, C3, and C2 respectively) and, in the case of the brain, arginine, presumably a constituent of the basic brain protein myelin. By measuring the relative intensities of the signals it is possible to assess the relative amounts of fatty and membrane material (Alger *et al.* 1981) and to determine the composition of the fatty acids which may be of importance in establishing nutritional fat deficiencies and for the study of diseases such as muscular dystrophy (Edwards *et al.* 1982).

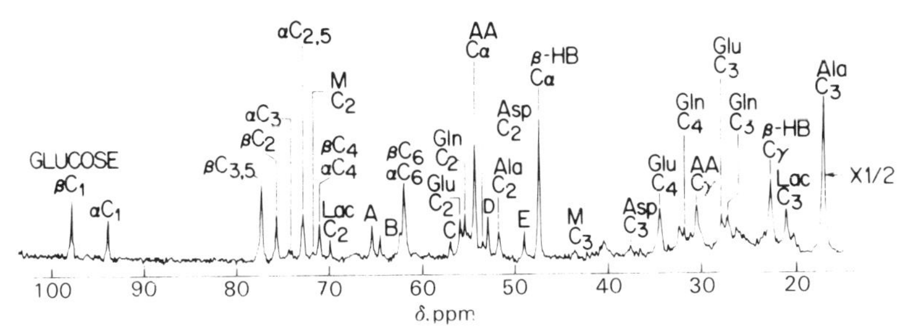

Fig. 2.31. ^{13}C n.m.r. spectra at 25°C. Hepatocytes from a euthyroid rat. This spectrum is the sum of 3000 scans accumulated over the interval 140–230 min following addition of 28 mM [3-^{13}C] alanine and 8 mM, D,L-β-hydroxybutyrate to the suspension of cells. Assignments: βC_1, αC_1–βC_6, and αC_6, carbons of glucose anomers; MC_2, malate C2; $LacC_2$, lactate C2; AACα, acetoacetate C2; β-HBCα, D-β-hydroxybutyrate C2; AACγ acetoacetate C4; β-HBCγ, D-β-hydroxybutyrate C4. Peaks A–E arise from the buffer. (From Cohen *et al.* 1981*a*).

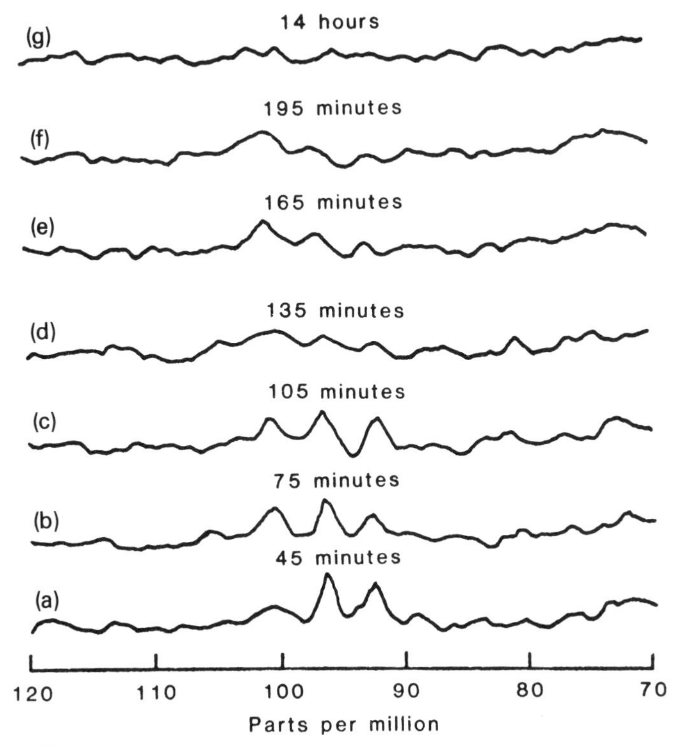

Fig. 2.32. ^{13}C spectra obtained *in vivo* from a rat abdomen following feeding with D-[1-^{13}C] glucose. Each spectrum took 0.5 h; times above spectra refer to interval between feeding and acquisition. Signals at 101, 96.8, and 42.3 p.p.m. arise from C1 carbons of glycogen and β and α anomers of D-glucose respectively. (From Alger *et al.* 1981. Copyright 1981 by the AAAS.)

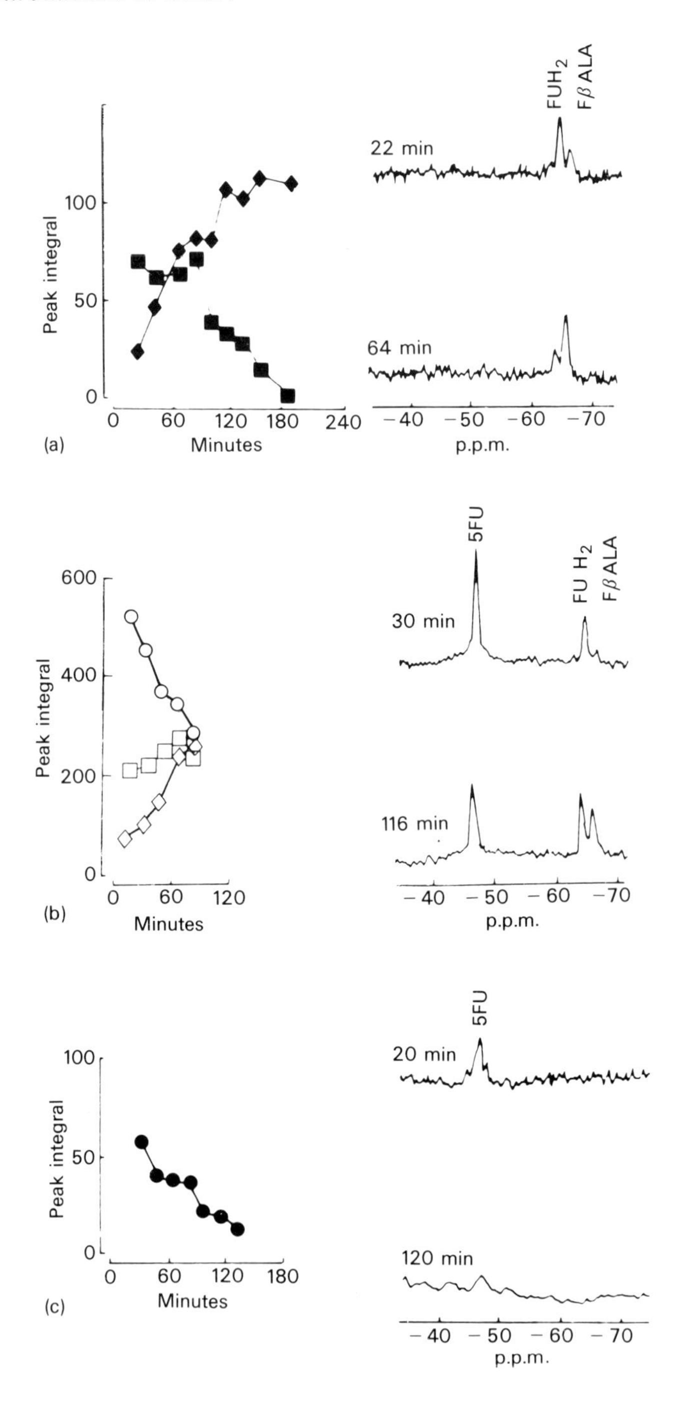

FUH2
FβALA
22 min
64 min
Peak integral
100
50
0
0 60 120 180 240
Minutes
(a)
−40 −50 −60 −70
p.p.m.
5FU
FU H2
FβALA
30 min
116 min
600
400
200
0
0 60 120
Minutes
(b)
−40 −50 −60 −70
p.p.m.
5FU
20 min
120 min
100
50
0
0 60 120 180
Minutes
(c)
−40 −50 −60 −70
p.p.m.

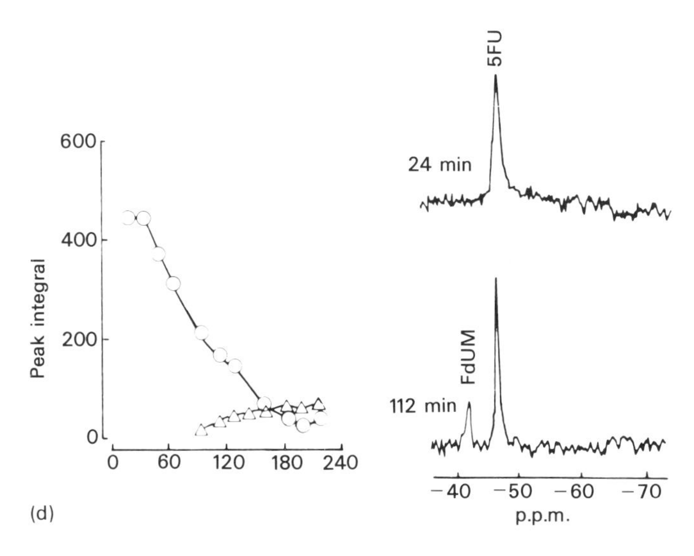

Fig. 2.33. Uptake of 5-fluorouracil (5FU) into liver (a, b) and implants of Lewis lung tumour (c, d) in C57 mice monitored by ^{19}F n.m.r. spectroscopy after i.v. injection of 30 mg/kg (a, c) and 180 mg/kg (b, d) of 5FU into the jugular vein. Times refer to interval following injection of 5FU. Probable peak assignments ●, ○ 5FU; ■, □ dihydrofluorouracil (FUH_2); ◆, ◇ fluoro-β-alanine (FβALA); ▲ 5-fluorodeoxyuridine; △ 5-fluorodeoxyuridine monophosphate (FduMP). (From Stevens *et al.* 1984.)

Of course, the important metabolites also give rise to ^{13}C resonances. However, since their concentrations are generally low (~5 mM or less) isotropic enrichment is normally required. Unfortunately, the high price of ^{13}C-labelled compounds (currently in excess of £500 per gram for ^{13}C-labelled glucose) has restricted their use, for the most part, to cellular systems. Since the ^{13}C chemical-shift range is large (~200 p.p.m.) the resonances are well resolved (one can easily separate the signals from the α- and β-anomers of glucose, for example) and it is possible to follow the metabolism of a labelled substance introduced into the system. By selectively labelling different carbon atoms, reaction mechanisms can be elucidated and fluxes through different branches of the metabolic pathways measured. The 'richness' inherent in ^{13}C spectra is well illustrated in Fig. 2.31, taken from the work of Cohen and co-workers (Cohen *et al.* 1981*a*). See the figure caption for assignments. Some exploratory ^{13}C-labelled studies of perfused organs, for example mouse liver (Cohen *et al.* 1979) have been performed and other *in vivo* experiments have also been reported. For example, Fig. 2.32 shows a small section from the ^{13}C spectrum of a rat abdomen recorded in half-hour blocks following the introduction of D-1-^{13}C glucose into the animal's stomach at time zero. The conversion of the glucose (two upfield signals corresponding to α (92.3 p.p.m.) and β (96.8 p.p.m.) anomers respectively) into glycogen (101 p.p.m.) is readily followed. Recently, it has been possible to observe natural abundance ^{13}C spectra *in vivo* from the livers of normal rats and rats with glycogen storage disease (a deficiency in liver phosphorylase resulting in an inability to metabolize glycogen). In the latter case, the glycogen concentration is some threefold higher (Stevens *et al.* 1982). ^{13}C and ^{31}P acquisition can be interleaved (Styles, Grathwohl, and Brown 1979). For a beautiful demonstration of the power of this approach see the study of perfused rat liver by Cohen (1983).

The routine use of ^{13}C-labelled components for human studies is an extremely attractive, but prohibitively expensive proposition. This is also true of ^{15}N, the low abundance (0.37 per cent) isotope of nitrogen which gives well-resolved resonances from the different amino acids, for example.

^{19}F is an interesting nucleus: it is the only naturally-occurring isotope of fluorine and has a sensitivity approaching that of the proton (see Table 2.1). It is not present in the body to any significant extent and is therefore an ideal candidate for tracer work—for example, lung ventilation studies, or imaging of the vascular system. It is also interesting from the spectroscopic viewpoint, since a number of important drugs contain fluorine. Figure 2.33, for example, shows a ^{19}F surface coil study of the uptake of 5-fluorouracil in liver (a, b) and Lewis lung tumours (c, d) implanted in C57 mice (Stevens, Morris, Iles, Sheldon, and Griffiths 1984). In the liver, only breakdown products are seen, whereas in the tumour, the anabolic pathway is manifest. The demonstration that one can monitor a drug at its site of action may ultimately prove to be of very considerable importance to the pharmaceutical industry. If the drug or metabolite in which one is interested does not contain fluorine then it is possible to produce a fluorine-labelled analogue for such studies.

Fluorine also occurs in an important class of anaesthetics of which halothane is a well-known example. Surface-coil techniques have been used to follow the time-course of its uptake and elimination in rabbit brain (Wyrwicz, Pszenny, Schofield, Tillman, Gordon, and Martin 1983). Retention times were found to be much longer than expected. Additionally, 'solvent' shifts were observed indicating that halothane is distributed in a number of membrane environments in the brain. Such studies have important bearing on the poorly-understood 'science' of anaesthesia.

^{19}F-labelled chelators are available which permit measurement of the intracellular concentrations of many of the biologically-important cations, for example H^+, Ca^{2+}, Mg^{2+}, Fe^{2+}, Zn^{2+}, Na^+ and K^+, through the chemical shift they induce on binding (Smith, Hesketh, Metcalfe, Feeney,

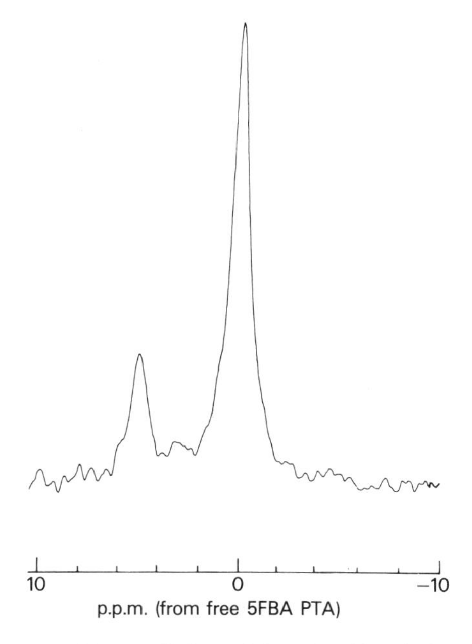

Fig. 2.34. Measurement of intracellular calcium level in pig lymphocytes using ^{19}F n.m.r. The larger peak is due to free 5-fluoro 1,2-bis(*o*-aminophenoxy)ethane-*N,N,N′,N′*-tetraacetic acid (5FBAPTA) and the smaller one is from calcium-bound 5FBAPTA. The intracellular free calcium concentration is determined from the ratio of the areas of the two peaks, and in this case is found to be 145 nM.

and Morris 1983). Methods for loading these chelators into cells and perfused organs have been developed and an example of intracellular calcium determination in pig lymphocytes is shown in Fig. 2.34. Of course, some of these cations are 'visible' directly by n.m.r. techniques, $^{23}Na^{+}$ for example, or, with isotopic enrichment $^{43}Ca^{2+}$ and $^{25}Mg^{2+}$ (see Fossel 1983). However, one then has the difficulty of distinguishing intra- from extracellular pools—a problem which can be attacked with the use of shift reagents which do not penetrate the cell membrane.

References

Abragam, A. (1961). *The principles of nuclear magnetism.* Clarendon Press, Oxford.

Ackerman, J. J. H., Bore, P. J., Gadian, D. G., Grove, T. H., and Radda, G. K. (1980). *Phil. Trans. R. Soc.* B **289,** 425.

———, Grove, T. H., Wong, G. G., Gadian, D. G., and Radda, G. K. (1980). *Nature, Lond.* **283,** 167.

Alger, J. R., Sillerud, L. O., Behar, K. L., Gillies, R. J., Shulman, R. G., Gordon, R. E., Shaw, D., and Hanley, P. E. (1981). *Science* **214,** 660.

Behar, K. L., Den Hollander, J. A., Stromski, M. E., Ogino, T., Shulman, R. G., Petroff, O. A. C., and Prichard, J. W. (1983). *Proc. natn. Acad. Sci. U.S.A.* **80** (16) 4945.

Bittoun, J., and Taquin, J. (1982). *C.R. Acad. Sci.* Ser. 2, **295,** 649.

Bore, P. J., Sehr, P. A., Cahn, L., Thulborn, K. R., Ross, B. D., and Radda, G. K. (1981). *Transplantatn Proc.* **13,** 707.

Bottomley, P. A., Hart, H. R., Edelstein, W. A., Schenck, J. F., Smith, L. S., Leue, W. M., Mueller, O. M., and Redington, R. W. (1983). *Lancet* **ii** (8344), 273.

Burt, C. T., Pluskal, M. G., and Sreter, F. A. (1982) *Biochim. biophys. Acta* **721,** 492.

Cady, E. B., Costello, A. M. de L., Dawson, M. J., Delpy, D. T., Reynolds, E. O. R., Tofts, P. S., and Wilkie, D. R. (1983) *Lancet* **i**, 1059.

Campbell, I. D., Dobson, C. M., Ratcliffe, R. G., and Williams, R. J. P. (1978). *J. magn. Reson.* **29,** 397.

Carr, H. Y., and Purcell, E. M. (1954). *Phys. Rev.* **94,** 630.

Chalovich, J. M., Burt, T. B., Danon, M. J., Glonek, T., and Barany, M. (1979). *Ann. N.Y. Acad. Sci.* **317,** 649.

Chan, L., French, M. E., Gadian, D. G., Morris, P. J., Radda, G. K., Bore, P. J., Ross, B. D., and Styles, P. (1981). In *Organ transplantation III.* (ed. D. E. Pegg, I. Jacobson and N. A. Halasz). MTP Press Ltd.

Chance, B., Eleff, S., Bank, W., Leigh, J. S., and Warnell, R. (1982). *Proc. natn. Acad. Sci. U.S.A.* **79,** 7714.

———, Younkin, D., Eleff, S., Warnell, R., and Deliveroria-Pappadopoulos, M. (1983). *Pediat. Res.* **17,** 307a.

Cohen, S. M. (1983). *J. biol. Chem.* **258** (23) 14294.

———, Glynn, P., and Shulman, R. G. (1981*a*). *Proc. natn. Acad. Sci. U.S.A.* **78,** 60.

———, Ogawa, S., and Shulman, R. G. (1979). *Proc. natn. Acad. Sci. U.S.A.* **76,** 1603.

———, Shulman, R. G., and McLaughlin, A. C. (1979). *Proc. natn. Acad. Sci. U.S.A.* **75,** 3742.

———, Rognstad, R., Shulman, R. G., and Katz, J. (1981*b*). *J. biol. Chem.* **256,** 3428.

Cox, D. W. G., Morris, P. G., Feeney, J., and Bachelard, H. S. (1983). *Biochem. J.*, **212,** 365.

Dawson, M. J. (1983). 'Nuclear magnetic resonance', in *Cardiac metabolism.* (ed. A. J. Drake and M. I. M. Noble). John Wiley, Chichester.

———, and Wilkie, D. R. (1983). 'Muscle and brain metabolism studied by ^{31}P nuclear magnetic resonance', in *Recent advances in physiology*, pp. 247–276. (ed. P. Baker). Churchill-Livingstone. Edinburgh.

Den Hollander, J. A., Brown, T. R., Ugurbil, K., and Shulman, R. G. (1981). *Proc. natn. Acad. Sci. U.S.A.* **75,** 1603.

Doyle, D. D., Chalovich, J. M., and Barany, M. (1981). *FEBS Lett.* **131,** 147.

Edwards, R. H. T., Dawson, M. J., Wilkie, D. R., Gordon, R. E., and Shaw, D. (1981). *Lancet* **i**, 725.
Farrar, T. C. and Becker, E. D. (1971). *Pulse and Fourier transform NMR.* Academic Press, New York.
Forsén, S., and Hoffman, R. A. (1963). *J. chem. Phys.* **39,** 2892.
Fossel, E. T. (1983). In 'Work in Progress', Society of Magnetic Resonance in Medicine, August meeting.
Gabillard, R. (1951). *C.R. Acad. Sci. (Paris)* **232,** 1551.
Gadian, D. G. (1982). *Nuclear magnetic resonance and its application to living systems.* Clarendon Press, Oxford.
———, and Radda, G. K. (1981). *A. Rev. Biochem.* **50,** 69.
———, ———, Ross, B. D., Hockaday, J., Bore, P. J., Taylor, D. J., and Styles, P. (1981). *Lancet* **ii,** 774.
Gordon, R. E., Hanley, P. E., and Shaw, D. (1982). *Prog. NMR Spectrosc.* **15,** 1.
Griffiths, J. R., Cady, E., Edwards, R. H. T., McCready, V. R., Wilkie, D. R., and Wiltshaw, E. (1983). *Lancet* **i,** 1435.
Gupta, R. J., and Moore, R. D. (1980). *J. biol. Chem.* **255,** 3987.
Haeberlen, U. (1976). High resolution NMR in solids: selective averaging, in *Advances in magnetic resonance,* Suppl. 1. Academic Press, New York.
Hahn, E. L. (1950). *Phys. Rev.* **77,** 297.
Hoult, D. I., and Richards, R. E. (1976). *J. magn. Reson.* **24,** 71.
———, Busby, S. J. W., Gadian, D. G., Radda, G. K., Richards, R. G., and Seeley, P. J. (1974). *Nature, Lond.* **252,** 285.
Iles, R. A., Stevens, A. N., and Griffiths, J. R. (1982). *Prog. NMR Spectrosc.* **15,** 49.
———, Griffiths, J. R., Gadian, D. G., and Porteous, R. (1980*a*). *Clin. Sci. mol. Med.* **58,** 2P.
———, ———, Stevens, A. N., Gadian, D. G., and Porteous, R. (1980*b*). *Biochem. J.* **192,** 191.
Ingwall, J. S. (1982). *Am. Physiol. Soc.* **242,** H729.
Mansfield, P., and Grannell, P. K. (1973). *J. Phys.* C **6,** L422.
———, and ——— (1975). *Phys. Rev.* **12,** 3618.
———, and Morris, P. G. (1982). 'NMR imaging in biomedicine', in *Advances in magnetic resonance,* suppl. 2 (ed. J. S. Waugh). Academic Press, New York.
Mansour, T. E., Morris, P. G., Feeney, J., and Roberts, G. C. K. (1982). *Biochim. biophys. Acta* **721,** 336.
Meiboom, S., and Gill, D. (1958). *Rev. scient. Instrum.* **29,** 688.
Moon, R. B., and Richards, J. H. (1973). *J. biol. Chem.* **248,** 7276.
Morris, P. G., Allen, D. G., and Orchard, C. H. (1985). In *Adv. Myocardiology* **5,** 27.
Newman, R. J., Bore, P. J., Chan, L., Gadian, D. G., Styles, P., Taylor, D., and Radda, G. K. (1982). *Brit. med. J.* **284,** 1072.
Ng, T. C., Evanochko, W. T., Hiramoto, R. N., Ghanta, V. K., Lilly, M. B., Lawson, A. J., Corbett, T. H., Durant, J. R., and Glickson, J. D. (1982). *J. magn. Reson.* **49,** 271.
Orchard, C. H., Allen, D. G., and Morris, P. G. (1985). In *Adv. Myocardiology,* **5,** 417.

Radda, G. K., Bore, P. J., Gadian, D. G., Ross, B. D., Styles, P., Taylor, D. J., and Morgan-Hughes, J. (1981). *New Engl. J. Med.* **304,** 1338.

Ross, B. D., Radda, G. K., Gadian, D. G., Rocker, G., Esiri, M., and Falconer-Smith, J. (1981). *New Engl. J. Med.* **304,** 1338.

Smith, G. A., Hesketh, R. T., Metcalfe, J. C., Feeney, J., and Morris, P. G. (1983). *Proc. natn Acad. Sci. USA* **80,** 7178.

Stevens, A. N., Iles, R. A., Morris, P. G., and Griffiths, J. R. (1982). *FEBS Lett.* **150,** 489.

———, Lutaya, G. L., Morris, P. G., Iles, R. A., and Griffiths, J. R. (1983). *Biochem. Soc. Trans.* **11,** 92.

———, Morris, P. G., Iles, R. A., Sheldon, P. W., and Griffiths, J. R. (1984). *Brit. J. Cancer,* **50,** 113.

Styles, P., Grathwohl, C., and Brown, F. F. (1979). *J. magn. Reson.* **35,** 329.

Ugurbil, K., Brown, T. R., Den Hollander, J. A., Glynn, P., and Shulman, R. G. (1978). *Proc. natn Acad. Sci. U.S.A.* **75,** 3742.

Wind, R. A., and Yannoni, C. S. (1979). *J. magn. Reson.* **36,** 269.

Wyrwicz, A. M., Pszenny, M. H., Schofield, J. C., Tillman, P. C., Gordon, R. E., and Martin, P. A. (1983). *Science* **222** (4622) 428.

3. Point and line imaging methods

3.1. Introduction

Since the first publications in 1973, the field of n.m.r. imaging has expanded rapidly to the extent that over twenty different techniques have now been reported in the literature. We shall attempt to at least mention most of these. However, the principal emphasis will be on those techniques, such as Fourier imaging, which are currently finding application in commercial systems or which offer the promise of further extending the range of n.m.r. imaging, for example by improving the speed and resolution or by allowing the measurement of further n.m.r. parameters. See Ljunggren (1983*a*) for a simple graphical description of a number of n.m.r. imaging methods.

The techniques will be divided into four categories: point, line, planar, and full three-dimensional imaging, according to the manner in which the image data points are acquired. In a point method, for example, each individual n.m.r. signal or f.i.d. carries information about a single image point, whereas in a line method the signal contains information about all the points making up an image line. Planar and three-dimensional methods are defined by obvious extension. However, the planar methods can often be derived from their corresponding three-dimensional method (usually by the simple expedient of replacing one stage of the imaging process by a selective excitation) and we therefore treat the two cases together in the subsequent chapter.

It should be clear from their ability to receive signal simultaneously from the entire region of interest that planar and three-dimensional techniques will be more efficient than a point or line method in terms of image S/N per unit time and will therefore be preferred in most situations. Point methods do, however, have the advantage of experimental simplicity: the sensitive point method, for example, was originally implemented without the benefit of computer or microprocessor control, a situation almost unthinkable in the case of echo-planar imaging. As well as requiring less hardware, making the imaging system potentially cheaper, point methods also make less stringent demands on magnet design. The homogeneity need only be good over a very small region corresponding to an image point. Indeed, as we shall see below, the f.o.n.a.r. type methods require that the field be homogeneous *only* at the image point. Topical magnetic resonance (t.m.r.) a method developed in

1980 by Oxford Research Systems, takes advantage of the greater homogeneity available over a restricted volume to measure *in vivo* ^{1}H, ^{13}C, ^{19}F, ^{31}P and other chemical shifts, enabling the major metabolites to be identified and measured. Although this facility can be incorporated into the more efficient imaging techniques, the overall homogeneity currently available from whole-body magnets (see Chapter 5) is barely sufficient for this purpose.

3.2. Point methods

3.2.1. F.o.n.a.r. and related methods

F.o.n.a.r. or *f*ield f*o*cused *n*uclear m*a*gnetic *r*esonance was the name given by Damadian and colleagues to their (point) method of n.m.r. imaging (Damadian, Minkoff, Goldsmith, Stanford, and Koutcher 1976*a*, *b*). It relies on the use of shaped magnetic fields and r.f. pulses to isolate a small signal-producing region or 'resonance aperture' within an extended sample.

The principle of the method will be familiar to any n.m.r. spectroscopist who has worked with instruments in which proper sample alignment is the responsibility of the operator. Such a person will know that any movement of the sample away from the central homogeneous region of the magnet causes the f.i.d. to be curtailed. This is because different regions of the sample are then located in different magnetic fields. The n.m.r. signal therefore contains a wide range of frequency components which rapidly dephase giving a shortened f.i.d. and a correspondingly broad lineshape on Fourier transformation. A sample which is sufficiently large to extend beyond the homogeneous region will give an n.m.r. signal which, to a first approximation, consists of two components—one which is long-lived (i.e. long T_{2e}) from the central homogeneous region, and one which is short-lived from the surrounding non-homogeneous region. It is possible to separate these two signals by means of a convolution difference method (Campbell, Dobson, Williams, and Xavier 1973) or similar procedure (Gordon, Hanley, and Shaw 1982). Alternatively, one can use a selective pulse (see § 3.3.1) with a sufficiently narrow bandwidth to excite only those spins lying within the resonance aperture. Either way one effectively isolates these spins and an n.m.r. signal can be selectively recorded from them. By relocating the resonance aperture or, more simply, by moving the sample relative to the magnet, an image can be built up point by point. This is the essence of the f.o.n.a.r. technique.

Whilst one can create rather crude resonance apertures by relying on the basic inhomogeneity of the magnet, finer definition can be obtained through the addition of field-profiling coils, which have the additional advantage that the spatial resolution can then be varied by altering the current passing through them. Thus, with a prototype instrument used for animal studies and based on a conventional spectrometer operating at 10 MHz for ^{1}H, Damadian and colleagues (1976*a*, *b*) were able to vary the size of the aperture over the range 0–15 mm.

In order to produce a f.o.n.a.r. scan, the subject undergoes a series of lateral displacements as illustrated in Fig. 3.1(a). At each location a number of f.i.d.s are accumulated to achieve a satisfactory S/N. After completion of an image line, the subject is displaced in an orthogonal direction and a new image line is scanned (Fig. 3.1(b)). Such mechanical movement of the subject can be a time-consuming operation. The small-scale animal studies took in the region of 4 h to complete and the first human scans some 4.75 h (Damadian, Goldsmith, and Minkoff 1977; Damadian, Minkoff, Goldsmith, and Koutcher 1978). With improvements in the mechanical scanning procedure, this time was subsequently reduced to a clinically more acceptable 30 min (Damadian 1980). Of course, the time necessary to reposition the patient can be virtually eliminated if the resonance aperture is scanned electronically by altering the current balance in the profiling coils. Technically this is a more difficult procedure, and one then runs into the more fundamental problem of the long signal averaging time necessary to achieve good S/N at reasonable resolution with a point technique.

Figure 3.2(a) shows a field contour plot for the f.o.n.a.r. magnet and coil system used to produce the first ever whole-body scan shown in Fig. 3.3 (Damadian *et al.* 1977, 1978). The 53-inch bore superconducting magnet (Goldsmith, Damadian, Stanford, and Lipkowitz 1977) and associated dewar (Minkoff, Damadian, Thomas, Hu, Goldsmith, Koutcher, and Stanford 1977) was constructed by Damadian and his colleagues who had no previous experience in superconducting technology—a truly remarkable feat. Although originally conceived as one of a Helmholtz pair, this single coil magnet has been successfully used for f.o.n.a.r. studies at 50.8 mT (508 G; 2.18 MHz for ^{1}H).

The horizontal displacements of Fig. 3.2(a) relate to the axial (z) and

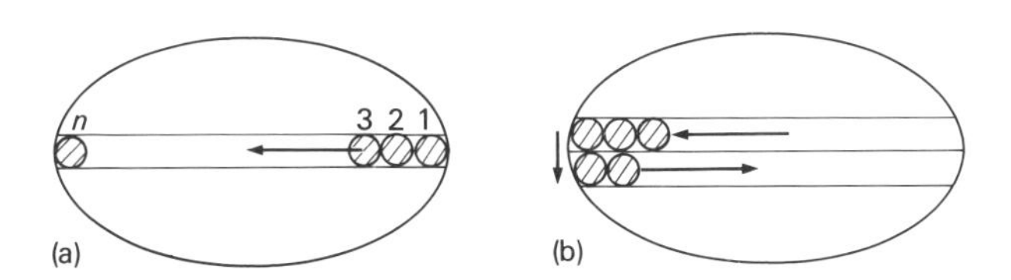

Fig. 3.1. The scanning principle of the f.o.n.a.r. experiment. (a) successive positions of the resonance aperture (shaded) as the patient is moved laterally along an image line; (b) orthogonal displacment of patient to start new image line.

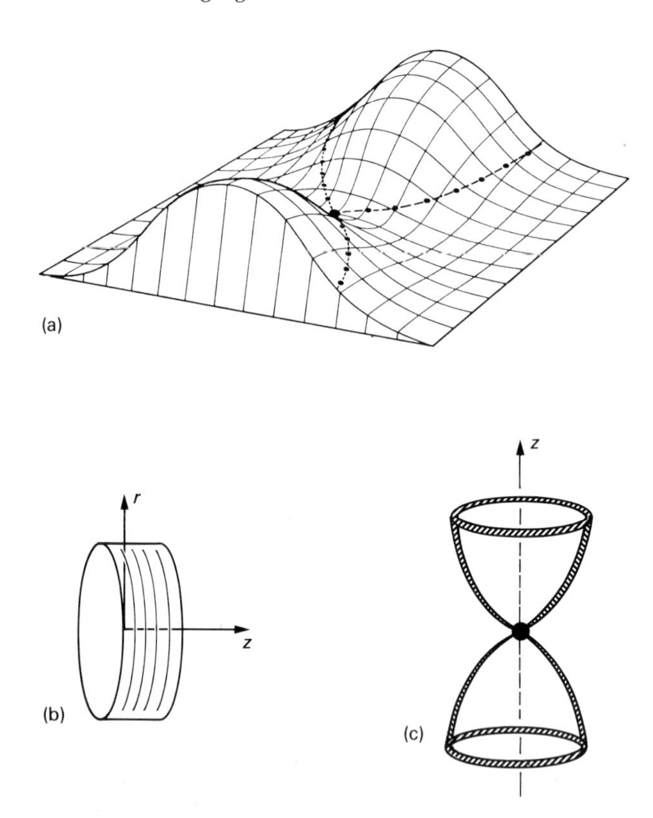

Fig. 3.2. The f.o.n.a.r. resonance aperture. (a) Field contour plot showing the isomagnetic field plane passing through the resonance aperture (from Damadian 1980). (b) Magnet co-ordinate definition. (c) The three-dimensional sensitive region.

radial (r) co-ordinates of the magnet coil (see Fig. 3.2(b)), and the vertical displacement gives the field amplitude in the z-direction $B_z(r, z)$. The field profile is thus seen to be saddle-shaped, with the resonance aperture at the saddle point corresponding to $\partial B_z/\partial r = \partial B_z/\partial z = 0$.

A selective r.f. pulse chosen to interact with spins at the resonance aperture will also excite those other spins which lie in the same magnetic field. These occur at the intersection of the (isomagnetic) field plane passing through the resonance aperture with the field surface, i.e. along the broken lines shown in Fig. 3.2(a). The full three-dimensional extent of the signal-producing region can be obtained by rotating the resonance aperture and lines of intersection about the z-axis of the coil as generator. The resulting 'egg-timer' shape is sketched in Fig. 3.2(c). The number of spins which contribute to the signal from the two cone-shaped

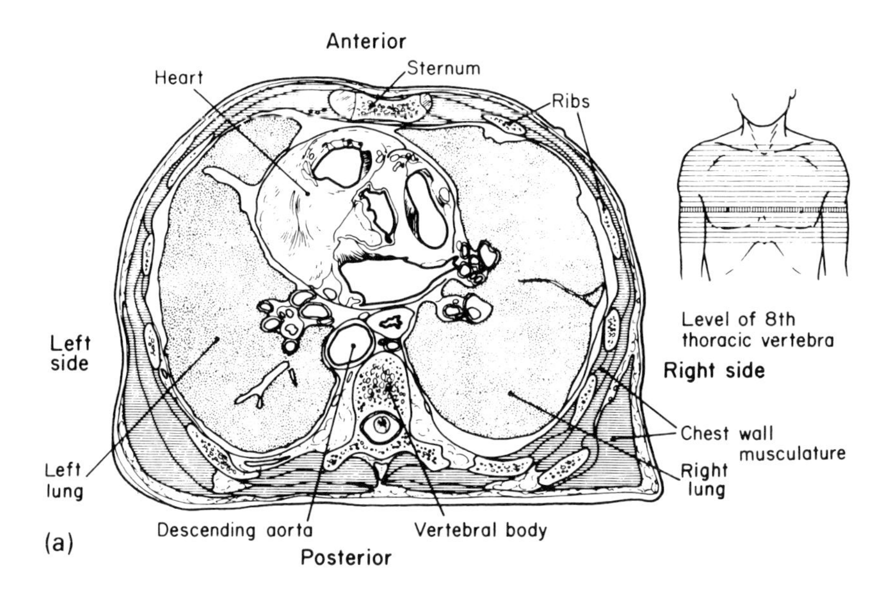

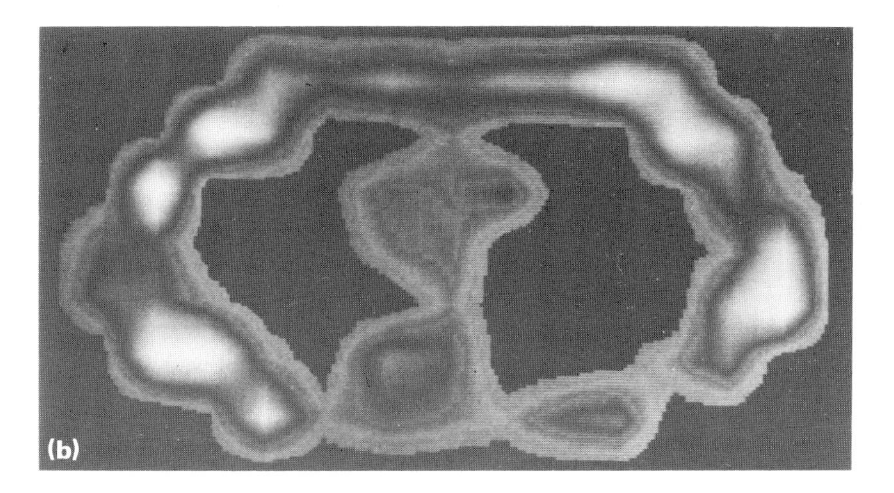

Fig. 3.3. The first ever whole-body n.m.r. scan. (a) Schematic of human chest at the level of the eighth thoracic vertebra. (b) F.o.n.a.r. cross-section of the live human chest at this level. Top of image is anterior boundary of chest wall, left area is left side of chest. Proton signal intensity is coded with black assigned to zero amplitude. (From Damadian, Minkoff, Goldsmith, and Koutcher 1978.)

regions is determined by the thickness of the cone walls. This in turn is controlled by the bandwidth of the selective pulse and the curvature of the field profile. Clearly, one should choose values which make the signal contributions from this unwanted source negligibly small. It is, however, difficult to eliminate them entirely and rather poor spatial resolution is a feature of this method. As an alternative, or in addition to the selective pulse, it is possible to use a simple filter in the receiver circuit. For example, for the human f.o.n.a.r. scan shown in Fig. 3.3 the signal following a 60 μs r.f. pulse was recorded 5 kHz off resonance: a 5 kHz narrow band filter applied following signal detection (removal of the r.f.) would therefore achieve the desired selectivity.

The image intensity is derived from the strength of the signal; either the integral of the f.i.d. envelope or, in the example shown, as the maximum peak to peak signal amplitude. Since signal averaging is normally necessary and the delay between r.f. pulses is generally insufficient to allow complete recovery of the spin system, the effective mobile spin density ρ' includes T_1 effects. Thus

$$\rho' = \rho_0\{1-\exp(-\mathrm{TR}/T_1)\}, \tag{3.1}$$

where ρ_0 is the actual spin density and TR is the repetition interval. We shall discuss the relative importance of ρ_0 and T_1 in tissue discrimination in Chapter 6.

A number of patients with advanced cancers (primary and metastatic adenocarcinomas, and alveolar and oat cell carcinomas) have been studied by Damadian and colleagues (Damadian, Goldsmith, and Minkoff 1978; Damadian 1980) using this method. It was also the basis of the Fonar Corporation's[1] first n.m.r. imaging system, the QED80. A preliminary clinical evaluation was reported by Partain, James, Watson, Price, Coulam, and Rolls (1980) and the product was released in 1981, the first n.m.r. imaging system to be made commercially available. In spite of the rather poor imaging times and the inherent difficulties with spatial resolution, this instrument had the advantage of being relatively inexpensive (roughly one-quarter the price of the much more advanced projection reconstruction and Fourier imaging systems currently being marketed.) The f.o.n.a.r. method no longer finds direct application in commercial imaging systems—the Fonar Corporation's products, e.g. their model QED β3000 scanner, now operate on the Fourier imaging principle. However, the method is still retained for region of interest T_1 measurements and could be used for spectroscopic studies at a suitably high field strength ($\gtrsim$1.2 T).

Variants of the f.o.n.a.r. method have been published by Abe, Tanaka,

[1] Fonar Corporation, 110 Marcus Drive, Melville, NY 11747, USA.

and co-workers (Abe, Tanaka, Hotta, and Imai 1974; Tanaka, Yamada, Shimizu, Sano, and Abe 1974; Tanaka, Yamada, Yamamoto, and Abe 1978) and by Crooks, Grover, Kaufman, and Singer (1978). In the method of Crooks *et al.* (1978), ferromagnetic shims were placed on the poles of a conventional electromagnet to achieve spatial localization. This instrument has been used to study small animals, but lacks the versatility of the f.o.n.a.r. method in that new shims have to be fitted in order to alter the size of the resonance apertures. There is also no possibility of electronic scanning in the manner discussed above.

3.2.2. *Topical magnetic resonance and surface coil methods*

Topical magnetic resonance or t.m.r. is another technique which aims to observe signals from a small isolated region (Greek τοποσ = place) within a larger sample. It was developed by Oxford Research Systems[2] in close collaboration with the Department of Biochemistry, Oxford, and achieves its spatial selectivity in a similar manner to f.o.n.a.r. by degrading the magnetic field homogeneity everywhere outside the desired signal-producing region. However, from the start, its object has been the observation of high-resolution spectra which enable important metabolites to be identified and measured within intact animal or human samples. Although the original application (Gordon, Hanley, Shaw, Gadian, Radda, Styles, and Chan 1980) was to phosphorylated metabolites such as phosphocreatine, inorganic phosphate, ADP, ATP, and NAD which occur in living systems at concentrations of <30 mM, ^{1}H and ^{13}C (Gordon 1981) and more recently ^{19}F (Stevens, Morris, Sheldon, and Griffiths 1984) studies are now also contributing to our understanding of *in vivo* metabolism. See § 2.6 for a brief discussion of high-resolution spectra and § 2.7 for examples of t.m.r. studies.

In order to resolve the chemical shifts of the different metabolites and make useful measurements of the internal pH (see Fig. 2.23) the basic homogeneity of the magnet needs to be very good (0.1 p.p.m. or better) over the central region in which measurements are made. The resonance aperture is defined through the use of an axially symmetric profile coil system which gives gradients up to fourth-order. The magnetic field $B(r, \theta)$ generated by such a coil is given by Hanley and Gordon (1981):

$$B(r, \theta) = \sum_{n=1}^{4} B_n r^n P_n(\cos\theta), \qquad (3.2)$$

where $P_n(\cos\theta)$ are the nth order Legendre polynomials and r, θ are the polar co-ordinates relative to the magnet centre. The axial variation is

[2] Oxford Research Systems Ltd, Nuffield Way, Abingdon, Oxon OX14 1RY, UK.

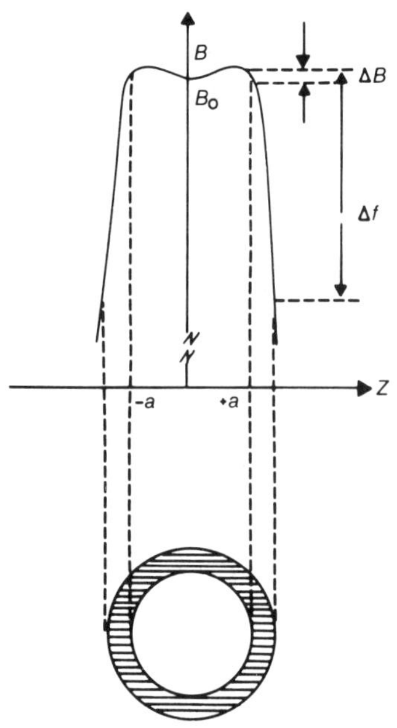

Fig. 3.4. The profile of the magnetic field along the magnet axis in t.m.r. The inhomogeneity ΔB within the central region shown schematically in the projection below the profile is constrained to remain less than the typical ^{31}P linewidths. In the shaded region the ^{31}P lines will be broadened by the field gradients. (From Gordon *et al.* 1980. Reprinted with permission from *Nature* **287,** 736, (1980). Copyright © 1980 Macmillan Journals Limited.)

illustrated in Fig. 3.4 for the case of $B_0 = B_1 = B_3 = 0$, $B_4 = -B_2$ (remember, however, that it is superimposed on the much larger uniform static field). The inhomogeneity ΔB over a central region of width $2a$ is given by

$$\Delta B = B_2^2/B_4, \tag{3.3}$$

where:

$$2a = 3.1(B_2/B_4)^{\frac{1}{2}}. \tag{3.4}$$

The required resolution ΔB can be achieved over the resonance aperture by appropriate variation of the ratio $B_2 : B_4$. The achievement of small sensitive volumes becomes more difficult as the object size increases since the dimensions of the profiling coils must increase in proportion and much larger currents are then required. (Efficient water cooling is therefore essential.) The coils for t.m.r. systems are in consequence built with a specific sensitive volume in mind, though an approximately threefold variation is possible through current adjustment.

In contrast to f.o.n.a.r., no shaping of the r.f. pulse is employed. The n.m.r. signal therefore contains the high-resolution information from the central homogeneous region together with a broad component from the surrounding inhomogeneous one. Several techniques exist for the separation of these narrow and broad components which can arise in other contexts, for example from the mobile and immobile parts of a macromolecule. One simple (and rather crude) method is the removal of the first portion of the f.i.d. (setting to zero of the first n points in a sampled f.i.d.). The signals contributing to the unwanted broad component dephase very rapidly and so contribute little to the f.i.d. at times greater than the reciprocal of the linewidth. The high-resolution components persist well beyond this time and are therefore preserved in the subsequent Fourier analysis. This method has several drawbacks; it introduces a first-order phase shift proportional to the number of points removed, it degrades the S/N since some of the required signal is discarded with the unwanted component, and it alters the relative intensities of the remaining resonances if they have different linewidths. This last effect is particularly invidious if one is attempting to make quantitative comparisons. Another technique is the so-called convolution difference method (Campbell *et al.* 1973). This involves the multiplication of the f.i.d. by an exponential function to 'smear out' the high-resolution component. This modified f.i.d. is then subtracted from the original one and the difference is Fourier transformed. Though much superior to the previously described method, convolution difference still suffers from a number of artefacts; the characteristic negative wings on the sides of large peaks and the more disturbing distortion of relative intensities. Many other more sophisticated methods are now available; see Lindon and Ferrige (1980) and Gordon *et al.* (1982) for examples.

If conventional r.f. coils are used with a t.m.r. system there is a considerable S/N problem, originating from the fact that all the material (for example, a whole animal) contained within the receiver coil generates noise, whereas the signal originates only from the sensitive volume. Even if sample lossiness is not the dominant source of noise, the problem still remains since large coils require long lengths of conductor which have greater electrical resistance and hence generate more noise. If the organ of interest lies deep within the body of the subject then one must either accept this fact or else resort to minor surgery—a practice which rather negates the potentially non-invasive nature of the method. However, if the organ is peripheral then one can make use of surface coils. These were originally developed for n.m.r. flow measurement (Morse and Singer 1970) but were subsequently also shown to be of value for biomedical studies by Ackerman, Grove, Wong, Gadian, and Radda (1980) who used them to obtain ^{31}P spectra from the muscle and brain of an anaesthetized rat. These coils can be placed alongside the organ of interest and pick up signal (and noise) only from the adjacent tissue. The extent of the excited region can to some extent be controlled by the coil radius and the length of the applied r.f. pulse. Surface coils therefore provide a means of spatial selection in their own right and do not suffer the disadvantage of receiving noise from non-excited regions. They have been widely used for *in vivo* high-resolution studies with or (more commonly nowadays) without t.m.r. field profiling. For a circular surface coil the region from which signal is detected corresponds very roughly to a disc-shaped volume with a cross-sectional area equal to that of the coil and a depth equal to its radius. However, the B_1 field, and hence also the receiver response function, is in reality very non-uniform (see § 5.1.1 where the full expressions are derived in connection with magnet design). It is therefore difficult to be certain from which region the signal originates. Better definition of the 'sensitive volume' of a surface coil can be obtained through the use of 'depth' pulse sequences (Bendall and Gordon 1983). These use the phase-cycling methods of Bodenhausen, Freeman, and Turner (1977) to average out the signal from regions where the pulse angle deviates from 90°. A typical scheme would use a 2θ; $\theta[\pm x]$; $(2\theta[\pm x, \pm y])_2$ sequence, where θ is the pulse angle at any sample point and $\pm x$, $\pm y$ denotes the phase-cycling. By increasing the pulse length the region corresponding to the 90° condition is moved progressively further into the sample. Bendall and Aue (1983) have shown that this scheme works well for the axial case. However, to remove off-axis contributions one has to indulge in further sophistications of the method. Surface coils have been used for T_1 measurements (Evelhoch and Ackerman 1983). They can also be used in conjunction with full (slice or volume) ^{1}H-imaging techniques to achieve better spatial resolution, for region of interest studies, of the eye, ear, or knee for example.

3.2.3. *The sensitive point method*

In 1974 Hinshaw (*a*, *b*) introduced another method for isolating a small signal-producing volume within an extended object. The technique was known as the sensitive point method and was based on the use of three sinusoidal oscillating orthogonal field gradients. The effect of a gradient in the x-direction, $G_x(t) = G_x \cos \Omega t$ where Ω is the frequency of the oscillating gradient, is to make the magnetic field time-dependent for all points which lie off the null plane defined by $x = 0$. Two such orthogonal gradients define a sensitive line $x = 0$, $y = 0$ and a third reduces the time-invariant region to a sensitive point $x = 0$, $y = 0$, $z = 0$. By altering the current balance between gradient coil pairs this sensitive point can be electronically scanned through the object to build up a two- or three-dimensional image.

The signal from the sensitive point is extracted by virtue of its time-invariance. This can be achieved in a number of ways. For example, one can integrate and co-add successive f.i.d.s so that the unwanted signals average out. In such a case one can show that the real component of the signal $S_R(\mathbf{r})$ in the presence of a single oscillating gradient $G_x(t)$ is given by (Mansfield and Morris 1982):

$$S_R(\mathbf{r}) = \frac{M_0(\mathbf{r})}{\mathrm{TR}} \sum_k J_k^2\left(\frac{\gamma G_x x}{\Omega}\right) \frac{T_2(\mathbf{r})}{1+(\Omega k/T_2(\mathbf{r}))^2}, \tag{3.5}$$

where TR is the pulse repetition period, $M_0(\mathbf{r})$ the equilibrium magnetization, $J_k(\gamma G_x x/\Omega)$ are Bessel functions of the first kind of order k and the expression is valid in the limit $\mathrm{TR} \gg T_2(\mathbf{r})$. If in addition $\Omega \gg 1/T_2(\mathbf{r})$, eqn (3.5) can be simplified to:

$$S_R(\mathbf{r}) \sim M_0(\mathbf{r}) \frac{T_2(\mathbf{r})}{TR} J_0^2\left(\frac{\gamma G_x x}{\Omega}\right), \tag{3.6}$$

which is plotted for various values of Ω in Fig. 3.5. An improved spatial response can be obtained by using non-sinusoidal functions for the gradients, for example a square wave (Bottomley 1978; Ljunggren 1983*b*).

A particularly convenient way in which to collect the signal is to use the steady state free precession (s.f.p.) method (Carr 1958). This involves the application of a string of phase alternated pulses with an interpulse interval $\mathrm{TR} < T_2(\mathbf{r})$. The magnetization soon reaches a steady state in which the evolution between pulses is constant. For 90° pulses and in the limit $\mathrm{TR} \ll T_1, T_2$, the amplitude of the resultant signal is given by $M_0(\mathbf{r})/2(1+T_1/T_2)$. Once again one can integrate the signal (over a whole number of complete gradient cycles) to obtain an average signal which originates predominantly from the sensitive point (Mansfield and Morris 1982; Meiere and Thatcher 1979). Again, if $\Omega \gg 1/T_2(\mathbf{r})$ only the $k=0$ Bessel term is important (see eqn (3.5)). However, if, as is often the case, this condition is not fulfilled, then the higher-order terms give rise to successively smaller spinning-sideband-like lobes at frequencies which are discrete multiples of Ω, corresponding to unwanted signal from planes at $x = \pm\Omega/\gamma G_x$, $\pm 2\Omega/\gamma G_x$, etc.

In the original experiments of Hinshaw (1974*a*, *b*, 1976) the sensitive point was scanned over the image plane using potentiometers driven by clockwork motors. The signals were obtained with the s.f.p. method and, after removal of the sign alternation (a useful feature of the s.f.p. method, giving the spectrometer the stability of a chopper amplifier,

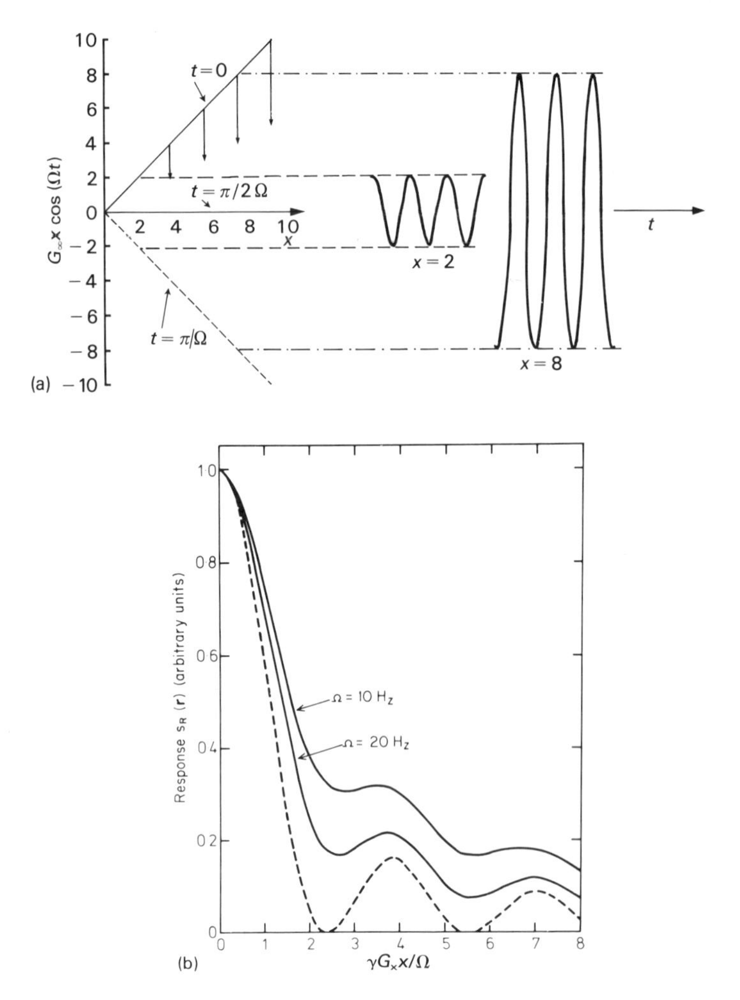

Fig. 3.5. The application of a field gradient $G_x \cos \Omega t$ to a sample: (a) spatial and time dependence of magnetic field due to the gradient; (b) spatial response for integrated f.i.d. sensitive point method. Gradient frequencies 10 and 20 Hz are illustrated. Corresponding curves for 100 and 500 Hz are indistinguishable from $J_0^2(\gamma G_x x/\Omega)$ shown as a broken line.

eliminating d.c. bias, coherent logic noise, etc.), were averaged using a simple low-pass filter with a time constant in the range 0.1–1.0 s. The images were output directly to an *xy* plotter driven in synchrony with the scan, the d.c. signal from the low pass filter being fed to the *y*-channel to give an additional deflection in this direction. Figure 3.6 is an example of an image obtained with this primitive imaging system. It shows a thin

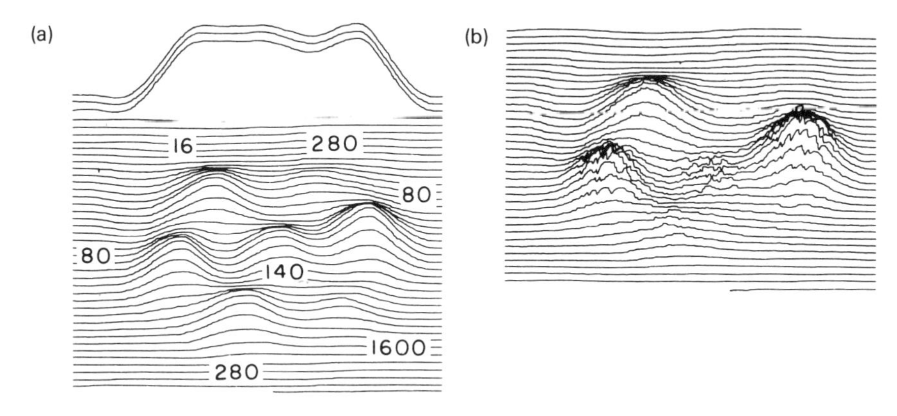

Fig. 3.6. (a) Cross-section of seven glass tubes each filled with water. The numbers are the T_1 values in ms; (b) shows the effect of adding a 180° pulse every 100 ms during the scan. Note that the effect of these saturating pulses depends upon the T_1 of the sample. The three traces above the top of the image represent a projection of the image and were obtained by turning off the y (vertical) gradient. (From Hinshaw 1976.)

cross-section through seven glass tubes filled with water. The spatial resolution in the xy-plane, estimated as the distance over which the output changes from 20 to 80 per cent of the maximum amplitude, is about 0.5 mm. The r.f. pulses were 2 μs in length corresponding to 30° nutations and had a repetition rate of 6 kHz. The z-gradient had a strength of 20 mT/m (2 G/cm) and a frequency of 400 Hz whereas the x- and y-gradients had strengths of about 24 mT/m (2.4 G/cm) and a frequency of 57 Hz. Each x sweep took 64 s giving a total imaging time of about 20 min.

The merit of the sensitive point method lies both in the simplicity with which it can be implemented and in the lack of dependence on gradient linearity (the image of Fig. 3.6 has no detectable spatial distortion, though there is some intensity variation arising from changes in the size of the sensitive volume). Its disadvantage is its inefficiency resulting in inordinately long imaging times (in some cases several hours). However, as noted by Hinshaw in his early papers, the method allows one to perform localized n.m.r. measurements—there is no necessity to form an image. Thus, for example, one could use it to measure T_1 (see Fig. 3.6b for an example of a T_1-dependent image) or record a high-resolution spectrum at the sensitive point. In the latter case the f.i.d.s from the s.f.p. echo train are averaged (but not integrated) and Fourier transformed to give the high-resolution spectrum. Bottomley (1982) has recently published examples of ^{31}P spectra obtained by this method. Other groups are actively pursuing similar aims (see, for example, Fitzsimmons 1982; Scott, Brooker, Fitzsimmons, Bennett, and Mick 1982). It is also possible to use selective excitation methods to isolate a defined volume from which high-resolution spectra can be recorded (see § 3.3.1).

3.3. Line-scan methods

We now turn to a discussion of line-scan imaging methods of which there are many variants. The basic idea is to isolate a line (or more correctly a rod since it will have a finite cross-section) within a three-dimensional object and subsequently to distinguish between signals emanating from different points along the selected line. We shall discuss line selection in the next section before going on to describe a number of the line-scanning techniques which have been proposed. In the remainder of this section, however, attention will be directed to the discrimination between image points along a line. This process, as the reader who has worked his way through Chapter 2 may surmise, is accomplished through the application of a magnetic field gradient (the 'read' gradient) along the direction of the selected line. If this is in the x-direction the n.m.r. signal from a small region of length δx will be, following phase sensitive detection and in the absence of relaxation effects, proportional to $\cos(\gamma G_x x t)\rho(x)\delta x$ where G_x is the field gradient and $\rho(x)$ the (line) spin density. This is just a simple harmonic signal with frequency proportional to the distance along the line and amplitude proportional to the spin density. Fourier transformation thus immediately yields the spin density along the image line. Consider as an example of this process the object illustrated in Fig. 3.7, consisting of two parallel water-filled tubes of square cross-section, and suppose that we have selected for observation the line of spins designated α. If we apply our read gradient parallel to this line then the signal or f.i.d. emanating from the line segment within the tube of larger cross-section will be a sinc function with a width proportional to $1/a$ (see § 2.4.2.4), modulated by an offset frequency $\gamma G_x x_1$ corresponding to the centre of this line segment (see Fig. 3.7(c)). Similarly, the signal from the water within the tube of smaller cross-section will also be a sinc function but with a width proportional to $1/b$ and modulated at a frequency of $\gamma G_x x_2$ (see Fig. 3.7(d)). The net signal is thus the sum of the two components illustrated in (c) and (d). Fourier transformation gives the spin density along the selected line as shown in (e). Note that the Fourier amplitudes are the same for both tubes since we are measuring the spin density along a line rather than of all material lying on a plane normal to it (see § 2.2.2). Thus, if we were to move the line to position β (Fig. 3.7(b)), we would no longer observe the smaller tube. The f.i.d. would then simply be as shown in (c) and the Fourier transform as shown in (f). Finally, if we were to reduce the density of mobile spins in the larger tube, for example by filling it with small glass beads so that water only remained in the interstitial spaces, the amplitude of the component from the large tube would be reduced proportionately and the line profile corresponding to position α would be as shown in (g).

It is useful to enquire at this stage as to how the information from a set of lines is best presented. Clearly a collection of graphs like Fig. 3.7(e) and (f) would make interpretation of an object of any complexity a tedious business. A simple alternative is to use a television-type display in which line profiles are used to intensity modulate the electron beam as it is swept across the corresponding line on the monitor screen. In this manner a full two-dimensional image can be built up with, for example, bright areas corresponding to regions of high spin density, dark ones to low, and grey areas to intermediate spin density. See § 5.6 for a fuller discussion of image display.

Line-scan methods are in general reasonably insensitive to magnetic field inhomogeneities, though to a lesser degree than point techniques. For example, Hutchison, Sutherland, and Mallard (1978) have demonstrated that images can be recovered when field distortions of up to 50 per cent of the maximum field difference generated across the sample by the applied gradient are present if a reference uniform (flood) sample is scanned under similar conditions. Provided a difference method is not employed (see § 3.3.4), the images are less susceptible to motional artefacts than with planar or three-dimensional methods, in the sense that motion occurring during the scanning of a particular line only destroys information corresponding to that line, leaving the remainder of the image unaffected. Such line artefacts can be readily observed in, for example, the multiple sensitive point image of *in vivo* samples such as the wrist of Fig. 6.16.

3.3.1. *Selective irradiation*

In most applications of conventional n.m.r. spectroscopy a non-selective pulse is used to uniformly excite all the resonances which are of interest. In other words, the bandwidth of the excitation pulse is large compared with the spectral width. However, if the aim is to excite a limited region of the spectrum, then a selective pulse or train of pulses (such as the d.a.n.t.e. sequence, see Morris and Freeman 1978) is required. These have been widely used in high-resolution n.m.r. as aids to the assignment of complex spectra and for inversion transfer experiments which permit studies of reaction kinetics (Campbell, Dobson, Ratcliffe, and Williams 1978). Selective irradiation has also been of major importance in n.m.r. imaging (Garroway, Grannell, and Mansfield 1974) since the application

(a)

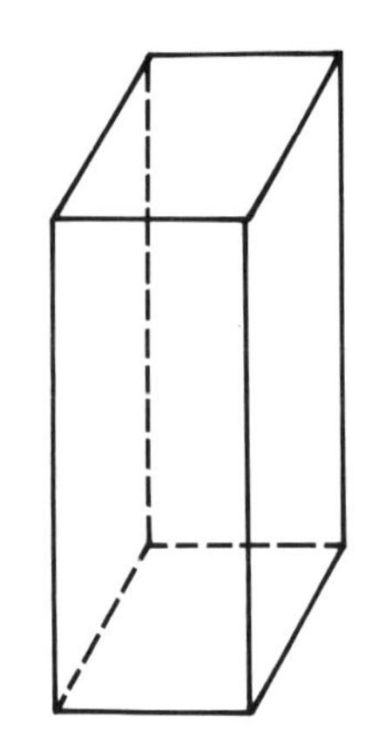

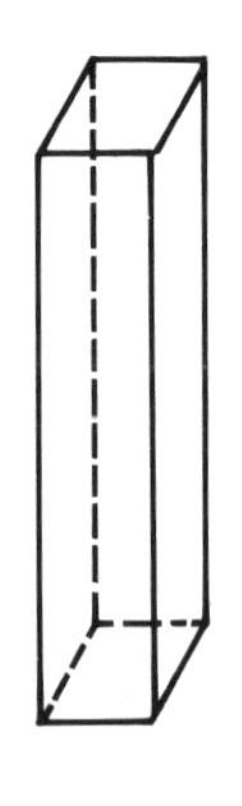

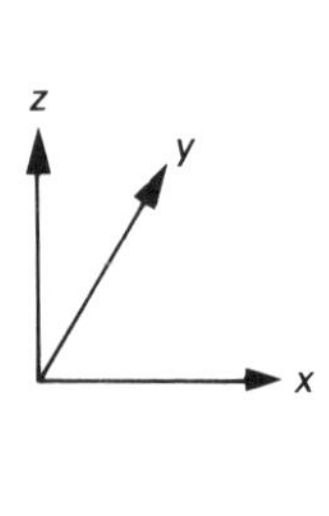

(b)

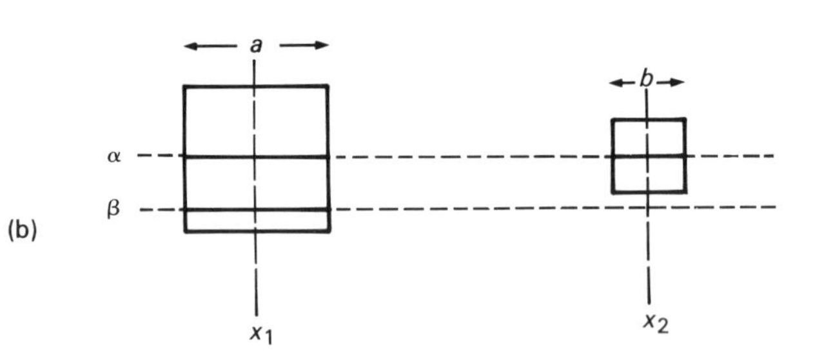

(c)

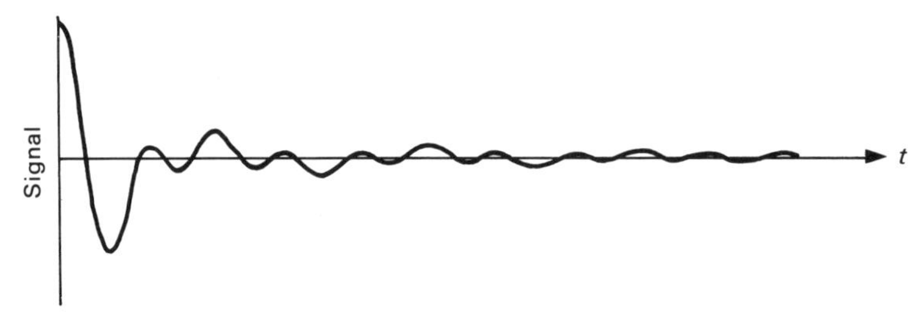

(d)

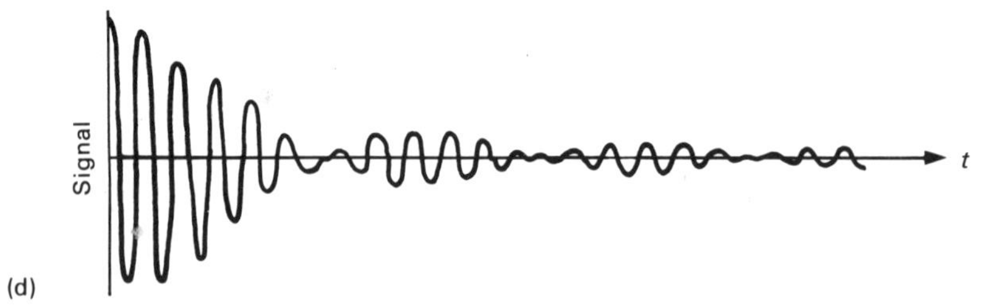

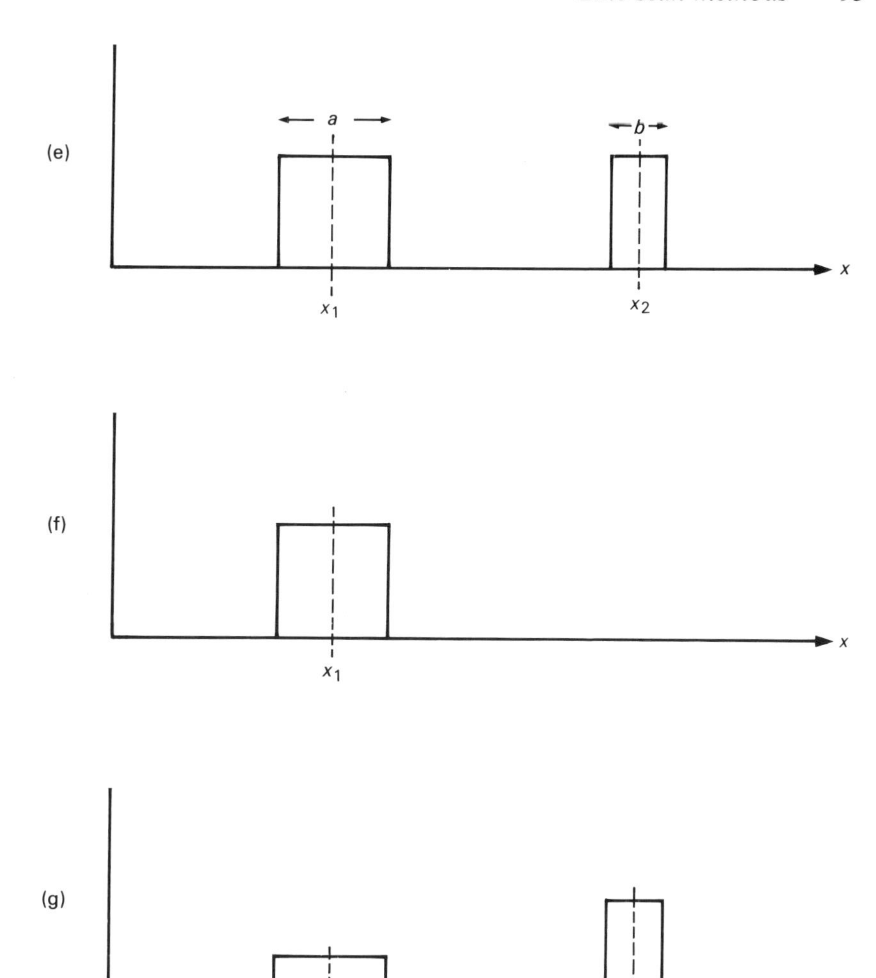

Fig. 3.7. (a) Phantom consisting of two square-sectioned tubes of side a and b, oriented parallel to the z-axis and located at $x = x_1$ and $x = x_2$ respectively; (b) xy-section through tubes showing positions of sensitive lines α and β corresponding to profiles shown in e and f; (c) signal arising from tube of larger cross-section in x-gradient; (d) signal arising from tube of smaller cross-section in x-gradient; (e) line profile through both tubes (α) arising from Fourier transformation of (c) + (d); (f) line profile (β) arising from Fourier transformation of (c); (g) line profile of similar phantom with spin density in larger tube reduced by presence of glass beads.

of a field gradient to the sample means that irradiation over a limited bandwidth corresponds to the excitation of a restricted spatial region. We shall discuss the details of selective pulse generation shortly, but let us suppose for the moment that we can generate selective pulses with any desired frequency profile. In particular, consider the application to our sample of a selective pulse which excites the spin system uniformly over a narrow frequency range of width $\Delta\omega$ (the 'top hat' frequency profile illustrated in Fig. 3.8(a)). In the presence of a linear field gradient G_y, the spins which are excited will lie in a plane of thickness $\Delta y = \Delta\omega/\gamma G_y$ normal to y and centred at $y = (\omega - \gamma B_0)/\gamma G_y$ (see Fig. 3.8(b)). The isolation of a plane corresponds to a onefold reduction in the dimension of the sample (from three to two) and this is all the preparation needed for the two-dimensional variants of the imaging methods such as projection reconstruction, Fourier and echo-planar imaging. In the case of line-scanning however, a further onefold reduction in dimensions is required so that a second selective process is necessary. This is not quite as straightforward as one might at first imagine. If only a single selective pulse is required, one normally arranges that it nutates the magnetization away from its equilibrium direction (z) through an angle of 90° into the $x'y'$-plane of the rotating frame, where it can be observed as a f.i.d. (see § 2.4). However, a second selective pulse would take the magnetization out of this plane replacing it with magnetization from different sample regions. Various schemes have been devised to overcome this difficulty and account for some of the diversity in line-scanning methods. Some of

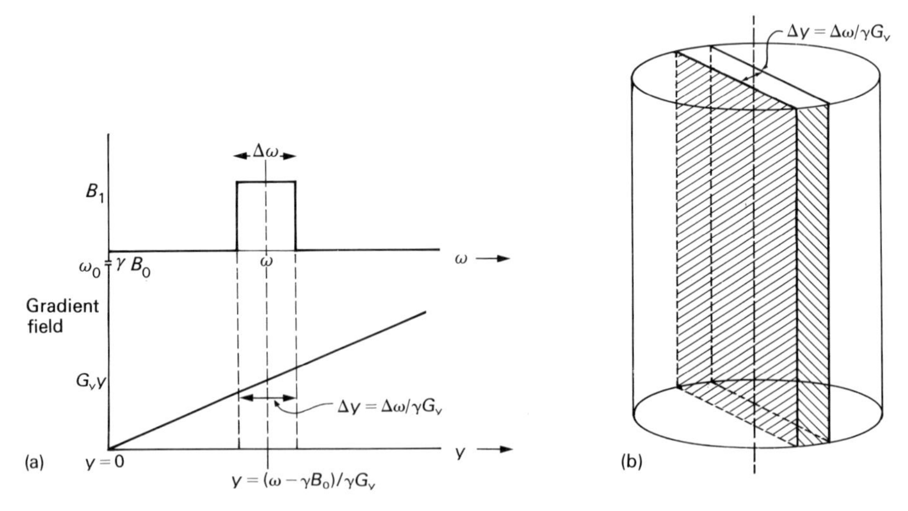

Fig. 3.8. Selective excitation of a plane of spins: (a) r.f. and gradient fields; (b) excited volume corresponding to irradiation profile of (a).

the more important variants are discussed below. The first line-scan experiments either avoided the problem by relying on the spatial response of the receiver coil, which gives a rather crude plane definition, or else they used selective saturation. As a simple example of the latter method, consider a selective pulse with the frequency profile illustrated in Fig. 3.9. If this is applied to a sample in the presence of a field gradient G_z, all z-magnetization can be destroyed (the spin system is saturated) apart from that which is situated in a plane of width $\Delta z = \Delta\omega/\gamma G_z$. A selective excitation pulse of the type shown in Fig. 3.8, applied in the presence of an orthogonal field gradient, will then nutate a line of spins from this pre-selected slice into the $x'y'$-plane where the magnetization can be observed. Although this method has been used successfully, it suffers from a number of difficulties. First, high-power r.f. pulses are required with good dynamic range, and secondly, any spins outside the selected line which are not totally saturated (those lying at the periphery of the sample, for instance) will contribute to the signal following the selective 90° pulse. It should be mentioned at this point that the other schemes also suffer from a number of drawbacks so that the two phase selective process is in many ways rather unsatisfactory.

A similar, but rather more efficient method than that of selective saturation has recently been proposed (Aue, Müller, Cross, and Seelig 1984). It uses a combination of selective and non-selective pulses (a 45° selective, 90° broadband, 45° selective sandwich) in the presence of an applied field gradient to preserve the z-magnetization in a plane of spins normal to the gradient direction whilst generating transverse magnetization elsewhere. The transverse magnetization rapidly dephases in the applied gradient, leaving an isolated plane of unexcited spins. This procedure can be used in combination with standard selective excitation methods as discussed above for the selective saturation method, or can itself be repeated to achieve further localization. Three such composite pulses have been used with x, y, and z field gradients to isolate a small volume, whose size and location can be altered by varying the applied

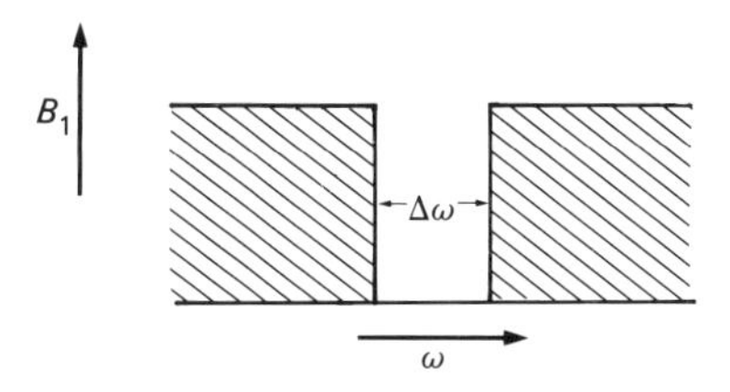

Fig. 3.9. Selective saturation irradiation profile. All spins are irradiated apart from those in a narrow frequency band $\Delta\omega$.

field gradients. A broadband 90° pulse allows the magnetization from this volume to be inspected, enabling regional chemical-shift studies to be performed (Aue *et al.* 1984).

3.3.2. *Selective pulse theory*

In Chapter 2 the inverse relationship was seen which exists between the extent of a function in the frequency and time domains (see Fig. 2.13). Thus, narrow n.m.r. linewidths are associated with prolonged f.i.d.s and broad linewidths with foreshortened ones. Similarly, if a r.f. pulse with a large bandwidth is required, it must be of very short duration. For high-resolution spectroscopy 90° pulse widths of 5–100 μs are typical; for solids where n.m.r. linewidths are potentially much greater (see § 2.5.2) pulse widths <1 μs are desirable. In the case of selective pulses for n.m.r. imaging, where narrow bandwidths are the order, pulsewidths are typically a few milliseconds.

In general, the selective pulse will need to be 'tailored' or shaped to excite a clean spatial region. However, for the sake of simplicity in our initial treatment, we shall consider the case of a simple rectangular top hat pulse of duration t_w and amplitude B_1 (see Fig. 3.10(a)) which we should expect to excite a region defined by a sinc function of width $4\pi/t_w$. Let us take as our sample a cylinder of spins with uniform equilibrium magnetization per unit length along the z-axis m_0 as shown in Fig. 3.10(b). Suppose that, in the presence of a field gradient G_z those spins lying in the plane $z = z_0$ satisfy the resonance condition for the r.f. carrier frequency ω, i.e.

$$\omega = \gamma(B_0 + G_z z_0). \tag{3.7}$$

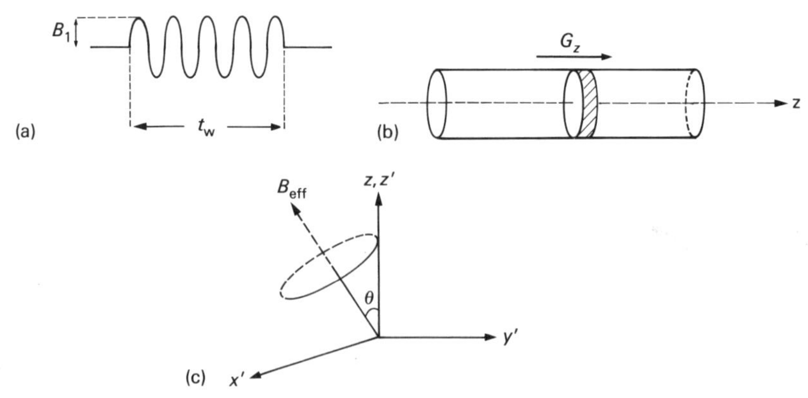

Fig. 3.10. Selective excitation. (a) Square selective pulse; (b) cylindrical sample with uniform magnetization per unit length, m_0; axis parallel to z; (c) precession of elemental magnetization in effective field $\boldsymbol{B}_{eff}$.

The effective field in the rotating reference frame for any other plane of spins then consists of two components: a static magnetic field deriving from the field gradient, $G_z(z-z_0)$, and an r.f. field B_1 directed along the x'-axis of the rotating frame. The total effective field in a general plane normal to z is therefore given by a vector of magnitude

$$B_{\text{eff}}=[B_1^2+G_z^2(z-z_0)^2]^{\frac{1}{2}}, \tag{3.8}$$

oriented at an angle

$$\theta=\tan^{-1}\{B_1/G_z(z-z_0)\} \tag{3.9}$$

to the z-axis in the $x'z$-plane (see Fig. 3.10(c)). The motion of the spins in this plane is a precession of the elemental magnetization about the effective field direction shown as a broken line in Fig. 3.10(c). After a time $t<t_w$ the distribution of the magnetization in the x'-, y'-, and z-directions, in the absence of relaxation effects, is given by (Mansfield, Maudsley, Morris, Pykett 1979):

$$m_{x'}=m_0 \sin\theta \cos\theta\{1-\cos(\gamma B_{\text{eff}}t)\} \tag{3.10a}$$

$$m_{y'}=m_0 \sin\theta \sin(\gamma B_{\text{eff}}t) \tag{3.10b}$$

$$m_z=m_0\{\cos^2\theta+\sin^2\theta \cos(\gamma B_{\text{eff}}t)\}. \tag{3.10c}$$

Note that these equations are not independent, since $m_{x'}^2+m_{y'}^2+m_{z'}^2=m_0^2$. We can combine the expressions for $m_{x'}$ and $m_{y'}$ to obtain the distribution for the net magnetization $m_{x'y'}$ in the $x'y'$ frame. Thus:

$$m_{x'y'}=[m_{x'}^2+m_{y'}^2]^{\frac{1}{2}} \tag{3.11a}$$

$$=m_0 \sin\theta\{\cos^2\theta[1-\cos(\gamma B_{\text{eff}}t)]^2+\sin^2(\gamma B_{\text{eff}}t)\}^{\frac{1}{2}}. \tag{3.11b}$$

The phase angle of this magnetization, relative to the y'-direction is given by

$$\phi=\tan^{-1}(m_{x'}/m_{y'}) \tag{3.12a}$$

$$=\tan^{-1}\frac{\cos\theta\{1-\cos(\gamma B_{\text{eff}}t)\}}{\sin(\gamma B_{\text{eff}}t)}. \tag{3.12b}$$

It is also instructive to study the total magnetization in the $x'y'$-plane. The x' and y' components $M_{x'}$ and $M_{y'}$ are given by the integrals of the respective distributions (eqns (3.10a) and (3.10b)) over z. Thus:

$$M_{x'}(t)=m_0B_1G_z\int_{-\infty}^{\infty}\frac{(z-z_0)\{1-\cos[\gamma G_z(z-z_0)t]\}}{B_1^2+G_z^2(z-z_0)^2}\,dz, \tag{3.13a}$$

$$M_{y'}(t)=m_0B_1\int_{-\infty}^{\infty}\frac{\sin[\gamma G_z(z-z_0)t]}{[B_1^2+G_z^2(z-z_0)^2]^{\frac{1}{2}}}\,dz. \tag{3.13b}$$

Since, for our uniform sample, $m_{x'}$ is an antisymmetric distribution, $M_{x'} = 0$ for all t. The $M_{y'}$ integral can be evaluated analytically, giving,

$$M_{y'}(t) = \frac{\pi B_1 m_0}{G_z} J_0(\gamma B_1 t), \tag{3.14}$$

where J_0 is a zeroth order Bessel function of the first kind. Equation (3.14) is illustrated in Fig. 3.11 for two values of r.f. field. An interesting and at first sight rather paradoxical feature of this result is the appearance at time zero of the maximum magnetization. This can be understood if we consider the limiting behaviour as the pulse width is made smaller and smaller. In this case, although the angle of nutation tends to zero, the bandwidth of the excitation pulse increases so that the region of the sample over which spins are excited is extended. Furthermore, signal contributions from all these spins will be in phase at the time origin, accounting for the theoretical maximum there. Of course, in practice, neither the sample nor the field gradients are of infinite extent so that the actual response is an initial rapid rise shown schematically as broken lines in Fig. 3.11.

We now move on to consider the time evolution of the magnetization distributions given by eqns (3.10)–(3.12). This is illustrated in Fig. 3.12

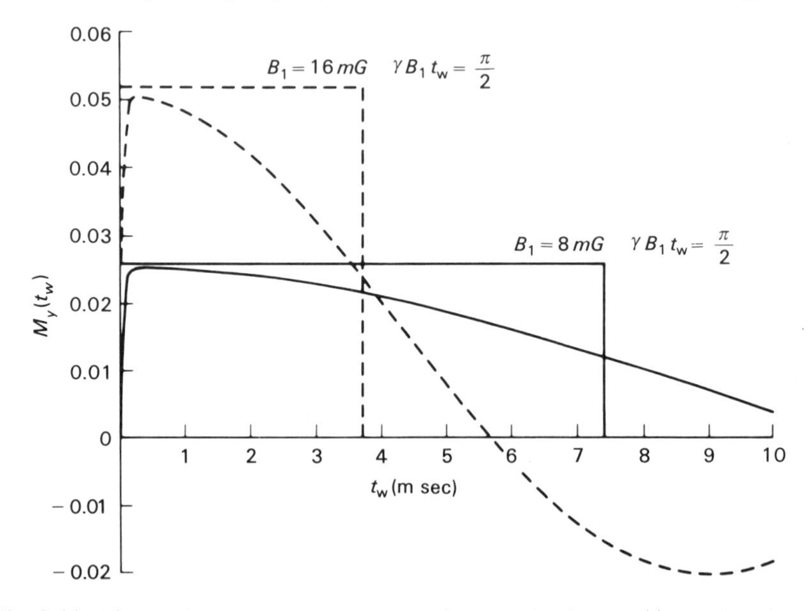

Fig. 3.11. The total transverse component of magnetization $M_{y'}(t)$ as a function of time for $B_1 = 1.6\ \mu T$ (broken lines) and $B_1 = 0.8\ \mu T$ (solid lines). The corresponding 90° rectangular pulses (for on-resonance spins) are also shown with $t_w = 3.7$ ms (1.6 μT) and 5.6 ms (0.8 μT) respectively. (From Mansfield, Maudsley, Morris and Pykett 1979.)

which shows $m_{x'y'}$, ϕ and m_z at two successive times corresponding to 30° and 90° pulses respectively for the resonant spins ($z = z_0$). At the earlier time, the central maximum which accounts for a large percentage of the total $x'y'$-magnetization is broad corresponding to excitation over a large region of the sample, whereas at the later time it has narrowed (in inverse proportion to the total elapsed time) corresponding to a sharpening of the spatial response. Inspection of eqn (3.10b) indicates the resemblance of $m_{y'}$ to the sinc function $\text{sinc}(\gamma G_z(z-z_0)t)$ for offsets $G_z(z-z_0) > B_1$. This is just what we should expect, namely that the distribution of excited magnetization corresponds to the Fourier transform of the selective excitation pulse. The (small) departure of the actual distribution from a true sinc function, which is more apparent in the case of larger pulse angles, is a measure of the non-linearity in the response of a spin system to an applied r.f. pulse. We shall return to this point later. Of course, a sinc-like distribution, in which a high proportion of the total $x'y'$-magnetization comes from the wings of the function in regions remote from the central excitation maximum, is far from ideal. The solution to this problem is to approach it from the reverse direction, namely to start with the desired excitation distribution and calculate, within the limits of linear response theory, the required selective pulse shape by (inverse) Fourier transformation. Thus a Gaussian profile of half-width Δz can be obtained with a Gaussian pulse of width $\Delta t = 1/\gamma G_z \Delta z$. The spatial response can be further sharpened by using as an excitation pulse a Gaussian modulated by a sinc function (Hutchison, Sutherland, and Mallard 1978). The Fourier transform of this pulse is a Gaussian convoluted with a top hat function as illustrated in Fig. 3.13(a). The response of the spin system $m_{x'y'}(t)$, $\phi(t)$, and $m_z(t)$ can be evaluated by numerical integration of the Bloch equations (Locher 1980) and is illustrated in Fig. 3.13(b). Another possible excitation pulse is the simple sinc function which gives rise to an approximate square wave response. This was the method used in the early line-scan imaging experiments of Mansfield, Maudsley, and Baines 1976; Mansfield and Maudsley 1976*a*; Mansfield and Maudsley 1977*a*). Thus far in this description nothing has been said about the phase distribution $\phi(z, t)$. In fact, this presents something of a problem, since, although spins can be excited over any defined region of the sample, in general, we cannot then observe the magnetization directly as the different spin isochromats have accumulated different phases. Their contributions to the induced signal are therefore out of phase, reducing the magnitude of the total received signal. In fact, at the end of the selective pulse when the r.f. amplitude falls to zero, the signal itself is also zero at least to a first approximation. One way around this difficulty, which can be applied in the case of the rectangular pulse discussed above, is to turn off the G_z gradient at the end of the pulse

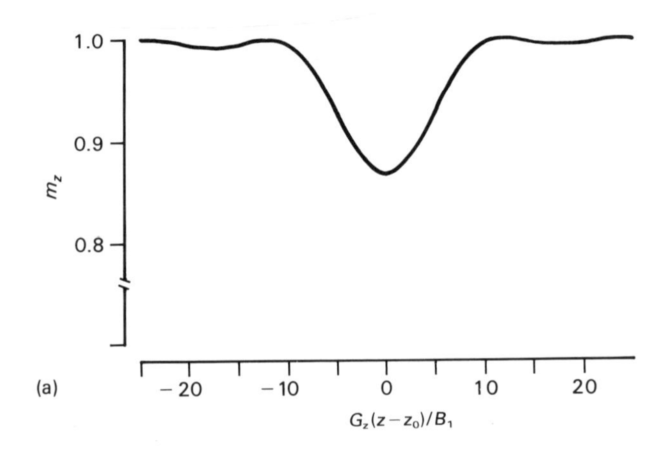
1.0
0.9
0.8
m_z
(a)
−20
−10
0
10
20
$G_z(z-z_0)/B_1$

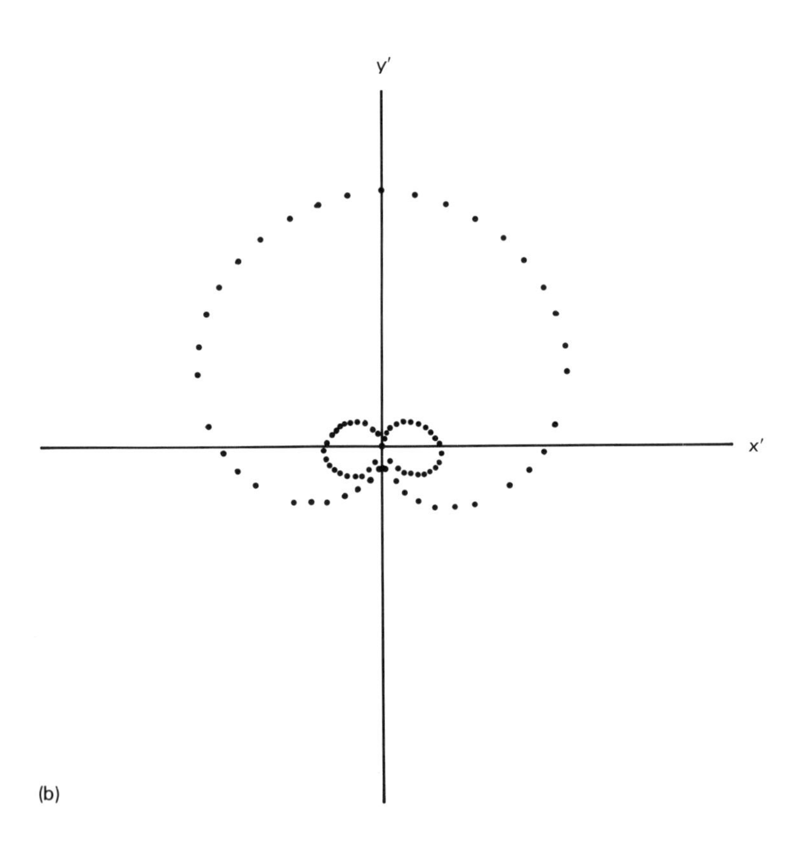
y′
x′
(b)

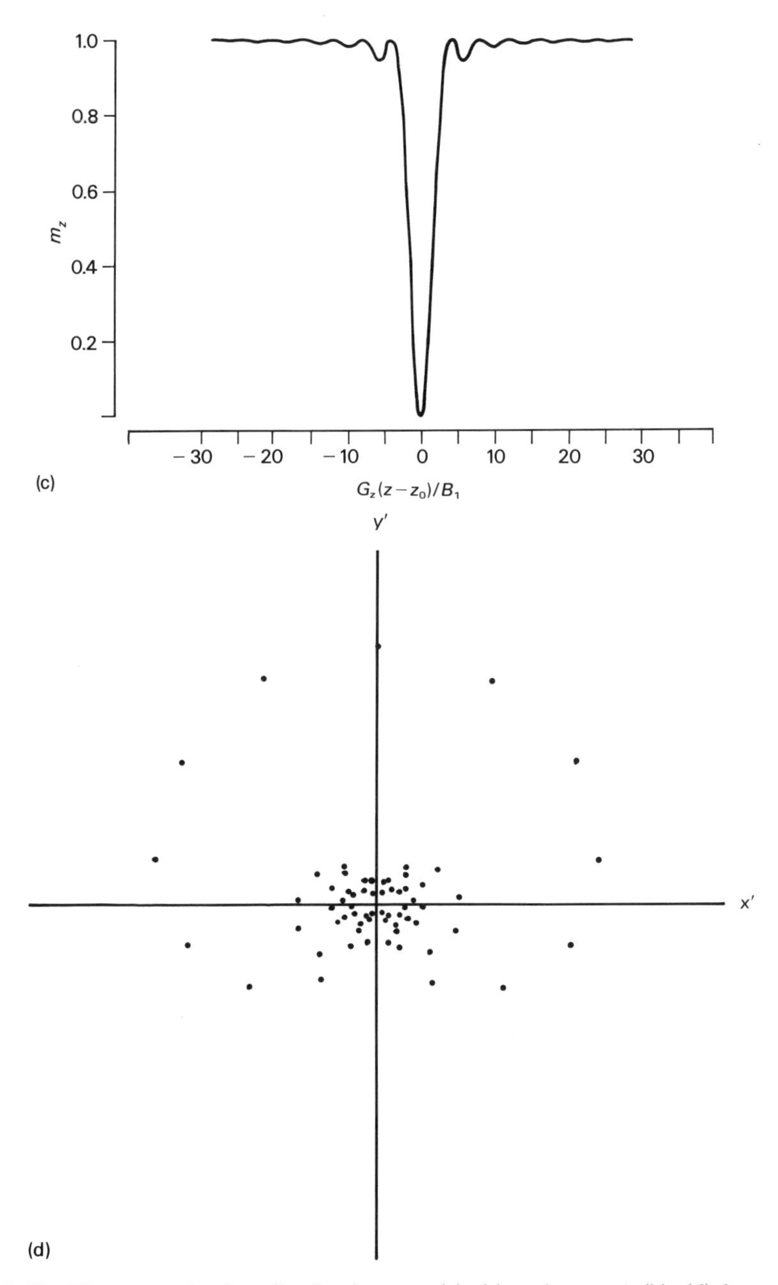

Fig. 3.12. The magnetization distributions m_z (a), (c) and $m_{x'y'}$, ϕ (b), (d) for two values of t_w, corresponding respectively to 30° (a), (b) and 90° (c), (d) pulses for on-resonance spins. Each successive point in the phase diagrams corresponds to an additional displacement of $B_1/2G_z$ along the z axis.

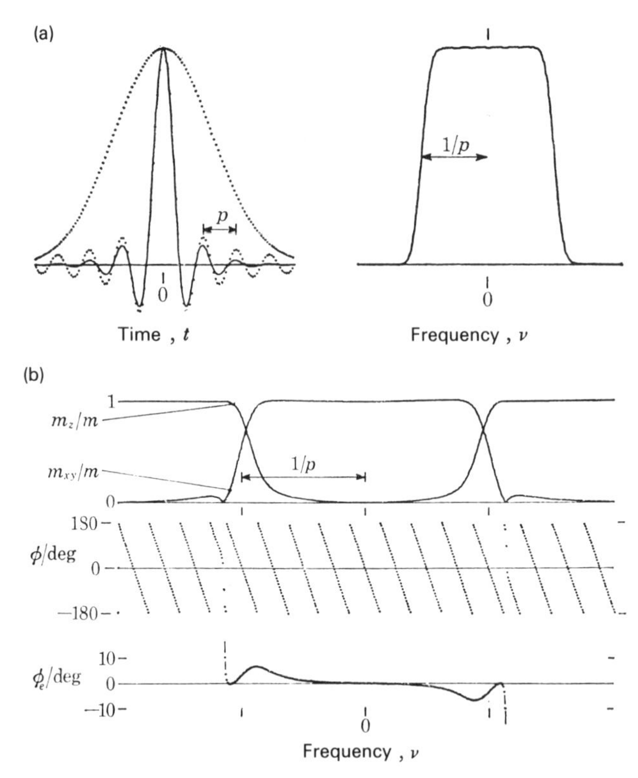

Fig. 3.13. (a) Tailored excitation function (left) and its Fourier transform (right). The dotted lines give the sinc and the Gaussian function of which the excitation function is the product. (b) Magnetization at the end of the selective pulse shown in (a) for $\theta = 90°$. The lower curve gives the phase error at the centre of the echo. (From Locher 1980.)

(Mansfield *et al.* 1979*a*). The $m_{x'}$ component, as we have seen, is antisymmetric for a uniform sample and cannot therefore contribute to the ensuing signal. The $m_{y'}$ component has a spatial distribution which looks like the expected sinc function for $G_z(z - z_0) \gg B_1$. When the sample is non-uniform over the width of the excited plane or line, the antiphase components of $m_{x'}$ do not cancel and give rise to phase and intensity anomalies. This effect is particularly pronounced near the periphery of a sample and many of the early line scan images show evidence of such 'edge artefacts'. A much more satisfactory solution to the above pattern is suggested by a consideration of the origin of the phase shifts. These arise from the relative precession of the spin isochromats in the effective field (see Fig. 3.10(a)). For field offsets which are large compared with the amplitude, i.e. for $\gamma G_z(z - z_0) \gg B_1$, the phase shift introduced is proportional to the frequency offset $\gamma G_z(z-z_0)$ and hence to the linear displacement from the selected line/plane. See Fig. 3.12 and eqn (3.12*b*) which in the above limit reduces to $\phi=\tan^{-1}\{\cot[\gamma G_z(z-z_0)t/2]\}$ so that the accumulated relative phase is $\gamma G_z(z-z_0)t_w/2$. It is therefore possible largely to eliminate these phase shifts by the simple expedient of field-gradient reversal. This was an integral part of the line-scanning technique described by Sutherland and Hutchison (1978; see also Hutchison, Edelstein, and Johnson 1980) and was the subject of a paper by Hoult (1979). Figure 3.14 reproduces some of the results from the latter work and shows the experimentally-observed magnetization in the presence of a field gradient following 180° (a), 90° (b), and 30° (c) rectangular selective pulses of duration 10 ms (magnetization is not observable during and immediately following the pulse due to receiver saturation and recovery). The signal following the 30° pulse is virtually zero due to the distribution of the phase angles for the different isochromats. These phase differences can be removed by switching the direction of the field gradient or, as in this case, by the application of a non-selective 180° pulse, phase shifted by 90° with respect to the selective pulse. The result is an echo centred at $t=t_w/2+\tau$, where τ is the time between the end of the selective pulse of width t_w and the field gradient reversal or 180° non-selective pulse (remember the phase difference builds up as if it started at the centre of the excitation pulse—hence the $t_w/2$, see above). Having achieved more or less complete rephasing of the isochromats (see Fig. 3.13(b) for the phase distribution of the isochromats at the echo maximum resulting from the field gradient reversal following a sinc-modulated Gaussian pulse) the field gradient can be removed to lock in the constant (zero) phase. Subsequently, the f.i.d. can be recorded in an orthogonal read gradient, as in the case of line-scan and projection reconstruction methods, or else in the orthogonal switched gradient of an echo-planar experiment.

Notice that in the 30° case the echo closely mimics the envelope of the applied selection pulse. In fact, the departure from the true rectangular distribution is another manifestation of the non-linearity of the spin system's response to r.f. fields (see Hoult 1977, 1979). This is more apparent with the 90° selective pulse and even more so with the 180° one, where the signal following the pulse is far from zero and the echo itself is anything but rectangular. The evolution in a field gradient following a selective pulse can be easily calculated by Fourier transformation of the distributions $m_{x'}(t_w)$, $m_{y'}(t_w)$ which can themselves be determined analytically as above or by numerical integration of the Bloch equations (see Locher 1980). For example, Fig. 3.14(d, e, f) shows the analytical results for the situation corresponding to Hoult's experiments (Mansfield and Morris 1982). Agreement between theory and experiment is excellent.

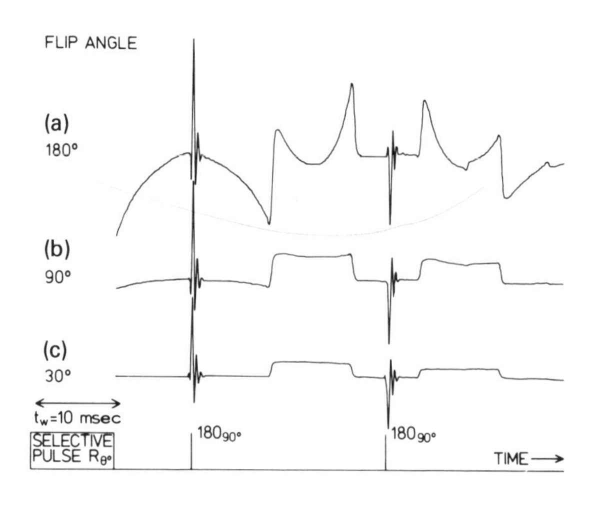

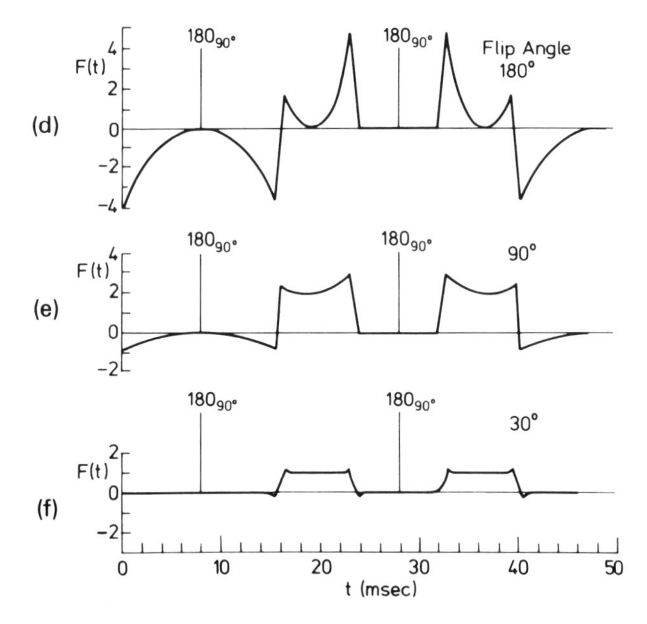

Fig. 3.14. Response of uniform spin distribution in a linear gradient to rectangular selective pulses of length 10 ms, followed by two non-selective 180° refocusing pulses. (a) On-resonance 180° selective pulse; (b) 90° selective pulse; (c) 30° selective pulse (Hoult 1979); (d), (e), (f) Theoretical responses for conditions corresponding to (a), (b), (c) respectively.

Note, however, that the experimental echoes following the second reversal (non-selective 180° pulse) are diminished in amplitude as a result of T_2-relaxation effects which were neglected in the theoretical treatment. The non-linear behaviour manifest in these experiments, gives rise to non-perfect refocusing, i.e. a phase distribution at the echo maximum which is not perfectly flat. Although this may correspond to only a few degrees variation over the excited region (see Fig. 3.13(b)), It is nevertheless important to minimize these effects in order to eliminate the small phase and intensity anomalies that would otherwise ensue. Some progress can be made in this direction by feeding back the measured errors into phase adjustments of the r.f. pulse in an iterative process (Ordidge 1981; see also Silver, Joseph, and Hoult 1983). This corresponds to making successive subtle alterations to the pulse envelope. For an example of what can be achieved in this manner, see Fig. 3 of Sutherland and Hutchison (1978). Hoult (1979) has developed a perturbation treatment which describes these non-linear effects. This may well be of help in the approach to excitation pulses which give truly flat phase responses.

The selective pulses described so far will excite a region centred about $z = z_0$, i.e. those spins lying in a magnetic field whose resonance condition corresponds to the frequency of the r.f. carrier wave. There are a number of alternative methods by which this plane or line can be scanned through the object. Least attractive is the physical movement of the object relative to the magnet in the manner of a f.o.n.a.r. experiment. Another approach would be to alter the current balance in the G_z gradient coils so as to move the resonance field through the sample. Alternatively, the same result could be achieved by altering the frequency of the r.f. carrier wave. Finally, it is possible to harmonically modulate the selective pulse which has the effect of moving the centre of the excitation spectrum by an amount equal to the (low, usually audio) modulation frequency. This frequency can then be changed to step the selected region through the object. In the method of Mansfield *et al* (1976) this modulation is applied automatically by adding the appropriate offset to the desired excitation profile prior to the Fourier transform which determines the pulse envelope. Unless single sideband (s.s.b.) modulation methods (see § 5.3.2.2) are used, however, two excitation bands are obtained, centred at $\omega \pm \omega_m$ where ω and ω_m are the carrier and modulation frequencies respectively. In order to limit excitation to a single region, it has to be arranged that when one area of the sample is excited by a particular sideband, no other region lying within the receptivity of the receiver coil is resonant at the other sideband frequency. In practice, this is easily ensured by working at offsets which are sufficiently large for the condition

$$\omega_m > \gamma G_z L_z/2, \tag{3.15}$$

to be fulfilled. (L_z is the extent of the object in the z-direction of the imaging field.) Note also that, in the absence of s.s.b. modulation, half of the r.f. power is 'wasted' in the unwanted sideband. We should add that these problems apply *a fortiori* to the d.a.n.t.e. sequence and the tailored excitations of Freeman and Hill (1971), relying as they do on pulse-width modulation techniques.

3.3.3. Fast scan imaging

Following the early selective irradiation experiments of Garroway *et al.* (1974), and Lauterbur, Dulcey, Lai, Feiler, House, Kramer, Chen, and Dias (1974), Mansfield *et al.* (1976) published details of a more efficient line-scanning method which they called 'fast scan imaging'. It was used extensively for small scale proton imaging (Mansfield and Maudsley 1976*a*, *b*, 1977*a*; Mansfield and Pykett 1978; Pykett and Mansfield 1978) with a home-converted spectrometer operating at 15 MHz. The same scheme was later used for human studies at 4 MHz (Mansfield, Pykett, Morris, and Coupland 1978; Mansfield, Morris, Ordidge, Coupland, Bishop, and Blamey 1979*b*; Morris, Mansfield, Pykett, Ordidge, and Coupland 1979; Pykett, Mansfield, Morris, Ordidge, and Bangert 1980; Mansfield, Morris, Ordidge, Pykett, Bangert, and Coupland 1980). Their basic sequence is shown in Fig. 3.15. In the full version of the experiment a field gradient G_z is applied to the sample which is then subjected to a selective saturation pulse (see § 3.3.1). This saturates all the spins apart from those lying in a narrow plane normal to the z-direction. (The position and width of this plane are respectively controlled by the offset and profile of the saturating pulse.) In the next phase the z-gradient is replaced by one in y-direction followed by the application of a 90° selective excitation pulse (§ 3.3.1). In the absence of any initial saturation

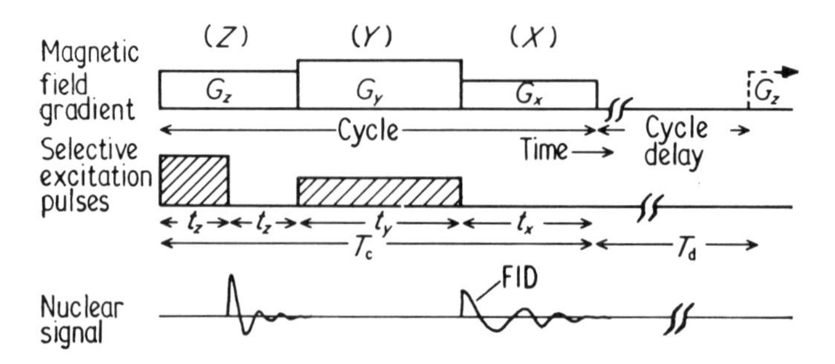

Fig. 3.15. Pulse sequence for fast scan imaging'. Only the f.i.d. in the final (G_x) gradient is sampled. (From Mansfield, Maudsley, and Baines 1976, copyright Institute of Physics 1976.)

the effect of the selective excitation would be to nutate all the spins lying in a narrow plane normal to y into the $x'y'$-plane of the rotating frame. However, the effect of the selective saturation is to ensure that the only magnetization which can be excited by the 90° pulse lies in a narrow strip or line at the intersection of the two planes normal to y and z. In the final read phase of the cycle the field gradient orientation is again switched, this time to the x-direction, and the f.i.d. is recorded.

Under most circumstances it is necessary to signal average by co-adding a number of f.i.d.s prior to Fourier transformation. It is then possible to achieve greater imaging speed by incrementing the line position after each scan. When one has accumulated a single f.i.d. for every line in the final image the procedure is repeated n times to build up the required S/N. Moving the line after each f.i.d. has the advantage that one is continually exciting fresh magnetization whereas averaging on a single line generally requires a delay between scans in order to allow the z magnetization to recover via a T_1 relaxation process (see § 2.5.1). If the complete image can be scanned in a time which is less than T_1 then the efficiency of the method in terms of image S/N per unit time, approaches that of the planar techniques. (A similar argument applies to planar methods where it is possible to excite a number of planes (typically up to 15) during the recovery period.) Note that such a method presupposes that one can change the line offset quickly. In the original implementation in which each line irradiation pattern was computed separately, this would not have been possible. However, with the inexpensive computer memory now available all the patterns can be prestored for immediate use. Alternatively, they could be calculated rapidly with the aid of an array processor or else a hardware approach could be adopted using a voltage controlled oscillator to change the operating frequency. In the case of long T_1 values an additional time-saving can be achieved if it is possible to collect a number of f.i.d.s before reapplying the selective saturation pulse.

Usually one is also interested in introducing T_1 effects since, as we shall see in Chapter 6, these give rise to much greater tissue contrast than spin density information alone. The simplest way in which this can be achieved is to average a single line with an interpulse interval TR (between successive 90° selective irradiation pulses) which is short compared with T_1. In this manner one obtains an effective spin density ρ' related to the real spin density ρ_0 through the expression (Mansfield and Maudsley 1976*b*)

$$\rho' = \rho_0\{1 - \exp(-\mathrm{TR}/T_1)\}. \tag{3.16}$$

(Note that this is the same expression as that applying for the f.o.n.a.r. method, eqn (3.1).) If the interpulse delay $\mathrm{TR} \gg T_1$ there is complete recovery and $\rho' \simeq \rho_0$. On the other hand, as TR is reduced below T_1, relaxation effects become dominant. Such a scheme is rather like the progressive saturation method for measuring T_1 values sometimes used in high-resolution spectroscopy (see § 2.5.1.1).

If such a method is used to try and obtain quantitative T_1 values in an imaging experiment, one has to give very careful consideration to r.f. pulse imperfections. For example, if the excitation function of the selective pulse is anything other than rectangular, the signal from the wings of the selected region will become increasingly important as the repetition rate is increased and the central region becomes progressively saturated. Effectively, this amounts to a variation in the dimension of the excited region with interpulse interval. Exactly the same problem arise in the case of those planar methods which utilize selective excitation.

In all the line-scanning experiments of Mansfield and co-workers the selective irradiation pulse has been determined for each line by digital Fourier transformation of the required excitation spectrum. This means that the sample is effectively irradiated at a discrete set of frequencies and the analysis of § 3.3.2 has to be modified to take this into account (Mansfield *et al.* 1979*a*; Mansfield and Morris 1982). In practice, the best way to calculate the response of the spin system is to use the actual pulse modulation function applied to the r.f. The stepped nature lends itself particularly well to a matrix treatment (Jaynes 1955) which allows the spatial response function to be optimized (Ordidge 1981).

If selective saturation pulses are used for plane definition the spatial response may not be clean for the reasons outlined in § 3.3.1. Most fast scan experiments therefore relied on the receiver coil geometry or the use of a filtering gradient for this purpose. Early experiments were also performed in the absence of any refocusing, a feature which would undoubtedly have been of benefit.

An extension of fast scan imaging, known as planar imaging, has been proposed by Mansfield and Maudsley (1976*b*, *c*, 1977*b*). It involves the simultaneous irradiation of several image lines followed by detection in the presence of a combined preparation and read gradient G_y and G_x (equivalent to a single gradient at an angle $\tan^{-1}(G_y/G_x)$ to the x-axis). In an even more ambitious scheme known as multiplanar imaging (Mansfield and Maudsley 1976*c*, 1977*b*) the authors envisaged extending the selective irradiation to sets of lines in distinct image planes coupled with observation in a triple gradient G_x, G_y, G_z. Whereas these techniques greatly increased the imaging speed they did so at the expense of imaging efficiency (image S/N per unit time), the reason being that the width of the lines and consequently the amount of sample contributing to each picture element, had to be reduced in order to allow discrimination between successive lines and planes (Brunner and Ernst 1979). The method was abandoned in favour of echo-planar imaging which also increases the imaging speed but retains imaging efficiency.

3.3.4. *Focused selective excitation*

In 1976, Hutchison proposed a method of line-scanning by selective irradiation based on the use of Gaussian-shaped pulses. Refocusing of the $x'y'$-magnetization in the manner of § 3.3.2 became an integral part of the procedure, which was refined by computer-aided analysis (Sutherland and Hutchison 1978) culminating in its use for human studies in 1979 (Mallard, Hutchison, Edelstein, Ling, Foster, and Johnson 1980; Hutchison *et al.* 1980). These experiments were carried out using a four-coil air-cored electromagnet at a field of 0.04 T corresponding to a frequency of 1.7 MHz for protons (see Fig. 5.8).

The imaging procedure is essentially a difference method, in the first stage of which the pulse sequence shown in Fig. 3.16 is applied to the subject. The combination of the 180° Gaussian pulse and the G_z gradient selectively inverts those spins lying in a narrow plane normal to z. The subsequent application of a refocused 90° selective pulse nutates all spins lying in a plane normal to y into the $x'y'$-plane of the rotating frame. However, those spins which occupy the region (line) defined by the intersection of this plane with the one normal to z will be 180° out of phase with the remainder. The signal S_a is recorded in the presence of an orthogonal read gradient G_x, starting at the echo maximum. It consists of

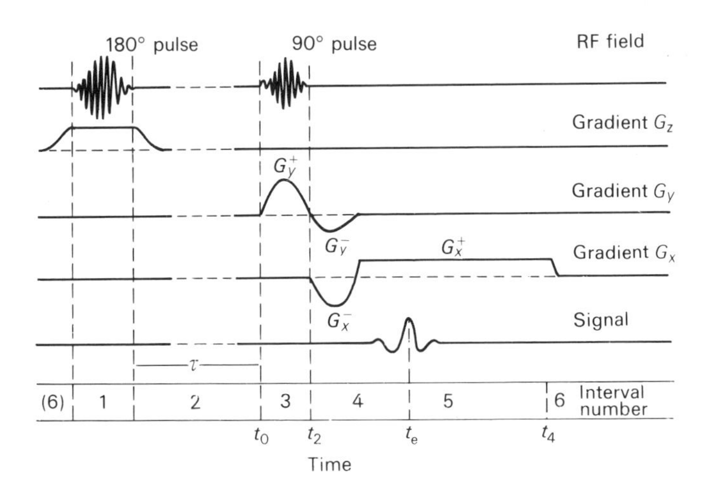

Fig. 3.16. Pulse scheme for 'focused selective excitation'. (From Hutchison *et al.* 1980, copyright Institute of Physics 1980.)

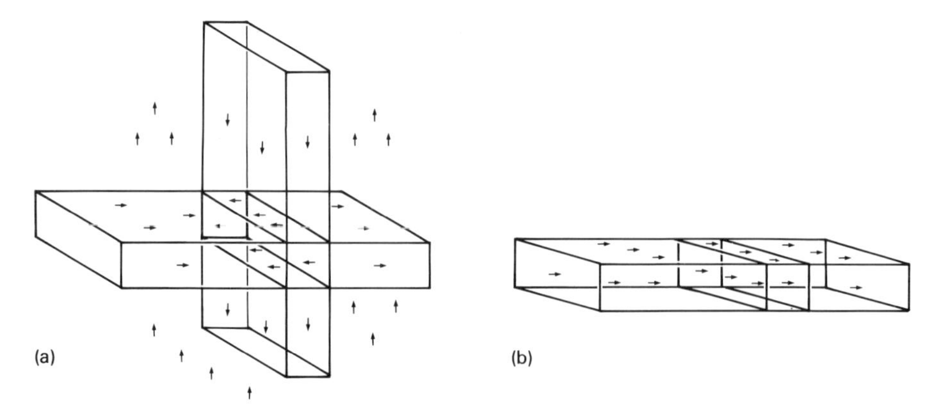

Fig. 3.17. Schematic diagram showing orientations of nuclear magnetization following the *a* (a) and *b* (b) phases of the focused selective excitation experiment.

two components in antiphase, S_1 and S_2, corresponding to the contributions from the selected line and the remainder of the selected plane respectively (see Fig. 3.17(a)). In the second stage of the experiment the 180° selective pulse is omitted. This time the two signal components S_1 and S_2 are in phase giving a total signal S_b (see Fig. 3.17(b)). Thus:

$$S_a = -S_1 + S_2 \tag{3.17a}$$

$$S_b = \quad S_1 + S_2. \tag{3.17b}$$

By subtracting the results of these two experiments one can thereby obtain the response from the selected line, S_1. Thus:

$$2S_1 = S_b - S_a. \tag{3.18}$$

This simple form of the experiment suffers from errors which arise from the imperfect nature of the 180° pulse which creates an unwanted component of magnetization in the $x'y'$-plane. This can be removed by improving the selectivity of the 180° pulse or by performing a third experiment in which its phase is shifted by 90° (Sutherland and Hutchison 1978). If the signal arising from this stage is denoted S_c then the line expression S_1 is given by

$$4S_1 = 2S_b - (S_a + S_c). \tag{3.19}$$

The line position was incremented to build up an image by changing the r.f. carrier frequency.

In the scheme discussed above the 90° pulse follows directly after the 180° one and, provided the delay between experiments is sufficiently large

($\sim$few T_1), the image reflects the spin-density distribution of the object. T_1 contrast can be introduced either by cycling more rapidly as in the case of fast scan imaging (§ 3.3.3), or more effectively by introducing a delay TI between the 180° and 90° pulses. The sequence then becomes very reminiscent of the non-selective 180°–TI–90° inversion recovery sequence, which is generally the method of choice for T_1 measurement in high-resolution spectroscopy (see § 2.5.1.1). By making measurements for a set of TI values it would be possible to determine accurately the T_1 distribution of the sample as well as its spin density. In practice, a single TI value of the order of T_1 is used and the sequence is repeated roughly once every second.

The method, whether used for T_1 measurement or simply for spin density mapping, suffers from a major drawback which is common to

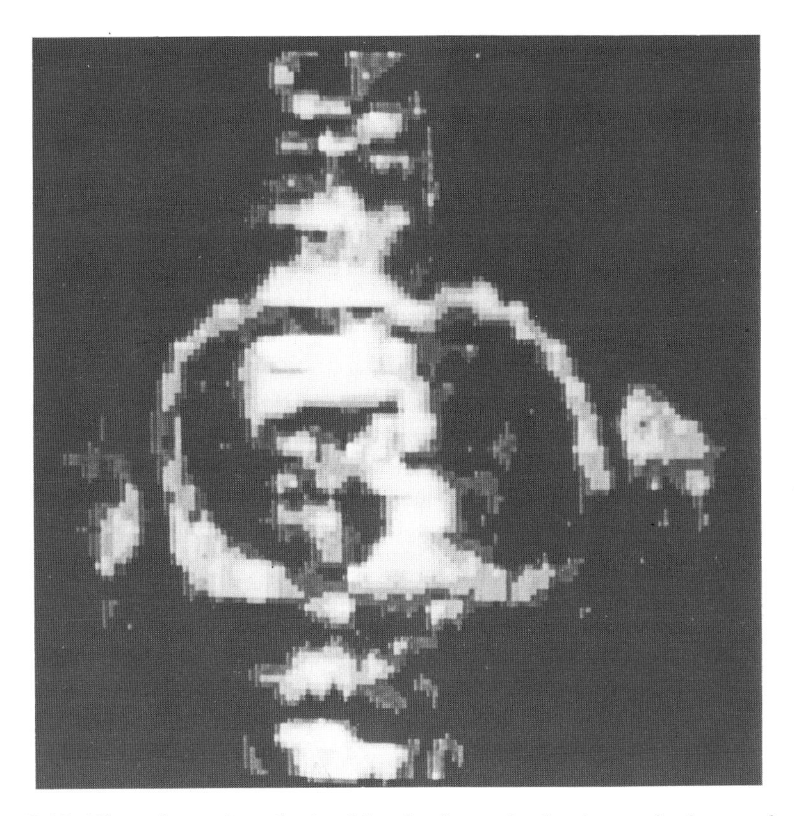

Fig. 3.18. Thoracic section obtained by the focused selective excitation method. Left and right lung cavities are clearly depicted, but the heart is obscured by an intense motional artefact (vertical bar). The arms are seen, but internal detail is lost due to the lower field homogeneity in the peripheral regions. (From Hutchison, Edelstein, and Johnson 1980, copyright Institute of Physics 1980.)

many difference methods, namely the presence of motional artefacts. These arise from the movement of the subject between the two (or three) stages of the imaging process. Thus, the images of test-tube phantoms, or the relatively static parts of the human anatomy, head and limbs for example, are of reasonably high quality and are free from motional artefacts (Mallard *et al.* 1980; Hutchison *et al.* 1980). The *in vivo* thoracic image reproduced in Fig. 3.18, on the other hand, shows an enormous artefact derived from the motion of the heart. In such cases it may prove possible to circumvent most of the problem by gating the scans from the e.c.g. However, there are other involuntary motions present which can give rise to difficulties, and one is forced to the conclusion that such difference methods are generally best avoided. This is the conclusion of the Aberdeen team, too, and they have since devoted their attention to a more efficient scheme known as spin-warp imaging. We discuss this in § 4.2.3.

3.3.5. *Selective spin-echo methods*

A line-scanning method which avoids both the problems of selective saturation and of difference measurements was proposed by Maudsley in 1980. It again makes use of selective 180° and 90° pulses but the order is reversed from Hutchison's focused selective excitation method. The scheme closely resembles the Carr–Purcell echo sequence used, with non-selective pulses, for the measurement of spin–spin relaxation times (see § 2.5.2.2). The basic scheme, illustrated in Fig. 3.19, commences

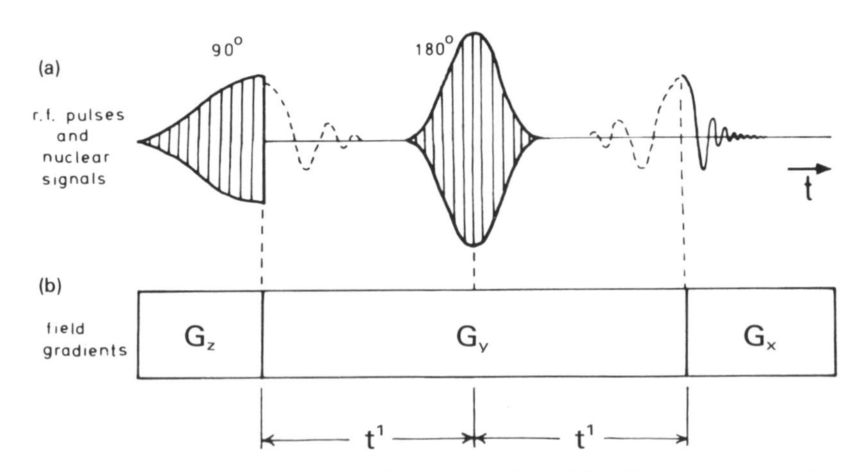

Fig. 3.19. The selective spin–echo imaging method. (a) Pulse sequence used for excitation of the volume $x\Delta y\Delta z$ shown in (b). (From Maudsley 1980.)

with a selective 90° pulse which nutates the spins in a plane normal to z into the $x'y'$-plane. The gradient direction is then switched from z to y and a time t^1 elapses during which the spin isochromats dephase. A selective 180° pulse along the y-direction then causes those spins lying at the intersection of the two selected planes to refocus after a further time t^1. Provided $t^1 > T_{2e}$ no transients remain from the rest of the sample and a clean line parallel to the x-direction is therefore defined. The spin density is observed in the usual manner by application of a field gradient G_x. Since the z-magnetization in the remainder of the imaging plane (i.e. excluding the selected line) has not been destroyed, merely inverted, there is no need for a relaxation recovery interval and one can proceed immediately to a new line offset. In this manner one can scan through the image plane taking note of the sign alternation deriving from the successive inversions of the remaining z-magnetization. When the full scan has been completed, the z-magnetization in the imaging plane is largely destroyed and a T_1 recovery period is then necessary. Alternatively, this can be integrated into the scanning cycle, either by introducing a delay between scans and averaging a single image line in a manner similar to that used for fast scan imaging (see eqn (3.16)) to give T_1 contrast, or as suggested by Maudsley (1980), by incrementing the line after each scan and introducing the delay following every second line.

As we should expect from a technique based on the Carr–Purcell sequence, the method is potentially sensitive to variations to T_2. By making t^1 short, one effectively eliminates T_2 effects and the image corresponds to a spin density map. However, as t^1 is increased and becomes comparable with the mean sample T_2, the image intensity becomes increasingly sensitive to variations of this parameter.

Maudsley analysed a number of possible pulse shapes for the selective 90° excitation. He came down in favour of a Hanning function $(0.5(1-\cos t/T))$ since it gives a smoothly varying in-phase magnetization distribution without refocusing. For the 180° selective pulse a symmetric function such as a Gaussian is suitable.

Crooks *et al.* (Crooks, Hoenninger, Arakawa, Kaufman, McRee, Watts, and Singer 1979, 1980; Crooks 1980) have used a very similar approach for their line-scanning method, choosing as a selective 90° pulse a curtailed sinc function. They have obtained good-quality proton density, T_1 and T_2 images of live rats using a Varian 30 cm electromagnet operating at 15 MHz for protons (Hansen, Crooks, Davis, De Groot, Herfkens, Margulis, Gooding, Kaufman, Hoenninger, Arakawa, McRee, and Watts 1980; Davis, Kaufman, Crooks, and Margulis 1982; Davis, Kaufman, and Crooks 1982). In general, of course, the image depends on all three of these parameters, their separation requiring a number of experiments (in this case four). Such a separation may well prove to be of importance in gaining an understanding of the origins of tissue contrast although, from a practical point of view, one would wish to use the parameters which optimized tissue contrast for a given clinical condition. See Chapter 6 for a fuller discussion of this point.

As with the other line-scanning methods, rapid movement of the subject leads to the loss of information in the line during which it occurred, leaving the remainder of the image unaffected. A crude visualization of blood flow can thus be derived from these effects (Crooks *et al.* 1980).

Maudsley (1980) has suggested ways of decreasing the imaging time of the spin-echo method. These involve the simultaneous irradiation of a number of image lines and possibly image planes. The excitation pulses are thus of a similar nature to those proposed by Mansfield and Maudsley (1976*b*, *c*; 1977*b*) for the planar and multiplanar imaging methods (see § 3.3.3). The difference lies in the field gradients used to record the f.i.d.s; rather than use a gradient which is inclined to the direction of the excited lines, causing resolution and/or S/N degradation, Maudsley proposed the retention of a simple co-linear read gradient. It is then not possible to distinguish between the various lines in a single experiment. However, suppose that m lines are simultaneously irradiated, and that in each of m experiments a different one of the m lines is excited with negative phase. It is then possible to extract the signals from any of the lines by simple algebraic manipulation. One thereby obtains an m-fold average for each line. Note that in essence this is a difference technique which is susceptible to motional artefacts. However, provided m is reasonably small, this will not present such severe problems as occur when the signal required is the small difference between two much larger quantities. Signal-to-noise can further be improved by recalling the magnetization as a series of echoes in the manner of a Carr–Purcell–Meiboom–Gill sequence (Meiboom and Gill 1958).

3.3.6. The multiple sensitive point method

The multiple sensitive point, or m.s.p., method differs from the sensitive point method (see § 3.2.3) in that it uses two rather than three oscillating field gradients. These serve to define an image line; discrimination along the line being achieved through the application of a linear field gradient parallel to the line. The basic approach is illustrated in Fig. 3.20. The technique was again developed by Hinshaw and colleagues at Nottingham who first used the method with an intermediate-sized imaging system, achieving excellent spatial resolution in objects up to 8 cm in diameter (see Andrew, Bottomley, Hinshaw, Holland, Moore, and Simaroj 1977; Hinshaw, Bottomley, and Holland 1977; Hinshaw, Andrew, Bottomley,

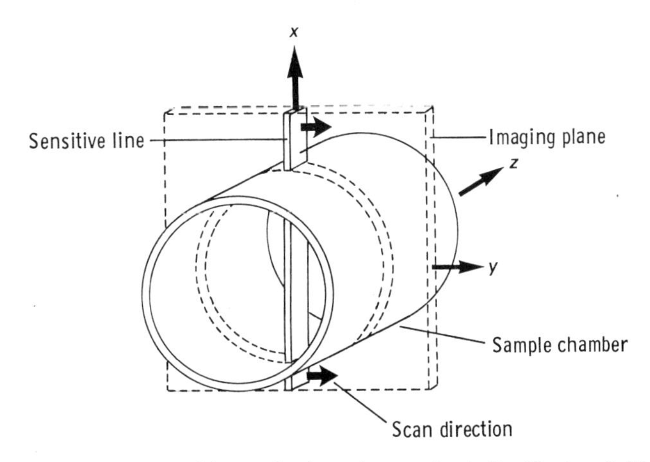

Fig. 3.20. The multiple sensitive point imaging method. Oscillating field gradients along the y- and z-directions generate a sensitive line parallel to x. Discrimination of image points along this line is achieved with a static field gradient G_x. (From Andrew *et al.* 1977, copyright Institute of Physics 1977.)

Holland, Moore, and Worthington 1978; Hinshaw and Holland 1978; Andrew, Bottomley, Hinshaw, Holland, Moore, Simaroj, and Worthington 1978; Bottomley 1978*a*, *b*; Hinshaw, Bottomley, and Holland 1979; Hinshaw, Andrew, Bottomley, Holland, Moore, and Worthington 1979; Bottomley 1979; Moore and Holland 1980*a*; Bottomley 1981). More recently, Moore and co-workers have developed a head-scanning system based on the same principle (Moore and Holland, 1980*a*, *b*; Holland, Moore, and Hawkes 1980*a*, *b*).

The technique again takes advantage of the steady-state free precession (s.f.p.) method but the signal has to be handled in a rather more sophisticated fashion than the simple integration used with the sensitive point method. The first difference arises from the digitization of the signal; typically some 128 (2^7) sample points are acquired between r.f. pulses. Successive sets of sample points are co-added, after removal of the sign alternation, in order to build up S/N and also to allow the oscillating gradients to achieve their spatial localization. For this purpose it is again necessary to arrange that the number of s.f.p. intervals which one averages to give a particular image-line coincides with an integral number of gradient cycles.

3.3.6.1. *Spatial resolution.* The two orthogonal oscillating field gradients used for line definition may be derived from separate asynchronous supplies or, more conveniently, from the quadrature outputs of a single supply. In the latter case the spatial response will have cylindrical symmetry about the sensitive line and depend only on the radial distance r from it. Meiere and Thatcher (1979) have analysed this response in depth and the interested reader is referred to their paper for details. The problem is a difficult one and the analytical solution, as obtained by Meiere and Thatcher, is unfortunately not separable into the product of two functions depending on r and x. This means that the cross-sectional area of the sensitive line depends on the distance along it. This is illustrated in Fig. 3.21, which shows the response as a function of the radial co-ordinate $\lambda = a\xi$ ($a = \gamma G_r T \operatorname{sinc}(\Omega T/2)$, where G_r is the strength and Ω the angular frequency of the radial oscillating gradient, and T is the interpulse interval) for various displacements along the line, $\xi = \gamma G_x Tx$ where G_x is the strength of the fixed gradient. The x-dependence is therefore seen not to be too severe, and Meiere and Thatcher have suggested an approximation which is separable, of the form

$$S(r, x) \sim \frac{\sin \xi}{\xi} \frac{1}{[1+(r/r_{\frac{1}{2}})^2]^{\frac{3}{2}}}, \tag{3.20}$$

where $r_{\frac{1}{2}} = 2.4\gamma G_r T$. This is shown as a broken line in Fig. 3.21. The x-dependence of eqn (3.20) is similar to that derived by Mansfield and Morris (1982) for the case of s.f.p. in the presence of a single time-independent field gradient. For 90° r.f. pulses, and in the limit $T < T_1$, T_2, they obtain an expression for the spatial sensitivity along the sensitive

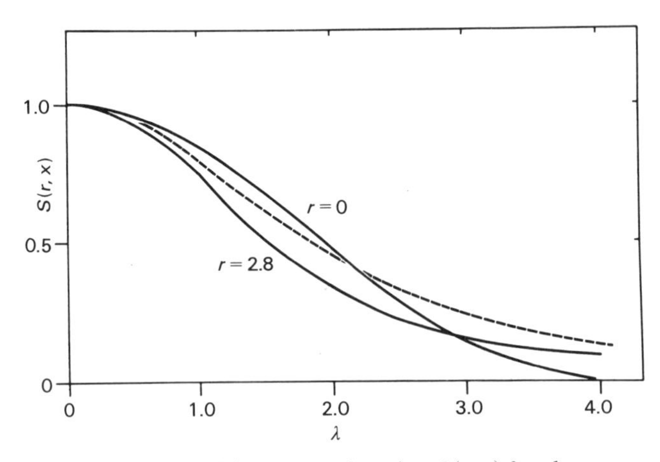

Fig. 3.21. The complete spatial response function $S(r, x)$ for the m.s.p. method. λ is proportional to r the distance from the sensitive line. The solid lines are extreme values depending on x. The broken line is an illustrative fit to $[1+(r/r_{\frac{1}{2}})^2]^{\frac{3}{2}}$ with $r_{\frac{1}{2}} = 2.4/\gamma GT$, the half-width of the magnetization. (From Meiere and Thatcher 1979.)

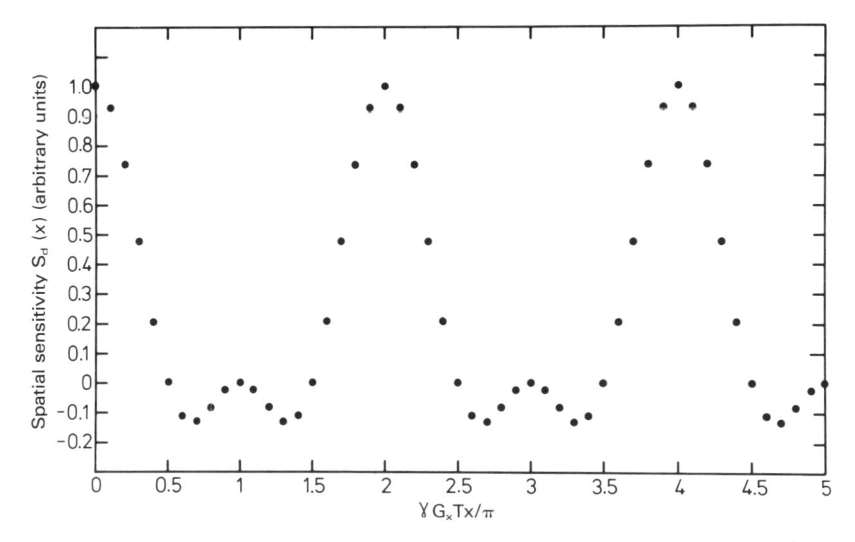

Fig. 3.22. The summed m.s.p. response $S_d(x)$ to an isochromat of spins. (From Mansfield and Morris 1982.)

line. This is illustrated in Fig. 3.22 and shows quite clearly the 'sensitive points' distributed along the sensitive line at intervals given by

$$x = 2\pi/\gamma G_x T. \tag{3.21}$$

Note that between these points there are regions which contribute no signal at all and others which contribute a (small) negative signal. This should not be construed as a particular failing of the m.s.p. method, however; all methods which use discrete Fourier transforms suffer from this problem in some form. Figure 3.22 shows the total signal $S_d(x)$ derived from a spin isochromat located at the indicated x-value; it gives no indication of where this signal is located in the frequency spectrum. This is illustrated in Fig. 3.23 for various values of x. For example, at a sensitive point the signal is seen to contribute only at the corresponding frequency point, $\omega = 2n\pi/T$, whereas an isochromat located exactly between two sensitive points does not contribute at all. Intermediate isochromats contribute to a range of perhaps 3–5 frequency points, and their contributions may be positive, negative, or a mixture of both. Note that the frequency response is symmetric about the midpoint of adjacent sensitive points i.e. $R(-x, \omega) = R(x, -\omega)$.

3.3.6.2. *Experimental considerations.* In their intermediate-sized m.s.p. imaging system operating at 30 MHz for protons, Hinshaw and colleagues used strings of phase-alternated pulses of approximately 20 μs duration

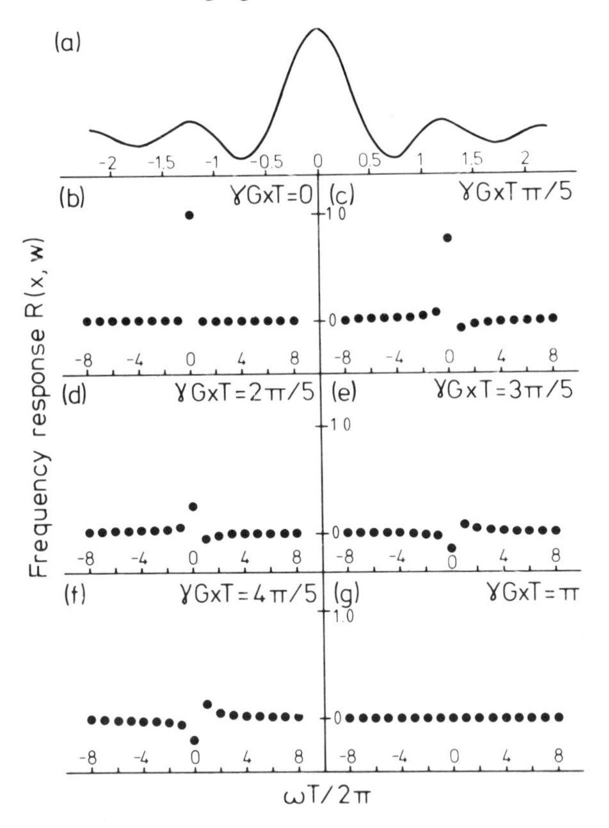

Fig. 3.23. The m.s.p. response to an isochromat of spins: (a) sinc function response $R(0, \omega)$ to an isochromat at $x=0$; (b) digital response to an isochromat at $x=0$, (c)–(g) digital responses to isochromats at (c) $x=\pi/5\gamma GT$; (d) $x=2\pi/5\gamma GT$; (e) $x=3\pi/5\gamma GT$; (f) $x=4\pi/5\gamma GT$; and (g) $x=\pi/\gamma GT$. (From Mansfield and Morris 1982.)

spaced some 2 ms apart ($T=2$ ms). Each interpulse interval was sampled 128 times ($N=128$) and somewhere in the region of 1000 intervals were averaged per image line. In this way 128 volume elements were made to span a $60\times0.5\times2$ mm 'line'.

It is important that T be made reasonably small in order that no significant T_2 decay occurs in the interpulse interval, i.e. to fulfil the condition $T \ll T_1, T_2$; solid-like regions for which $T_2<2$ ms will not therefore be observed. Even if T does satisfy this condition the signal response (for 90° r.f. pulses) will be proportional to $\rho_0/(1+T_1/T_2)$ and hence depends strongly on the ratio T_1/T_2. For liquids T_1 is often approximately equal to T_2 and the signal attains its optimum value, being proportional to $\rho_0/2$. However, for solids T_1/T_2 may easily be $\sim$10 000! Biological water corresponds to an intermediate case, with a ratio 10 being typical.

Another reason for making T small is that it determines the bandwidth per point in the image line. Thus a 2 ms interval gives a point bandwidth of 500 Hz and hence, for $N = 128$ sample points, a total signal bandwidth of 64 kHz. Such large bandwidths enable one to work in fields of modest homogeneity, for example 1 part in 10^4, equivalent to 400 Hz at a 4 MHz operating frequency, in the case of the whole-body system described by Moore and Holland (1980*a*, *b*).

Having chosen the number of sample points N, the static field gradient amplitude G_x must be adjusted so that the sample spans the image field. Thus, as we have seen (eqn (3.21)), the separation between adjacent sensitive points along the line is $2\pi/\gamma G_x T$. Ideally, we should like the sample to just occupy the full length of the sensitive line. If the extent of the sample in the x-direction is L_x, we then require that

$$x = 2\pi/\gamma G_x T = L_x/N, \tag{3.22a}$$

giving

$$G_x = 2\pi N/\gamma T L_x. \tag{3.22b}$$

Taking the case of $N = 128$, $T = 2$ ms and $L_x = 25$ cm (a head-scanning system) gives a value for G_x of $0.6\ \mathrm{G\,cm^{-1}}$. For a whole-body-sized system the gradient strength required is proportionately lower (provided N remains the same).

Since excitation is performed in the presence of the linear gradient, the r.f. pulses must be of sufficiently short duration to excite the full bandwidth of $2N\pi/T$ Hz (64 KHz for the parameters used in the above examples). For 90° pulses in a coil of volume V the r.f. power required to achieve this is given by (Moore and Holland 1980*a*)

$$P \simeq \frac{2\pi^3}{\mu_0 \gamma^2} \cdot \frac{VN^3}{T^3}. \tag{3.23}$$

Taking V as $0.01\ \mathrm{m^3}$, $T = 2$ ms and $N = 128$, we obtain a value of about 1.8 kW. This level of r.f. power can be obtained without too much difficulty, but important consideration should be given to the safety aspects (see § 6.8.3).

The s.f.p method continues to find some advocates in the imaging community. However, the m.s.p. method itself has been largely abandoned in favour of the more efficient planar and three-dimensional methods to be discussed in Chapter 4.

References

Abe, Z., Tanaka, K. Hotta, M., and Imai, M. (1974). In *Biologic and clinical effects of low frequency magnetic and electric fields*, pp 295–317. Thomas, Springfield, Illinois.

Ackerman, J. J. H., Grove, T. H., Wong, G. G., Gadian, D. G., and Radda, G. K. (1980). *Nature, Lond.* **283,** 167.

Andrew, E. R., Bottomley, P. A., Hinshaw, W. S., Holland, G. N., Moore, W. S., and Simaroj, C. (1977). *Phys. med. Biol.* **22,** 971.

———, Bottomley, P. A., Hinshaw, W. S., Holland, G. N., Moore, W. S., Simaroj, C., and Worthington, B. S. (1978). *Proc. 20th Congr. Ampere,* pp. 53–56. Tallinn, U.S.S.R.

Aue, W. P., Müller, S., Cross, T. A., and Seelig, J. (1984). *J. magn. Reson.* **56,** 350.

Bendall, M. R., and Aue, W. P. (1983). *J. magn. Reson.* **54,** 149.

———, and Gordon, R. E. (1983). *J. magn. Reson.* **53,** 365.

Bodenhausen, G., Freeman, R., and Turner, D. L. (1977). *J. magn. Reson.* **27,** 511.

Bottomley, P. A. (1978*a*). Ph.D. Thesis, University of Nottingham.

——— (1978*b*). *Cancer Topics* **2,** 4.

——— (1979). *Cancer Res.* **39,** 468.

——— (1981). *Experientia* **37,** 768.

——— (1982). *J. magn. Reson.* **50,** 335.

Brunner, P., and Ernst, R. R. (1979). *J. magn. Reson.* **33,** 83.

Campbell, I. D. Dobson, C. M., Williams, R. J. P., and Xavier, A. V. (1973). *J. magn. Reson.* **11,** 172.

———, ———, Ratcliffe, R. G., and Williams, R. J. P. (1978). *J. magn. Reson.* **29,** 397.

Carr, H. Y. (1958). *Phys. Rev.* **112,** 1693.

Crooks, L. E. (1980) *IEEE Trans.* N.S. **27,** 1239.

———, Grover, T. P., Kaufman, L., and Singer, J. R. (1978). *Invest. Radiol.* **13,** 63.

———, Hoenninger, J., Arakawa, M., Kaufman, L., McRee, R., Watts, J., and Singer, J. R. (1979). SPIE, *Recent and future developments in medical imaging II* **206,** 120.

———, ———, ———, ———, ———, ———, and ——— (1980). *Radiology* **136,** 701.

Damadian, R. (1980). *Phil. Trans. R. Soc.* B **289,** 379.

———, Goldsmith, M., and Minkoff, L. (1977). *Physiol. Chem. Phys.* **9,** 97.

———, ———, and ——— (1978). *Physiol. Chem. Phys.* **10,** 285.

———, Minkoff, L., Goldsmith, M., Stanford, M., and Koutcher, J. (1976*a*). *Physiol. Chem. Phys.* **8,** 61.

———, ———, ———, ———, and ——— (1976*b*). *Science* **194,** 1430.

———, ———, ———, and Koutcher, J. (1978). *Naturwissenschaften* **65,** 250.

Davis, P. L., Kaufman, L. and Crooks, L. E. (1982). *Proc. int. symp. NMR Imaging,* p. 101. Bowman Gray School of Medicine, Winston-Salem.

———, ———, ———, and Margulis, A. R. (1981). In *NMR imaging in medicine,* p. 71 (eds. Kaufman, L., Crooks, L. E. and Margulis, A. R.). Igaku-Shoin, New York.

Evelhoch, J. L., and Ackerman, J. J. H. (1983). *J. magn. Reson.* **53,** 52.

Fitzsimmons, J. R. (1982). *Rev. scient. Instrum.* **53,** 1338.

Freeman, R., and Hill, H. D. W. (1973). *J. magn. Reson.* **4,** 366.
Garroway, A. N., Grannell, P. K., and Mansfield, P. (1974). *J. Phys.* C **7,** L457.
Goldsmith, M., Damadian, R., Stanford, M., and Lipkowitz, M. (1977). *Physiol. Chem. Phys.* **9,** 105.
Gordon, R. E. (1981). *Phys. Bull.* **32,** 178.
———, Hanley, P. E., and Shaw, D. (1982). *Prog. NMR Spectrosc.* **15,** 1.
———, ———, ———, Gadian, D. G., Styles, P., Bore, P. J., and Chan, L. (1980). *Nature, Lond.* **287,** 367.
Hanley, P. E., and Gordon, R. E. (1981). *J. magn. Reson.* **45,** 520.
Hansen, G., Crooks, L. E., Davis, P. L., De Groot, J., Herfkins, R., Margulis, A. R., Gooding, C., Kaufman, L., Hoenninger, J., Arakawa, M., McRee, R., and Watts, J. (1980). *Radiology* **136,** 695.
Hinshaw, W. S. (1974*a*). *Phys. Lett.* **48A,** 87.
——— (1974*b*). *Proc. Ampere Congr., 18th,* Nottingham, p. 433, North-Holland, Amsterdam.
——— (1976). *J. appl. Phys.* **47,** 3709.
———, and Holland, G. N. (1978). *Trends biochem. Sci.* N84.
———, Bottomley, P. A., and Holland, G. N. (1977). *Nature, Lond.* **270,** 722.
———, ———, and ——— (1979). *Experientia* **35,** 1268.
———, Andrew, E. R., Bottomley, P. A., Holland, G. N., Moore, W. S., and Worthington, B. S. (1978). *Br. J. Radiol.* **51,** 273.
———, ———, ———, ———, ———, and ——— (1979). *Br. J. Radiol.* **52,** 36.
Holland, G. N., Moore, W. S., and Hawkes, R. C. (1980*a*). *Br. J. Radiol.* **53,** 253.
———, ———, and ——— (1980*b*). *J. comput. assist. Tomogr.* **4,** 1.
Hoult, D. I. (1977). *J. magn. Reson.* **26,** 165.
——— (1979). *J. magn. Reson.* **35,** 69.
Hutchison, J. M. S. (1976). In *Proc. 7th L. H. Gray Conf., Medical images: formation, perception and measurement* (ed. G. A. Hay). John Wiley, New York.
———, Edelstein, W. A., and Johnson, G. (1980). *J. Phys.* E: *Scient. Instrum.* **13,** 947.
———, Sutherland, R. J., and Mallard, J. R. (1978). *J. Phys.* E: *Scient. Instrum.* **11,** 217.
Jaynes, E. T. (1955). *Phys. Rev.* **98,** 1099.
Lauterbur, P. C., Dulcey, Jr., C. S., Lai, C-M., Feiler, M. A., House Jr. W. V., Kramer, D. M., Chen, C-N., and Dias, R. (1974). *Proc. Ampere Congr. 18th,* Nottingham, p. 27. North-Holland, Amsterdam.
Lindon, J. C. and Ferrige, A. G. (1980). *Prog. NMR Spectrosc.* **14,** 27.
Ljunggren, S. (1983*a*). *J. magn. Reson.* **54,** 338.
——— (1983*b*). *J. magn. Reson.* **54,** 165.
Locher, P. R. (1980). *Phil. Trans. R. Soc.* B **289,** 537.
Mallard, J. R., Hutchison, J. M. S., Edelstein, W. A., Ling, C. R., Foster, M. A., and Johnson, G. (1980). *Phil. Trans. R. Soc.* B **289,** 519.
Mansfield, P., and Maudsley, A. A. (1976*a*). *Phys. med. Biol.* **21,** 847.
———, and ——— (1976*b*). *Proc. Ampere Congr., 19th,* Heidelberg, p. 247. Groupment Ampere, Heidelberg.
———, and ——— (1976*c*). *J. Phys.* C **9,** L409.

———, and ——— (1977*a*). *Br. J. Radiol.* **50,** 188.

———, and ——— (1977*b*). *J. magn. Reson.* **27,** 101.

———, and Morris, P. G. (1982). *NMR imaging in biomedicine*, suppl, *2, Adv. Mag. Res.*, ed. J. S. Waugh, Academic Press, New York.

———, and Pykett, I. L. (1978). *J. magn. Reson.* **29,** 355.

———, Maudsley, A. A. and Baines, T. (1976). *J. Phys.* E: *Scient. Instrum.* **9,** 67.

———, ———, Morris, P. G., and Pykett, I. L. (1979*a*). *J. magn. Reson.* **33,** 261.

———, Morris, P. G., Ordidge, R. J., Coupland, R. E., Bishop, H. M., and Blamey, R. W. (1979*b*). *Br. J. Radiol.* **52,** 242.

———, ———, ———, Pykett, I. L., Bangert, V., and Coupland, R. E. (1980). *Phil. Trans. R. Soc.* B **289,** 503.

———, Pykett, I. L., Morris, P. G., and Coupland, R. E. (1978). *Br. J. Radiol.* **51,** 921.

Maudsley, A. A. (1980). *J. magn. Reson.* **41,** 112.

Meiboom, S., and Gill, D. (1958). *Rev. Scient. Instrum.* **29,** 688.

Meiere, F. T., and Thatcher, F. C. (1979). *J. appl. Phys.* **50,** 4491.

Minkoff, L., Damadian, R., Thomas, T. E., Hu, N., Goldsmith, M., Koutcher, J., and Stanford, M. (1977). *Physiol. Chem. Phys.* **9,** 101.

Moore, W. S., and Holland, G. N. (1980*a*). *Br. Med. Bull.* **36,** 297.

———, and ——— (1980*b*). *Phil. Trans. R. Soc.* B **289,** 381.

Morris, G. A., and Freeman, R. (1978). *J. magn. Reson.* **29,** 433.

Morris, P. G., Mansfield, P., Pykett, I. L., Ordidge, R. J., and Coupland, R. E. (1979). *IEEE Trans. nucl. Sci.* **26,** 2817.

Morse, O. C., and Singer, J. R. (1970). *Science* **170,** 440.

Ordidge, R. J. (1981). Ph.D. Thesis, University of Nottingham.

Partain, C. L., James, A. E., Watson, J. T., Price, R. R., Coulam, C. M., and Rolls, F. D. (1980). *Radiology* **136,** 767.

Pykett, I. L., and Mansfield, P. (1978). *Phys. med. Biol.* **23,** 961.

———, ———, Morris, P. G., Ordidge, R. J., and Bangert, V. (1980). In: Lecture notes in Physics 112, Imaging processes and coherence in physics; *Proc. Les Houches 1979* (eds. Schlenker, M., Fink, M., Goedgebuer, J. P., Malgrange, C., Vie-not, J. Ch., and Wade, R. H.). Springer-Verlag, Berlin and New York.

Scott, K. N., Brooker, H. R., Fitzsimmons, J. R., Bennett, H. F., and Mick, R. C. (1982). *J. magn. Reson.* **50,** 339.

Silver, M. S., Joseph, R. I., and Hoult, D. I. (1983). In 'Work in progress', Annual meeting of Society of Magnetic Resonance in Medicine, San Francisco, August.

Stevens, A. N. S., Morris, P. G., Sheldon, P. W., and Griffiths, J. R. (1984). *Br. J. Cancer* **50,** 113.

Sutherland, R. J., and Hutchison, J. M. S. (1978). *J. Phys.* E: *Scient. Instrum.* **11,** 79.

Tanaka, K., Yamada, Y., Shimizu, T., Sano, F., and Abe, Z. (1974). *Biotelemetry* **1,** 337.

———, ———, Yamamoto, E., and Abe, Z. (1978. *Proc. IEEE* **66,** 1582.

4. Two- and three-dimensional n.m.r. imaging methods

4.1. Projection reconstruction

4.1.1. Introduction

We have discussed in Chapter 2 how, as a consequence of the fundamental Larmor relationship ($\omega = \gamma B$), the application of a static magnetic field gradient to a sample causes all the spins lying in a plane normal to the gradient direction to be resonant at the same frequency and how that frequency is proportional to the displacement of the plane along the gradient direction. (Thus, for a gradient in the z-direction, centred at $z=0$, the frequency offset $\Delta\omega$ for a plane at z_0 is given by $\Delta\omega = \gamma G z_0$.) This is illustrated for three specific planes in the object of Fig. 4.1. The strength of the signal at any frequency will thus be directly proportional to the number of nuclei lying in the corresponding object plane

$$S(\Delta\omega) = \iint \rho(x, y, \Delta\omega/\gamma G)\, \mathrm{d}x\, \mathrm{d}y, \tag{4.1}$$

where $\rho(x, y, \Delta\omega/\gamma G)$ is the (two-dimensional) spin density in the plane $z = \Delta\omega/\gamma G$ and the n.m.r. frequency distribution or spectrum corresponds to a one-dimensional projection of the object normal to the field gradient direction. By altering this direction, projections or views of the object in different orientations are obtained, and, given a sufficient number, it is possible to reconstruct an image of the original object. This is the process of projection reconstruction. Although, as we shall see, the operation can be carried out in three dimensions yielding an $n \times n \times n$ matrix, two-dimensional images require proportionately fewer projections for reconstruction ($n \times n$) and are therefore to be preferred where imaging times are an important consideration (see below). In this case, one of the established techniques, such as selective irradiation, has to be used to isolate a plane of spins within the larger region defined by the active volume of the transmitter coil (see § 3.3).

The problem of determining a function such as $\rho(x, y, z)$ from the plane integrals (see eqn (4.1)) was first approached and solved by Radon (1917) in connection with gravitational theory. The first image reconstruction techniques, however, were not developed until 1956 when Bracewell used them to map the distribution of microwave emissions from the solar

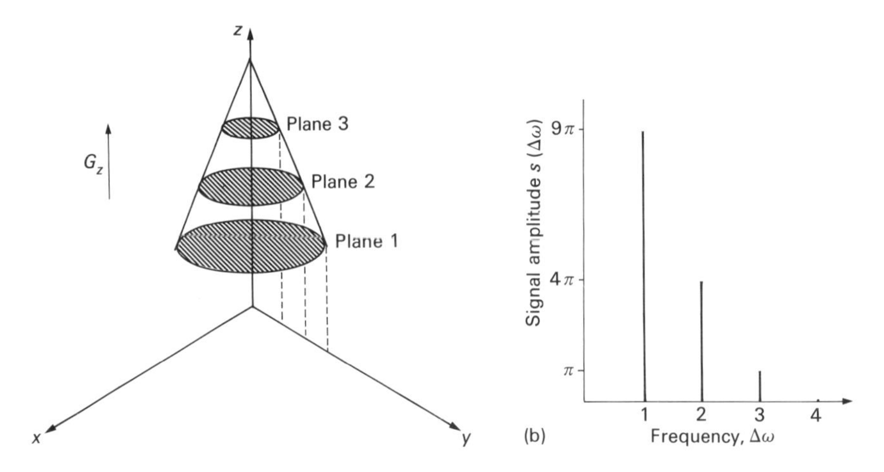

Fig. 4.1. (a) Application of a linear field gradient G_z to a uniform spin density cone of height 3 units and semi-vertical angle 45°. (b) Signal amplitude at frequencies γG_z, $2\gamma G_z$, $3\gamma G_z$ arising from the corresponding planes shown shaded in (a).

surface. Later, reconstruction methods were independently developed for the determination of molecular structure from transmission micrographs, and, again independently, for optical applications.

The first n.m.r. image was published in *Nature* by Paul Lauterbur in 1973 (see Fig. 4.2); it, too, was based on projection reconstruction techniques which, like their predecessors in the other scientific disciplines, evolved independently. Lauterbur christened his technique 'zeugmatography' from the Greek ξευγμα, 'that which is used for joining', referring to the coupling together of the field gradient and r.f. irradiation in the process of image formation. Other workers were also pursuing projection reconstruction methods at an early stage (Mansfield, Grannell, and Maudsley 1974; Hutchison, Mallard, and Goll 1974). Interest was rapidly kindled amongst the n.m.r. community and the number of groups active in the area proliferated, with researchers taking advantage of the great flexibility of n.m.r. to develop their own methods.

What really put projection reconstruction on the map was the release in 1972 of the EMI scanner, an X-ray projection-reconstruction system producing high-quality cross-sectional images, initially of the head and subsequently of the body—an invention which has revolutionized modern radiology and for which Hounsfield shared the 1979 Nobel prize for medicine (his co-recipient Cormack received recognition for the development of accurate reconstruction techniques, work which began in the late 1950s and came to fruition in the early 1960s (Cormack 1963, 1964)). It

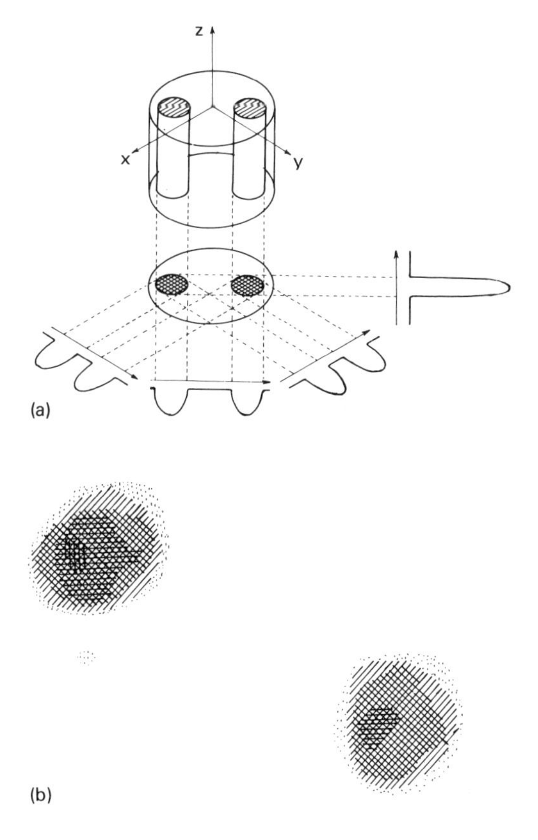

Fig. 4.2. (a) Relationship between a three-dimensional object, its two-dimensional projection along the z-axis and four one-dimensional projections at 45° intervals in the xy-plane. Arrows indicate gradient directions. (b) N.m.r. image reconstructed from the four projections shown in (a). (From Lauterbur 1973. Reprinted by permission from *Nature* **242,** 190 (1973).)

is interesting to note that Hounsfield devoted the latter part of his Nobel prize address to a discussion of the future of n.m.r. rather than X-ray CT methods (Hounsfield 1980).

The enormous fund of knowledge and experience gained on X-ray CT systems in the past decade made projection reconstruction a particularly attractive technique for manufacturers in the n.m.r. imaging field. It is an efficient technique which is relatively straightforward to implement. Commercial prototypes have already achieved excellent image quality using it and it featured prominently in first generation n.m.r. scanners. Currently, however, the trend is towards Fourier imaging-based systems which sample the spin density in a more uniform manner (§ 4.2).

There is a considerable literature on the subject of projection reconstruction. Gordon, Herman, and Johnson (1975) have given a very readable account of X-ray methods in a *Scientific American* article, whilst, for the more mathematically minded, the review article by Brooks and Di Chiro (1976) is recommended.

4.1.2. *Two-dimensional reconstruction algorithms*

4.1.2.1. Notation. Consider an image slice, as shown in Fig. 4.3, divided into a matrix of $n \times n$ image elements or pixels each of width w. The object of our reconstruction procedure will be to determine the correct value to assign to each pixel for the spin density ρ_{ij} (or any other n.m.r. parameter we choose to make the basis of our image—see § 4.1.7).

If a field gradient is applied to the sample at an orientation ϕ (see Fig. 4.4(a)), all spins lying on a particular line (or 'ray' to adopt the X-ray terminology) perpendicular to the gradient direction will resonate at a specific angular frequency offset $\Delta\omega$ (remember that the main frequency ω_0 is removed during phase sensitive detection—see § 2.4). This is given

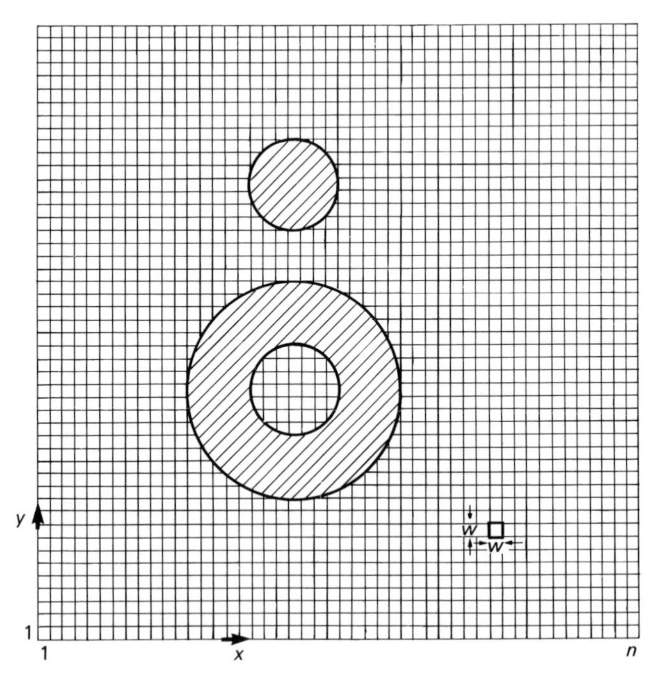

Fig. 4.3. Diagram showing superposition of a square grid of $n \times n$ pixels, each of side w, on a two-dimensional phantom consisting of a circle and an annulus with centres aligned along the y-axis.

by:

$$\Delta\omega(x, y) = \Delta\omega(r) = \gamma G_{\phi} r. \tag{4.2}$$

Here r is the perpendicular distance of the ray from the origin, itself defined as the gradient null point for which $\Delta\omega = 0$ for all orientations of G_{ϕ}. It is straightforward to show (see Fig. 4.4(b)) that the points x, y lying on a particular ray defined by (r, ϕ) satisfy the condition:

$$r = x \cos \phi + y \sin \phi, \tag{4.3}$$

a relationship we shall make use of later.

The total spin density contributing at a frequency $\Delta\omega$, or alternatively from eqn (4.2), to each ray (r, ϕ) is given by the line integral:

$$P(r, \phi) = \int_{r,\phi} \rho(x, y)\, \mathrm{d}s. \tag{4.4}$$

Here s is the distance measured along the ray direction. Each value of $P(r, \phi)$ is known as the 'ray sum' and the set of values for a given gradient orientation ϕ as the projection at angle ϕ (see Fig. 4.5). This is the function which we actually determine in a n.m.r. projection reconstruction experiment; it is simply the n.m.r. spectrum which is obtained by Fourier transforming the f.i.d. recorded in the gradient G_{ϕ}. By rotating the gradient angle, we collect a set of projections $P(r, \phi)$ which we use to reconstruct an image of the original object.

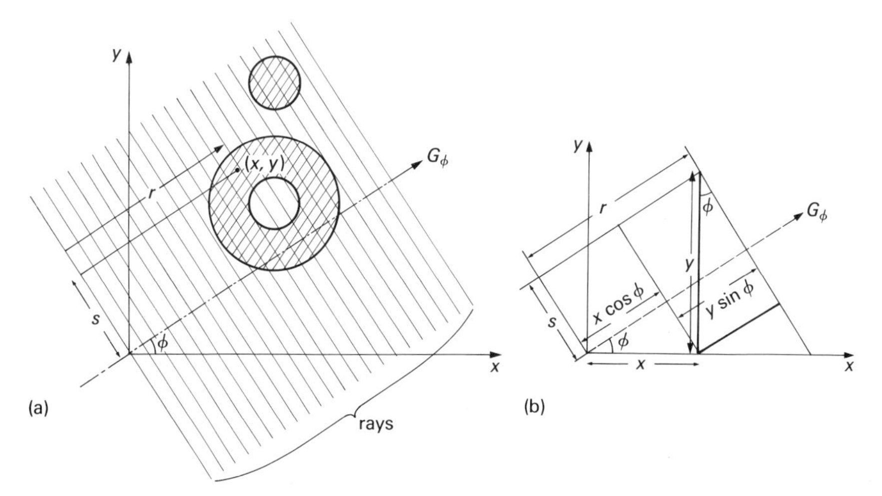

Fig. 4.4. (a) Application of a field gradient G_{ϕ} to the object of Fig. 4.3, showing rays and defining r, the location of the ray along the gradient direction, and s, the distance along the ray. (b) Demonstration of the relationship between x, y and r, s co-ordinate systems: viz. $r = x \cos \phi + y \sin \phi$.

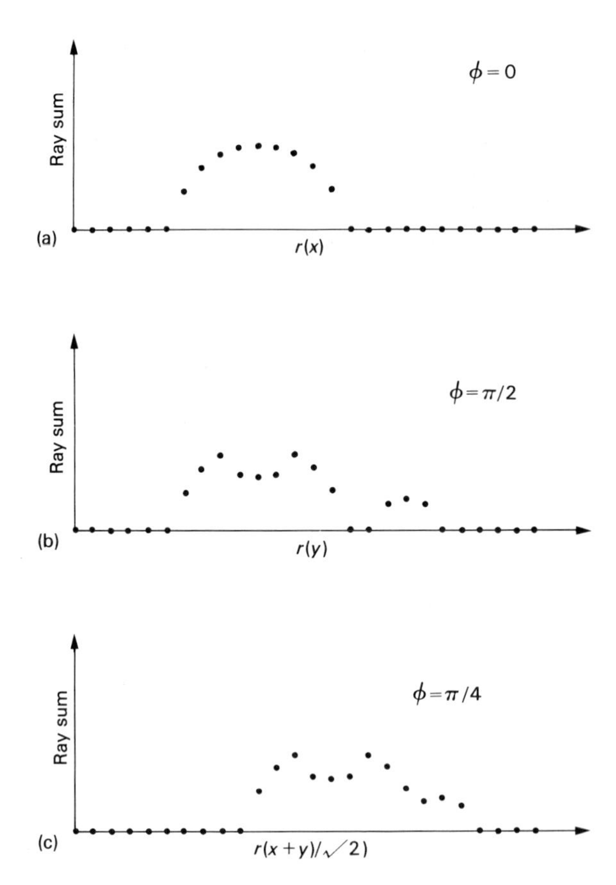

Fig. 4.5. Projections of the phantom of Fig. 4.3 in three directions, corresponding to $\phi = 0$ (a), $\phi = \pi/2$ (b) and $\phi = \pi/4$ (c). Each point gives the ray sum at the appropriate r value . For example, in (a) $r = 10$ corresponds to a summation of the spin density along a line parallel to y, passing through the centres of disc and annulus.

We now enquire as to the number of projections which are actually required. Suppose we wish to reconstruct onto a circular array with n pixels spanning its diameter. Then to determine an image properly we need to know the spin density at each of the $n^2\pi/4$ pixels lying within the circular boundary. If each projection consists of n ray sums then we require the number of projections m to be $n\pi/4$. For reconstruction onto a square matrix the total number of pixels is, of course, n^2, whence $m = n$. Note that for uniform resolution the projections should be uniformly spaced ($\Delta\phi = \text{constant}$).

4.1.2.2. Back projection. The simplest method of image reconstruction, and one widely used in the early development of X-ray CT, is the procedure of back-projection. This involves dividing each ray sum (i.e. each point on the projection profile) equally amongst all the pixels which lie along the corresponding ray. This 'smearing out' process is repeated for each of the projections in turn so that when the results are summed, reinforcement occurs in those image locations corresponding to high proton density in the object. Mathematically, the back-projection process can be described as:

$$\bar{\rho}(x, y) = \sum_{j=1}^{m} P(r, \phi_j)\, \Delta\phi, \tag{4.5}$$

where $\bar{\rho}(x, y)$ is the calculated spin density, $\phi_j = 1, m$ are the (m) different gradient orientations and $\Delta\phi$ is the angular increment between adjacent projections. Using the expression for r of eqn (4.3), this can be rewritten as:

$$\bar{\rho}(x, y) = \sum_{j=1}^{m} P(x \cos \phi_j + y \sin \phi_j, \phi_j)\, \Delta\phi. \tag{4.6}$$

A simple illustration is given in Fig. 4.6, which shows the reconstruction from eight projections (in practice, many more would be used) of a disc of uniform spin density. The back-projected rays overlap centrally to build up a disc-shaped image corresponding to the original object. However, even

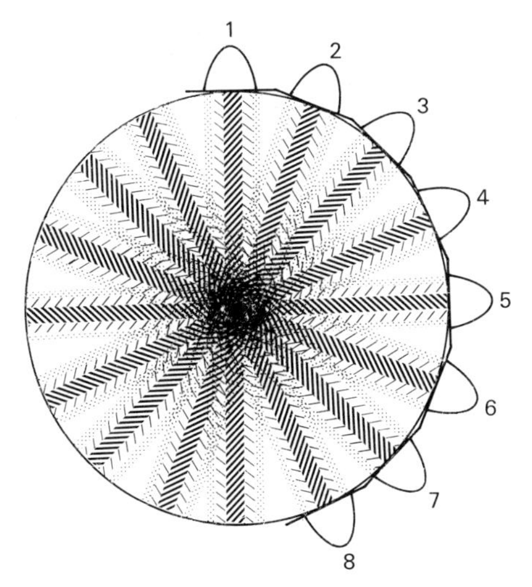

Fig. 4.6. Reconstruction of a disc from 8 projections by the method of back-projection.

the most cursory glance is sufficient to convince oneself that this method cannot be exact—the disc looks more like the hub of a bicycle wheel with the back-projected rays corresponding to the spokes. If a larger number of projections is employed, these spoke-like artefacts merge, giving rise to a uniform fogging of the background. A number of methods have evolved for the removal of such artefacts which are a common feature in back-projected images. However, as discussed in § 4.1.2.5, it is possible to correct or filter the projections prior to back-projection in a manner which makes the method analytic.

4.1.2.3. Iterative reconstruction. The iterative method of reconstruction, as the name suggests, involves making a series of approximations which (hopefully) converge to the true spin density in a small number of steps. From the starting approximation (often a uniform 'grey' matrix), the projections are calculated and compared with the measured ones. Appropriate corrections are then made and a second approximation derived. The various iterative procedures which have been employed differ in the type of correction used and in the manner in which it is applied. Thus, it is possible simultaneously to correct all the projections, applying a suitable damping factor since each pixel in the matrix is updated for each ray passing through it and would otherwise be grossly over-corrected. This is the iterative least squares technique or i.l.s.t. In a second method, known as the algebraic reconstruction technique or a.r.t., an individual projection is calculated and a correction applied. This updated information is then used in the calculation of the next projection. The method is a particularly efficient one and exists in a number of forms which differ in the way in which the correction is applied, for example multiplicative or additive a.r.t. One disadvantage is that it does not treat the projections equally; the later ones being more influential than the first. It is therefore important to ensure that projections are not processed in strict rotation. This fact was appreciated by Hounsfield who used the method for the early prototype EMI scanners.

Although such methods are not generally employed in modern X-ray CT scanning, they nevertheless still find frequent application with γ-ray systems since they allow the incorporation of beam attenuation factors in a fairly straightforward manner (Budinger and Gullberg 1974).

4.1.2.4. The central slice theorem. If the two-dimensional Fourier transform of $\rho(x, y)$ is denoted by $F(k_x, k_y)$ where k_x, k_y are the wavenumbers in x- and y-directions respectively, then:

$$F(k_x, k_y) = \int_{-\infty}^{\infty} \int_{-\infty}^{\infty} \rho(x, y) \exp[-2\pi i(k_x x + k_y y)]\, dx\, dy. \qquad (4.7)$$

This can be transformed from the (x, y) to the (r, s) co-ordinate system using the expression of eqn (4.3). Thus:

$$F(k_x, k_y) = \int_{-\infty}^{\infty} \int_{-\infty}^{\infty} \rho(x, y)\exp(-2\pi i k r)\, dr\, ds, \tag{4.8}$$

where $k = (k_x^2 + k_y^2)^{\frac{1}{2}}$ and ϕ, the angle relating (x, y) and (r, s) systems (see Fig. 4.4(a)), is given by $\phi = \tan^{-1}(k_y/k_x)$. Now the integral of $\rho(x, y)$ with respect to s is simply the projection $P(r, \phi)$ (see eqn (4.4)), so that

$$F(k_x, k_y) = \int_{-\infty}^{\infty} P(r, \phi)\exp(-2\pi i k r)\, dr \tag{4.9a}$$

$$= H(k, \phi) \tag{4.9b}$$

where $H(k, \phi)$ is the Fourier transform of $P(r, \phi)$ with respect to r. This is the central slice theorem which states that the Fourier coefficient of the spin density, $F(k_x, k_y)$, is equal to the Fourier transform of the projection taken in the direction of the Fourier wave defined by the wavenumbers k_x, k_y.

This provides the basis for the analytic method known as Fourier reconstruction in which one Fourier transforms the projections, and interpolates the Fourier coefficients thus obtained to achieve a square array in Fourier space before finally performing an inverse two-dimensional Fourier transform to get the spin density $\rho(x, y)$. Figure 4.7 illustrates the procedure for the case of our simple disc-shaped object. Whilst Fourier reconstruction appears an attractive technique in principle, in practice the two-dimensional interpolation proves to be a very time-consuming operation making the method less appealing than the now almost universally-employed filtered back-projection technique.

4.1.2.5. Filtered back-projection. Consider the expression for the back-projected spin density $\bar{\rho}(x, y)$, eqn (4.6), in its integral form:

$$\bar{\rho}(x, y) = \int_{0}^{\pi} P(x \cos \phi + y \sin \phi, \phi)\, d\phi. \tag{4.10}$$

If we replace $P(x \cos \phi + y \sin \phi, \phi)$ by its Fourier representation (see eqn (4.9)) this becomes

$$\bar{\rho}(x, y) = \int_{0}^{\pi} \int_{-\infty}^{\infty} \frac{H(k, \phi)}{|k|} \exp[2\pi i k(x \cos \phi + y \sin \phi)]\, |k|\, dk\, d\phi. \tag{4.11}$$

Taking the Fourier transform of $\bar{\rho}(x, y)$ we see that

$$\bar{F}(k_x, k_y) = H(k, \phi)/|k|, \tag{4.12}$$

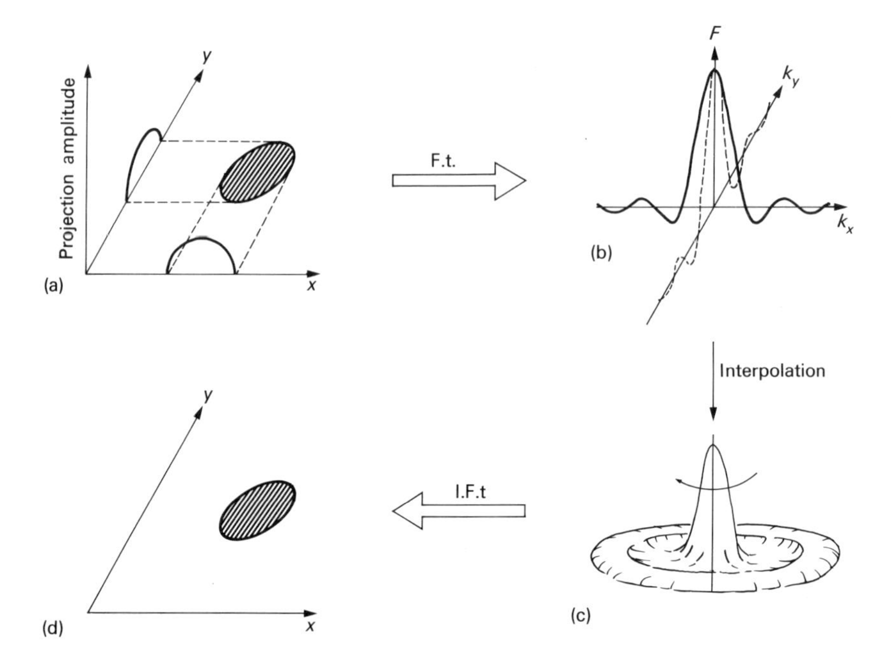

Fig. 4.7. Two-dimensional Fourier reconstruction of a disc phantom. (a) Projections along x- and y-axes. (b) Fourier transformation of the projections, giving Airy functions along k_x and k_y axes. (c) Full two-dimensional Fourier transform (corresponding to Airy ring diffraction pattern in Fourier optics). (d) Inverse Fourier transformation giving image of disc.

and, using the central slice theorem, that:

$$\bar{F}(k_x, k_y) = F(k_x, k_y)/|k|. \tag{4.13}$$

In other words, the actual Fourier coefficients $F(k_x, k_y)$ of the object can be obtained from those of the simple back-projected image by multiplying by the wave vector amplitude $|k|$. This suggests that back-projection can be made to work analytically provided the projections are appropriately modified prior to back-projection. To see how this can be achieved consider—

$$\rho(x, y) = \int_0^{\pi} \int_{-\infty}^{\infty} F(k, \phi)\exp(2\pi ikr)|k|\,\mathrm{d}k\,\mathrm{d}\phi, \tag{4.14}$$

which can be rewritten as:

$$\rho(x, y) = \int_0^{\pi} P^*(x\cos\phi + y\sin\phi, \phi)\,\mathrm{d}\phi, \tag{4.15}$$

where

$$P^*(r, \phi) = \int_{-\infty}^{\infty} |k|\, F(k, \phi) \exp(2\pi i k r)\, \mathrm{d}k. \tag{4.16}$$

Thus it is the modified projections $P^*(r, \phi)$ which we need to back project. They are obtained from the measured projections $P(r, \phi)$ through the relationship of eqn (4.16). This involves Fourier transforming $P(r, \phi)$ to obtain $H(k, \phi)(= F(k, \phi))$, multiplication by the wave-vector amplitude $|k|$ (a filtering operation, since it increases the Fourier amplitudes of the projection in proportion to the (spatial) frequency) followed finally by inverse Fourier transformation to get $P^*(r, \phi)$. One problem with this direct approach is that P^*, unlike P, is not spatially bounded and, although this difficulty can be overcome, the method has not found widespread application.

An alternative to this so-called Fourier filtering technique is the method of a Radon filtering, which uses the convolution theorem (see, for example, Brigham 1974, or Lighthill 1958) to replace the Fourier transform of the product of the two functions $F(k, \phi)$ and $|k|$ of eqn (4.16) by the convolution of their respective Fourier transforms, namely $P(r, \phi)$ and $-1/2\pi^2 r^2$. Thus:

$$P^*(r, \phi) = -\frac{1}{2\pi^2} \int_{-\infty}^{\infty} \frac{P(r', \phi)}{(r - r')^2}\, \mathrm{d}r'. \tag{4.17}$$

The quadratic singularity can be removed by integration by parts to give:

$$P^*(r, \phi) = \frac{1}{2\pi^2} \int_{-\infty}^{\infty} \frac{\partial P(r', \phi)/\partial r'}{r - r'}\, \mathrm{d}r', \tag{4.18}$$

which can be evaluated numerically, although care must be exercised with the digital representation since the integral still contains a singularity.

The quadratic singularity of eqn (4.17) arises from the Fourier transform of $|k|$ and can be removed if k is bounded, i.e. set to zero for all $|k| > k_m$. The Fourier transform of such a function is:

$$\int_{-k_m}^{+k_m} |k| \exp(2\pi i k r)\, \mathrm{d}k = \frac{k_m}{\pi r} \sin(2\pi k_m r) - \frac{\sin^2(\pi k_m r)}{\pi^2 r^2} \tag{4.19}$$

and this can be used to replace $-1/2\pi^2(r - r')^2$ in the integral of eqn (4.17) leading to:

$$P^*(r, \phi) = \int_{-\infty}^{\infty} P(r', \phi) \left\{ \frac{k_m \sin[2\pi k_m (r - r')]}{\pi (r - r')} - \frac{\sin^2[\pi k_m (r - r')]}{\pi^2 (r - r')^2} \right\} \mathrm{d}r'. \tag{4.20}$$

This can be further simplified to give finally

$$P^*(r,\phi)=\frac{P(r_\mathrm{i})}{4w}-\frac{1}{\pi^2 w}\sum_{\substack{\mathrm{j}=1\\ \mathrm{odd}}}^{n}\frac{P(r_\mathrm{j})}{(\mathrm{i}-\mathrm{j})^2}, \tag{4.21}$$

where $P(r_\mathrm{i})=P(\mathrm{i}w,\phi)$. This so-called convolution method is the basis of many reconstruction procedures and has been widely employed with X-ray CT systems. A simple FORTRAN program which implements this method is given in Brooks and Di Chiro (1975).

One problem with the back-projection process needed to complete image formation is that the algorithm (eqn (4.10) or (4.15)) requires the projection amplitude (ray sum) to be known at arbitrary values of r, whereas only n equally-spaced discrete values are available. It is possible to interpolate exactly though time-saving considerations generally dictate that a simpler procedure such as selection of the nearest available amplitude or a linear approximation is used. Even if an exact method were to be applied, inaccuracies could still arise from the fact that what is really required is an integral of the projection over the angular increment $\Delta\phi$. These errors manifest themselves through a reduction in the spatial resolution and in the appearance of streak artefacts at positions where the projection changes rapidly, namely edges such as occur at the skull boundary. Other difficulties arise at the filtering stage as a result of aliasing—the phenomenon whereby high (spatial) signals, if present, are folded back and appear as lower frequencies when the projection data are sampled (see § 5.4.3 for a discussion of aliasing in another context). The use of a sharp frequency cut-off, such as the square filter of the convolution method can also cause 'ringing' at high-contrast boundaries due to suppression of the naturally-present high frequencies. This is known as the Gibb's phenomenon, which, in X-ray CT studies, is most noticeable as a low intensity artefactual ring lying within the skull. Such problems can be alleviated by using improved filtering methods in which the frequency rolls off rather than being cut off prior to back-projection. A number of such filters are available which trade off artefact against spatial resolution to different degrees. Particularly successful is the Shepp–Logan filter (Shepp and Logan 1974).

Most of the work on projection reconstruction methods has been in association with X-ray CT and the development of the different filters has followed the need to suppress the artefacts common to such systems. However, in the case of n.m.r., the situation is considerably more favourable since the change in proton density on passing from tissue to bone is not nearly so great, in relation to the soft tissue contrast, as the corresponding change in X-ray absorption. This makes n.m.r. images inherently much less prone to contrast artefacts, enabling head scans to be obtained through the brain stem without interference from the posterior fossa, for example (see § 6.5.1).

4.1.3. Three-dimensional reconstruction algorithms

In the n.m.r. imaging experiment, unless one takes steps to isolate a particular image slice (see § 4.1.5, for example), the projections are integrals over planes of spins rather than along strips, and one has the option of performing a complete three-dimensional reconstruction in which every volume element (or voxel) within a defined region; the head, for example, is assigned a spin density.

One direct method for three-dimensional projection reconstruction is based on an extension of the filtered back-projection method described in § 4.1.2.5. The (three-dimensional) spin density $\rho(x, y, z)$ can be written in terms of its Fourier coefficients, $F(k_x, k_y, k_z)$:

$$\rho(x, y, z) = \int_{-\infty}^{\infty} \int_{-\infty}^{\infty} \int_{-\infty}^{\infty} F(k_x, k_y, k_z) \exp 2\pi i (k_x x + k_y y + k_z z) \, dk_x \, dk_y \, dk_z, \tag{4.22}$$

which, converting to spherical co-ordinates (k, θ, ϕ), becomes:

$$\rho(x, y, z) = \int_0^{\infty} \int_0^{2\pi} \int_0^{\pi} F(k, \theta, \phi) \exp(2\pi ikr) k^2 \sin\theta \, d\theta \, d\phi \, dk \tag{4.23}$$

$$= \int_0^{2\pi} \int_0^{\pi} P^*_{\theta\phi}(r) \sin\theta \, d\theta \, d\phi, \tag{4.24}$$

where

$$P^*_{\theta\phi}(r) = \int_0^{\infty} k^2 F(k, \theta, \phi) \exp(2\pi ikr) \, dk. \tag{4.25}$$

Usually a three-dimensional extension of the central slice theorem, namely that:

$$F(k, \theta, \phi) = \int_{-\infty}^{\infty} P_{\theta\phi}(r) \exp(2\pi ikr) \, dr = H_{\theta\phi}(k), \tag{4.26}$$

i.e. the projection in the direction θ, ϕ and the corresponding Fourier component of the spin density are a one-dimensional Fourier transform pair (see Lai 1981, or Shepp 1980, for a derivation) one can rewrite eqn (4.25) as:

$$P^*_{\theta\phi}(r) = \int_0^{\infty} k^2 H_{\theta\phi}(k) \exp(2\pi ikr) \, dk, \tag{4.27}$$

and the three-dimensional image is reconstructed by back-projecting the modified projections P^*. These are obtained from the measured projections by Fourier transformation, multiplication by k^2 and finally inverse Fourier transformation in similar fashion to the two-dimensional Fourier filtering method of reconstruction discussed in § 4.1.2.4.

The algorithm obtained by replacing the integral of eqn (4.24) with a discrete summation over the N projections is:

$$\rho(x, y, z) = \sum_{i=1}^{N} P^*_{\theta_i,\phi_i}(r) \sin\theta_i\, \Delta\theta_i\, \Delta\phi_i. \tag{4.28}$$

The solid angle corresponding to each projection is proportional to $\sin\theta_i\, \Delta\phi_i\, \Delta\theta_i$ and, for isotropic resolution, this would have the same value for all projection anglcs θ, ϕ. One way in which this can be achieved is by choosing $\Delta\theta_i = \text{constant}$, $\Delta\phi_i = \Delta\theta_i/\sin\theta$, as illustrated in Fig. 4.8. The algorithm of eqn (4.28) reduces in such a case to

$$\rho(x, y, z) = (\Delta\theta)^2 \sum_{i=1}^{N} P^*_{\theta\phi}(r)$$
$$= (\Delta\theta)^2 \sum_{i=1}^{N} P^*(x \sin\theta \cos\phi + y \sin\theta \sin\phi + z \cos\theta). \tag{4.29}$$

The expression for r in terms of the projection orientation θ, ϕ and the Cartesian co-ordinates x, y, z of the point at which the spin density is to be determined is analogous to the two-dimensional case of eqn (4.3); a geometric derivation is outlined in Fig. 4.9. If required, suitable filtering

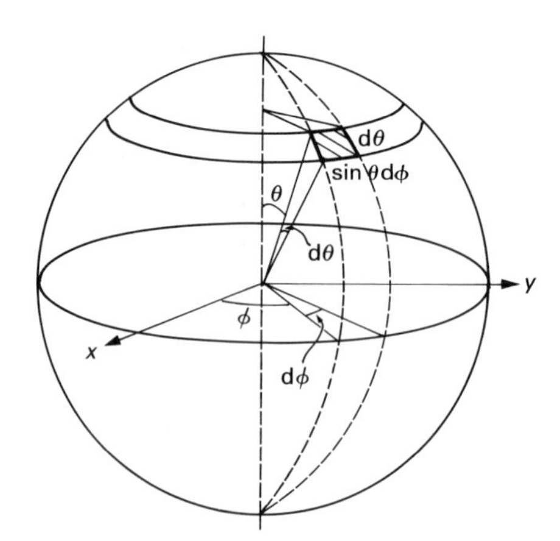

Fig. 4.8. Isotropic resolution in three-dimensional projection reconstruction is ensured by distributing the gradient orientations uniformly over the surface of a sphere. The area of the shaded element is $\sin\theta\, d\theta\, d\phi$ so θ could be incremented linearly and ϕ as $1/\sin\theta$.

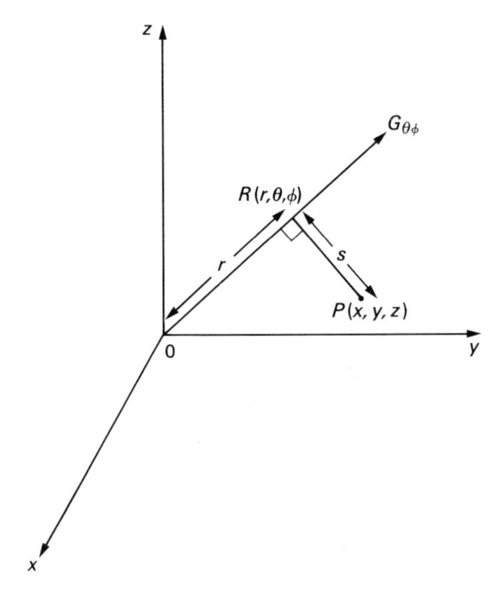

Fig. 4.9. Geometric derivation of the expression for r in terms of projection orientation θ, ϕ and Cartesian co-ordinates x, y, z. Position vector **OR** is $r\{\sin\theta\cos\phi\mathbf{i}+\sin\theta\sin\phi\mathbf{j}+\cos\theta\mathbf{k}\}$ and position vector **RP** is $(x-r\sin\theta\cos\phi)\mathbf{i}+(y-r\sin\theta\sin\phi)\mathbf{j}+(z-r\cos\theta)\mathbf{k}$. Since **OR** and **RP** are orthogonal $\mathbf{OR}\cdot\mathbf{RP}=0$. Thus $x\sin\theta\cos\phi+y\sin\theta\sin\phi+z\cos\theta-r\{\sin^2\theta\cos^2\phi+\sin^2\theta\sin^2\phi+\cos^2\theta\}=0$ and so $r=x\sin\theta\cos\phi+y\sin\theta\sin\phi+z\cos\theta$.

can easily be incorporated by introducing an additional factor $g(k)$ into the expression for the modified projection (eqn (4.27)). Lai (1981), for example, has used a function of the form: $g(k)=\frac{1}{2}(1+\cos k)$ in the range $|k|<k_{\mathrm{m}}$, $g(k)=0$ for $|k|>k_{\mathrm{m}}$. This is illustrated in Fig. 4.10.

This direct method of three-dimensional reconstruction has been called 'Fourier reconstruction zeugmatography' (F.r.z.). There is however an alternative approach which involves the application of two successive two-dimensional filtered back projection processes. The first of these leads to a reconstruction of the one-dimensional projections into a set of two-dimensional projections, and the second, using the same reconstruction algorithm, takes these two-dimensional projections and reconstructs them into a three-dimensional image as illustrated in Fig. 4.11. A formal derivation of the process has been given in a paper by Lai and Lauterbur (1980) which also describes a gradient reorientation device capable of generating the required set of projections either for the two-stage convolution method (see Lauterbur and Lai 1980, Lai and Lauterbur, 1981*a*, *b*, for examples of n.m.r. images obtained by this method) or for the three-dimensional direct algorithm discussed above. The two-stage method has the advantage that it involves fewer computational steps than its single stage counterpart ($\sim 2\times N^4$ as opposed to N^5—see Shepp, 1980).

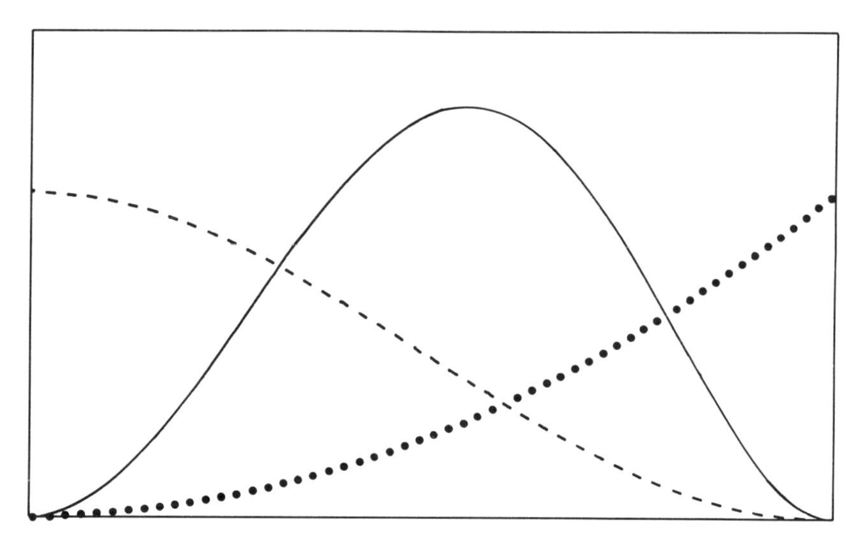

Fig. 4.10. The additional filtering function $g(k)=\frac{1}{2}(1+\cos k)$ shown as broken line, k^2 as the dotted line, and their product $\frac{1}{2}k^2(1+\cos k)$, the full filtering function, as the full line. (From Lai 1981.)

It also uses existing and well-tried reconstruction methods. However, the equal angular steps in both $\Delta\phi$ and $\Delta\theta$, which it requires, means that far more projections are collected near the poles (see Fig. 4.11(a)) and the spatial resolution will not be isotropic. We have seen that this is easily taken care of in the single-stage process, which does therefore yield uniform resolution. Another major advantage of this particular technique is that the image intensity depends only on those planes passing through the image point in question. It is therefore possible to reconstruct selectively a small region of the image.

4.1.4. *Application to n.m.r. imaging*

In the case of n.m.r. imaging, the signal digitized is not the projection itself but its Fourier transform, the free-induction decay $h_{\theta,\phi}(t)$. Thus the projection $P_{\theta,\phi}(r)$ is given by:

$$P_{\theta\phi}(r)=\int_0^{\infty} h_{\theta,\phi}(t)\exp(-\mathrm{i}\omega t)\,\mathrm{d}t. \tag{4.30}$$

The frequency is related to the displacement r through the field gradient G as expressed in eqn (4.2) so that eqn (4.30) can be rewritten as:

$$P_{\theta,\phi}(r)=\int_0^{\infty} h_{\theta,\phi}(t)\exp(-\mathrm{i}\gamma Grt)\,\mathrm{d}t. \tag{4.31}$$

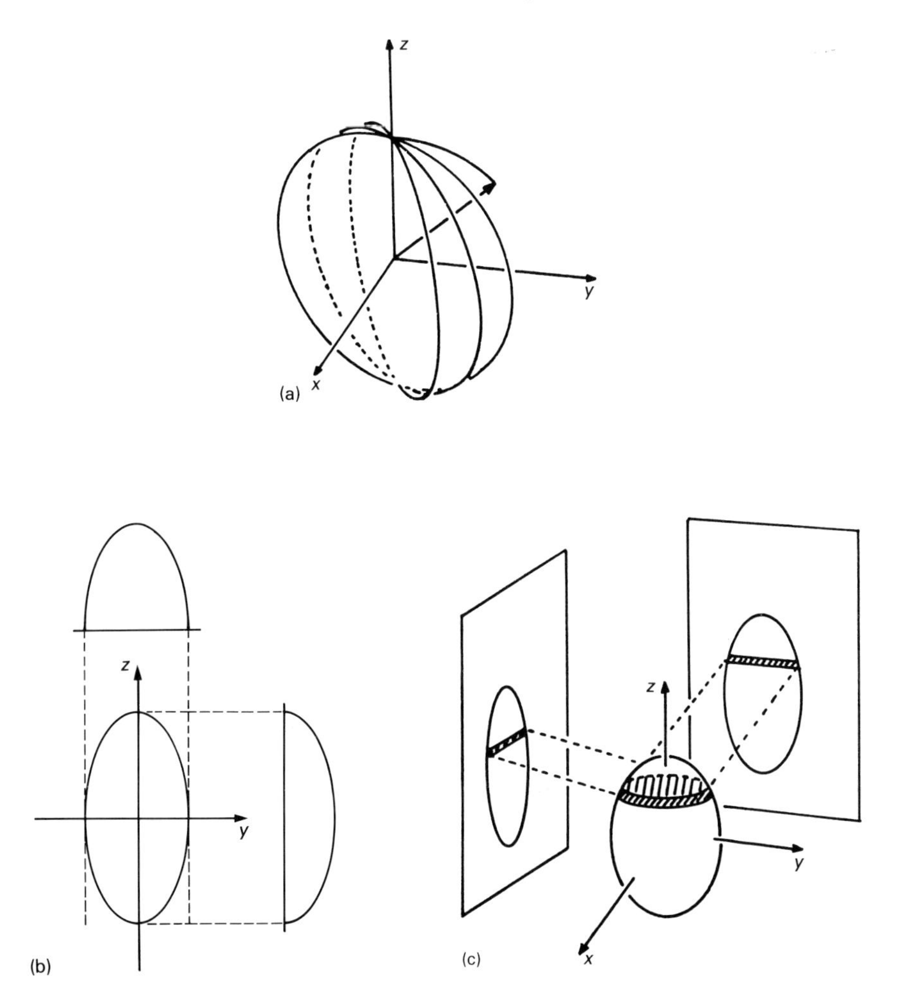

Fig. 4.11. Reconstruction of a uniform ellipsoid by the two-stage filtered back-projection method. (a) Gradient path. (b) Reconstruction of two-dimensional projections. (c) Reconstruction of three-dimensional image. (a, c from Lai and Lauterbur 1980. Copyright Institute of Physics 1980.)

The Fourier conjugate variables in the n.m.r. case are ω and t or, since ω is proportional to r, r and t. As the Fourier transform of the projection is obtained directly as the f.i.d. this allows some simplification of filtered back-projection and other methods which require Fourier transformation as a first step in the calculation of the modified projection. Thus in the case of two-dimensional filtered back-projection, the procedure for projection reconstruction would be as follows: take the f.i.d. and multiply by t (and any further filter function required), then take the inverse Fourier transform and back-project. In the three-dimensional case the product of the f.i.d. is taken with t^2 and any additional filter required ($\frac{1}{2}(1+\cos t)$ in the case of the F.r.z. method) before inverse Fourier transformation and back-projection.

There are a number of practical considerations which are of importance in the handling of n.m.r. projection data. First, there is the problem of correctly centering the projection. This can be overcome by using suitable markers (such as water-filled tubes) whose absolute positions are known. The position of these reference points can then be used to shift the projections appropriately prior to back projection. Alternatively, one can make use of the property that the centre of mass of any projection should always project to the same image point. Perhaps the simplest reference point to take, however, is the one corresponding to the null point of the field gradients (i.e. the frequency of resonance in the absence of any field gradient).

The errors in projection phase and baseline, which arise from instrumental imperfections such as finite r.f. pulse length, timing of data acquisition (see Fig. 4.12), audio filtering (see § 5.4.3), and reference phase instability, are more difficult to deal with. Whereas in the case

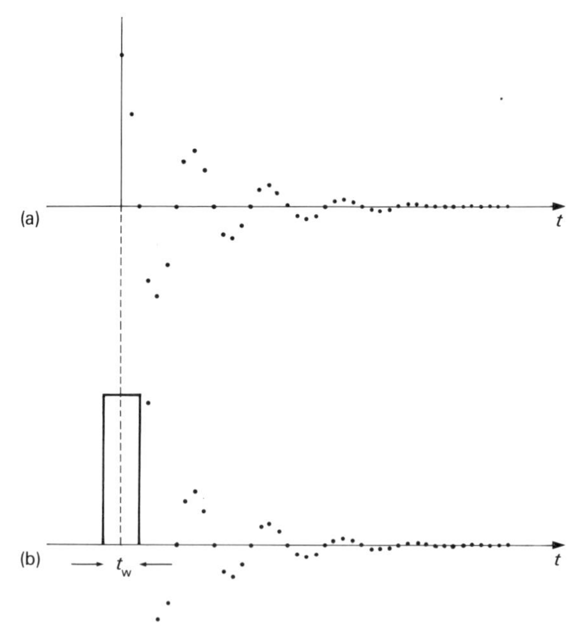

Fig. 4.12. Ideal (a) and actual (b) sampling schemes. In (b) the shift in time origin gives rise to a frequency-dependent phase error which can be largely removed by first-order phase correction. Individual points correspond to the times at which the signal is sampled by the analogue-to-digital converters.

of conventional n.m.r. spectroscopy these can be corrected manually in a quite satisfactory fashion (although even in this case automatic phase correction is becoming increasingly favoured by the major manufacturers), image projections are much more difficult to process and subjective judgements are clearly undesirable. Lai and Lauterbur (1981*c*) have analysed these errors in detail and have suggested automatic correction schemes for them. These involve the collection of f.i.d.s at different frequency offsets and in the absence of any field gradients, enabling the zeroth and first-order phase corrections to be determined and the information derived for projection centering. Baseline correction is achieved by fitting straight lines to the projection data immediately beyond the range of the object projection.

The projection-reconstruction method is susceptible to static field inhomogeneity effects, which give rise to characteristic streak artefacts, the problem being particularly acute at high field. Lai has considered such difficulties and has published details of modified projection reconstruction algorithms which compensate for static field inhomogeneity (Lai 1982, 1983*a*) and field gradient non-linearity (Lai 1983*a*, *b*). In the latter case it is shown that, by using a curvilinear co-ordinate system derived from the known field gradients, the image can be fully recovered. Spatial and intensity distortions are thereby eliminated although the resolution is then necessarily non-isotropic.

Timing is often an important consideration and there are two aspects to it, data acquisition and data processing. The data acquisition time will be determined by the number of projections required, for example 128 for a 128^2 two-dimensional array or 128^2 for a 128^3 three-dimensional one, and the cycle time (the time required to obtain a single projection including all relaxation delays) which can only be varied within narrow limits (typically 0.2–2 s) if loss of spatial resolution and/or contrast are not to be incurred. Some saving in total acquisition time can be made by giving careful consideration to the exact choice of projection angles used (Louis 1982). The data processing time, on the other hand, can vary enormously depending on the approach adopted. In the development stage (Lauterbur and Lai 1980) three-dimensional image processing often necessitated time-consuming data transfer from the system's minicomputer to a larger mainframe, possibly via a storage medium such as magnetic tape. The whole process for image generation was then an 'overnight' procedure. Of course, one would ideally like the reconstruction to be carried out in real time to enable preliminary image assessment to be conducted before release of the patient. This is especially important in the case of two-dimensional imaging where a judgement has to be exercised concerning the level and orientation of the image slice. It is less so in the case of three-dimensional imaging where, provided the data are good (i.e.

not marred by instrumental or motional problems) all slices are automatically available for later inspection. Some reduction in processing time can be achieved by overlapping it with acquisition. For example, in the case of the two-stage convolution reconstruction procedure one can reconstruct the first two-dimensional projection as soon as the set of projections corresponding to a particular azimuthal angle (ϕ) have been collected (see Lai and Lauterbur 1980). This fast back-projection algorithm can be employed in the presence of non-linear gradients, but correction for static field inhomogeneities still requires the direct three-dimensional method (Lai 1983). Greater improvements can be achieved by using array processors or, better still, hardware reconstructors such as the Analogic XAP-400 which can reconstruct a 128^2 array from 60 projections in just 150 ms. As n.m.r. imaging requirements become better defined we can expect to see great advances in this area.

4.1.5. Two-dimensional imaging

As mentioned in § 4.1.1, projection-reconstruction was the first method to be used for n.m.r. imaging. In these early experiments no effort was made to select the image plane, which consequently was defined in a rather rough and ready fashion by the active volume of the receiver and transmitter coils. This led to large plane thicknesses which were inappropriate to the study of most intact biological systems. For clinical application it is essential to have narrow and well-defined image planes and these can be achieved in a number of ways. Most popular is the method of selective excitation, in which irradiation with a low power r.f. pulse takes place in the presence of a field gradient oriented normal to the plane required. By suitable shaping of this pulse and by applying refocusing methods (see § 3.3.2), excellent plane definition can be achieved. An alternative approach involves the use of oscillating field gradients first introduced by Hinshaw in connection with his sensitive point and multiple sensitive point methods (see §§ 3.2.3 and 3.3.6). Their use in connection with projection reconstruction imaging was first discussed by Brooker and Hinshaw (1978) and later found application in the head and whole-body scanning systems of Moore and co-workers (Moore and Holland 1980; Hawkes, Holland, Moore, and Worthington 1980). More recently, a different use of oscillating field gradients in this regard has been reported by Taylor, May, and Jones (1981). In this application the spins lying along the direction normal to the plane required evolve in a time-varying field gradient. The time-dependence is changed from one cycle to the next so that, when the f.i.d.s are co-added, destructive interference occurs everywhere outside the null plane. It has the advantage over the straightforward oscillating field-gradient approach that multiple plane selection is possible and, in many ways, is more akin to the 'spin-warp' method (Edelstein, Hutchison, Johnson, and Redpath 1980) in which progressive variations in field-gradient amplitude along a particular direction form the basis for a two-dimensional Fourier transform type experiment (see § 4.2.3). Other methods of plane selection are available and have been described in detail (see Mansfield and Morris 1982, Chapter 4).

Two-dimensional projection reconstruction has found widespread application in commercial prototype and production instruments, being an efficient method which can be implemented in a particularly straightforward manner. Imaging times are largely determined by the need to acquire a sufficient number of projections to reconstruct an image with the required spatial resolution, i.e. the restriction is one of providing sufficient independent measurements to make the spin density determinate rather than of signal averaging to achieve adequate S/N. As an example, the time to acquire a single projection (including relaxation delays) is typically 1 s, so that for a 128^2 image matrix about two minutes of data acquisition is necessary.

4.1.6. Three-dimensional imaging

In the case of three-dimensional imaging, acquisition times are long: a 32^3 matrix (= 32 768 voxels) requiring typically $32^2 \times 1$ s or about 17 min. Larger matrices require prohibitively long collection times—a factor which is likely to prove a major restriction on the available spatial resolution. As a consequence of the signal averaging inherent in this method, the final image S/N is excellent, however.

Figure 4.13(c) shows a set of tomographic images selectively displayed from the three-dimensional image array obtained by projection reconstruction using a two-stage convolution method (§ 4.1.3). The original phantom is illustrated in Fig. 4.13(a), and Fig. 4.13(b) shows a set of two-dimensional projections obtained by reconstructing onto a 33×33 array using a filtered back-projection method. Each of these projections corresponds to a view of the object at a particular angle in the xy-plane; there are thirty such views taken at intervals of 12° over a full 360° rotation. Each view is itself reconstructed from 30 one-dimensional projections also obtained at 12° intervals over 360°. Note that rotation through either θ or ϕ has a two-fold redundancy since the projection obtained at an angle $\pi + \phi$ should simply be the mirror image of that at angle ϕ. That this is indeed so, can be seen by inspecting two-dimensional projections with numbers which differ by 15 (for example, Nos. 9 and 24). The two-dimensional projections can then be combined in the second stage of the three-dimensional convolution method to give the full image

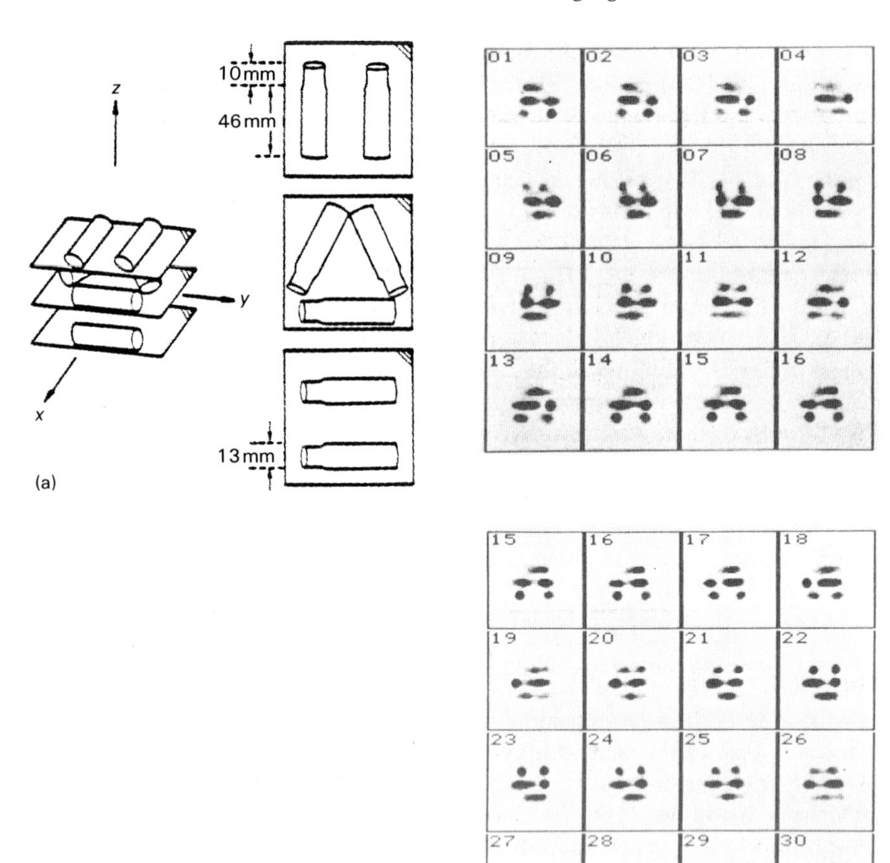

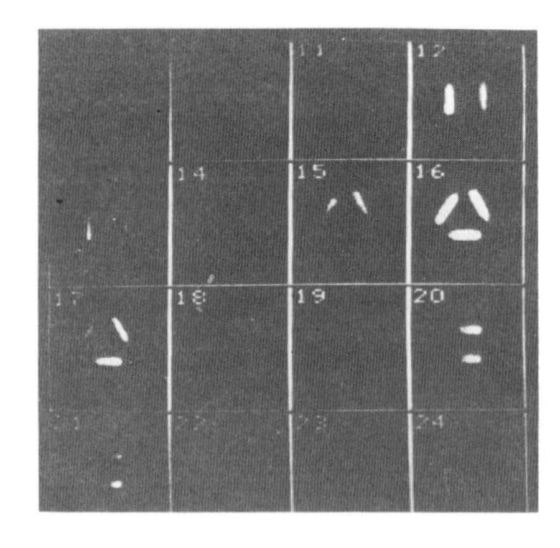

Fig. 4.13. (a) Sketch of a phantom consisting of seven vials filled with $NiCl_2$ to give a relaxation time of 50 ms and supported by glass plates in three layers as indicated. (b) Thirty two-dimensional views at 12° intervals of the phantom shown in (a). Each view is reconstructed from thirty one-dimensional projections measured over 360°. (c) Tomographic images reconstructed from the set of two-dimensional views shown in (b). (From Lauterbur and Lai 1980. Copyright © 1980 IEEE.)

matrix illustrated in Fig. 4.13(c). Thus, in this case, 900 projections were required to reconstruct the image. Total data acquisition time was ~12 min, corresponding to 800 ms per projection. The six tubes forming the phantom were filled with water doped with Ni^{2+} to give a short spin–lattice relaxation time of about 50 ms. It would therefore have been possible to increase the imaging speed by reducing the cycle time, for example to 100 ms. In addition, by removing the fourfold redundancy in the projection data the overall imaging time could have been reduced to ~22 s, corresponding to a scan time of less than 1 s per slice, which compares favourably with commercial X-ray CT systems (though not in terms of spatial resolution). Note, however, that in real biological tissues relaxation times are considerably longer than 50 ms, precluding the use of such short cycle times. Figure 4.14 illustrates another example of the use of the two-stage convolution method. It shows a three-dimensional array corresponding to a series of vertical cross-sectional slices through a coconut (Lai and Lauterbur 1981*b*). The fluid level of the milk is well visualized.

Figure 4.15 illustrates the use of the single stage F.r.z. method. It shows a series of cross-sections through an isolated empty pig heart (Lai 1981) and was obtained from 632 projections reconstructed onto a 33^3 matrix (again a fourfold redundancy in the projection data was required to allow correction of experimental imperfections). Total data collection time in this case was 8 min corresponding to 15 s per slice. A similar phantom to that illustrated in Fig. 4.13 was also imaged. It demonstrated, as expected, that the more isotropic distribution of projection angles required a shorter scanning time (about 40 per cent) to achieve a similar spatial resolution to that of the two-stage method. However, the field gradient controller used for the direct method needs much greater positional accuracy since the polar angle increments are not then constant (see § 4.1.6). In the device used by Lai and Lauterbur (1980) which was based on a Motorola 68000 micro-processor, the problem was overcome by choosing a small angular step and incrementing it a variable number of times between projections (some correction for the finite step size was, however, still necessary). For a more detailed discussion of gradient controllers see § 5.2.4.

Three-dimensional imaging methods have been used in human systems (see, for example, Lauterbur 1981; Pykett 1982) and are particularly well suited to head studies where motion does not pose such a serious problem.

4.1.7. Other experiments

4.1.7.1. Relaxation time measurements. The basic projection reconstruction experiment can be modified to make it more (or less) sensitive to the

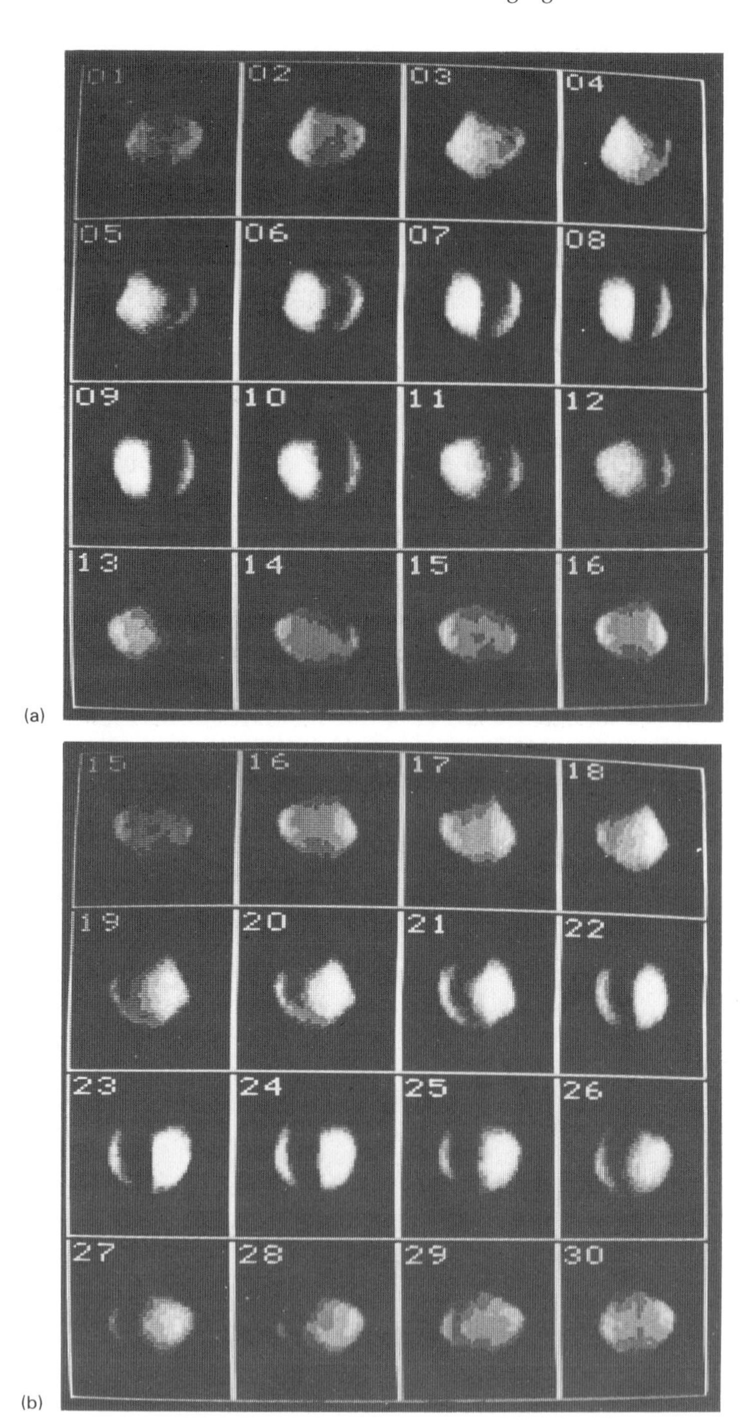
01
02
03
04
05
06
07
08
09
10
11
12
13
14
15
16
(a)
15
16
17
18
19
20
21
22
23
24
25
26
27
28
29
30
(b)

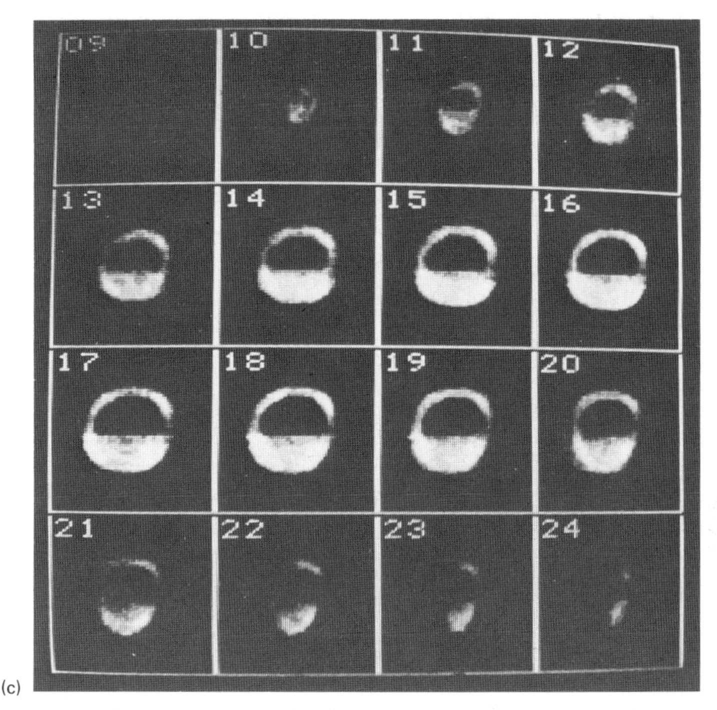

(c)

Fig. 4.14. (a), (b) Thirty two-dimensional projections of a coconut, each reconstructed from 30 one-dimensional projections. Note that, because a full 360° rotation was used, the second half of these projections should be mirror images of the first half. (c) The reconstructed three-dimensional image of a coconut, displayed as a set of vertical slices. The outer slices fade because there is appreciable B_1 and B_0 inhomogeneity in these regions. (From Lai and Lauterbur 1981. Copyright Institute of Physics 1981.)

relaxation times T_1 and T_2. Three commonly-used sequences are illustrated in Fig. 4.16 (two-dimensional versions). In the first of these (Fig. 4.16(a)), known as the repeated f.i.d., progressive saturation, or saturation recovery (SR) sequence, the $\pi/2$ pulse combines with the switched field gradient G_z to select a plane normal to z. The projection orientation itself is determined by the relative proportion of the G_x and G_y gradients applied during data acquisition (see § 5.2.4). Provided the time between the excitation pulse and the start of data acquisitions which is required as part of the refocusing procedure (see § 3.3.2), is kept short compared to T_2, the signal amplitude observed can be approximated by:

$$S_{\mathrm{SR}} = \rho[1-\exp(-\mathrm{TR}/T_1)], \tag{4.32}$$

so that by varying the repetition period TR it can be made more or less

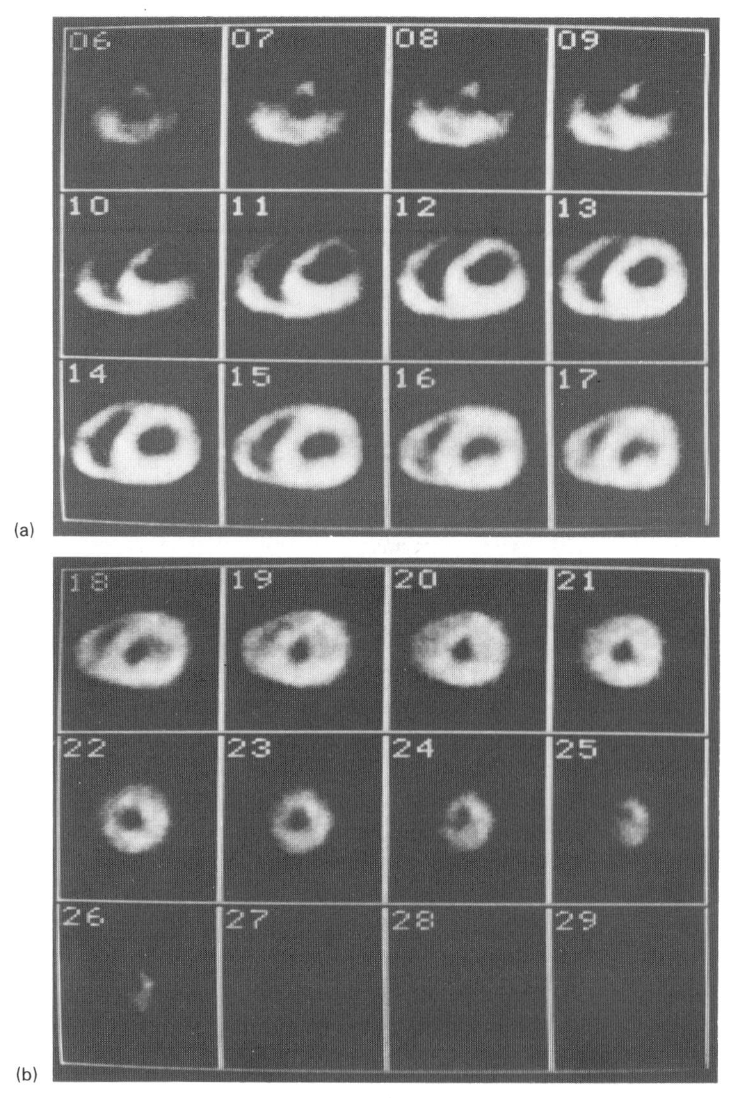

Fig. 4.15. Three-dimensional image obtained by the F.r.z. method from an isolated pig heart. Slices 6 to 17 are shown in (a) and slices 18 to 29 in (b). (From Lai 1981.)

sensitive to T_1. For example, if $\mathrm{TR} \gg T_1$, $S_{\mathrm{SR}} \sim \rho$, and the dependence is removed; for $\mathrm{TR} \sim T_1$ the dependence is especially strong. The method is similar to the conventional progressive saturation sequence often used for the measurement of T_1 (see § 2.5.1.1). For a three-dimensional imaging experiment the refocusing period is unnecessary and the projection

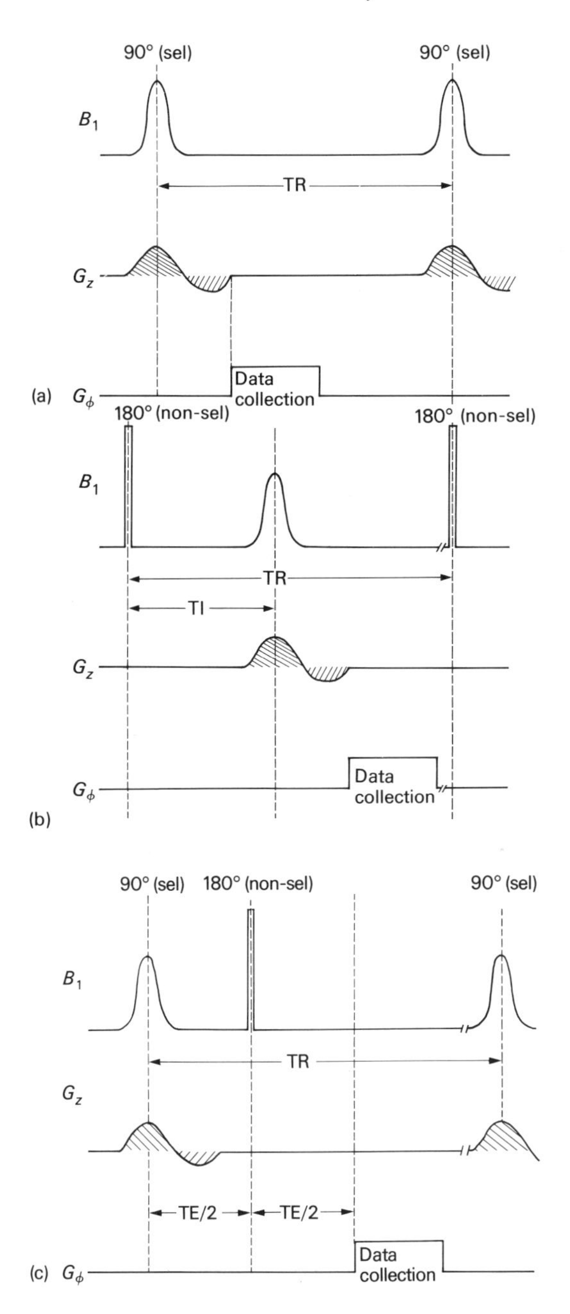

Fig. 4.16. Projection reconstruction experiments designed to measure, or be sensitive to, T_1 and T_2 relaxation times. (a) Saturation recovery (SR), (b) inversion recovery (IR), and (c) spin-echo (SE) sequences.

gradient is applied immediately following a simple non-selective $\pi/2$ pulse.

In the sequence of Fig. 4.16(b) an inversion recovery (IR) method is used in which a non-selective π pulse inverts the magnetization which is then allowed to recover before being inspected by the subsequent selective (two-dimensional experiment) or non-selective (three-dimensional experiment) $\pi/2$ pulse. Again, this technique is analogous to a conventional n.m.r. T_1 measuring method, the 180–TI–90 sequence (see § 2.5.1.1). The signal amplitude is given by:

$$S_{\mathrm{IR}} = \rho[1-2\exp(-\mathrm{TI}/T_1)+\exp\{(-\mathrm{TI}+\mathrm{TR})/T_1)\}]. \qquad (4.33)$$

The third of these schemes, known as the spin-echo (SE) sequence and shown in Fig. 4.16(c), is designed to be sensitive to the spin–spin relaxation time T_2. It is akin to the conventional Carr–Purcell spin-echo experiment discussed in § 2.5.2. The signal amplitude in this case is given by:

$$S_{\mathrm{SE}} = \rho[1-2\exp\{-(\mathrm{TR}+\mathrm{TE}/2)/T_1\}+\exp\{-(\mathrm{TE}+\mathrm{TR})/T_1\}]. \qquad (4.34)$$

It has become common practice to employ a spin-echo even for SR and IR experiments. Although the echo delay (TE) is kept short (except in the case of hybrid schemes), this nevertheless introduces an additional T_2-dependence. The expressions for this case are given in § 6.3.1 where consideration is also given to the tissue contrast generated by such sequences.

4.1.7.2. High-resolution spectra. Much interest has recently been expressed in combining n.m.r. imaging with the analytical power of conventional high-resolution n.m.r. (see Chapter 2 for a brief description of high-resolution n.m.r. and for a discussion of its application to biological systems). This can be achieved through a number of different modifications of the basic projection reconstruction experiment. The simplest approach would be to selectively excite a single resonance in the high-resolution spectrum. This throws away the multiplex advantage which would accrue if all shift components were simultaneously imaged. However, this can be partially offset by using the relaxation delay period to excite the other components. If all the remaining resonances can be interrogated in a time short compared with T_1, then the efficiency of this cyclical excitation method approaches that of a true chemical-shift imaging experiment.

Another approach, proposed by Lauterbur and colleagues (Lauterbur, Kramer, House, and Chen 1975) early in the development of n.m.r. imaging, uses a different combination of the projection reconstruction and selective irradiation procedures. It involves the selection of a single strip of spins within the sample from which a high-resolution spectrum is recorded (in the absence of any field gradients). This is repeated for each of the remaining parallel strips and the set of spectra so obtained yields a single projection for each of the n.m.r. resolvable lines in the high-resolution spectrum. The same procedure is then repeated for different orientations of the parallel selected strips to give other projections which can be used to reconstruct an image of each of the chemical shift components. Such a procedure, whilst achieving the desired result, is generally less efficient than one based on the excitation of a single resonance as described above.

One method which is both simple and efficient uses a straightforward projection reconstruction experiment. The difference lies in the fact that

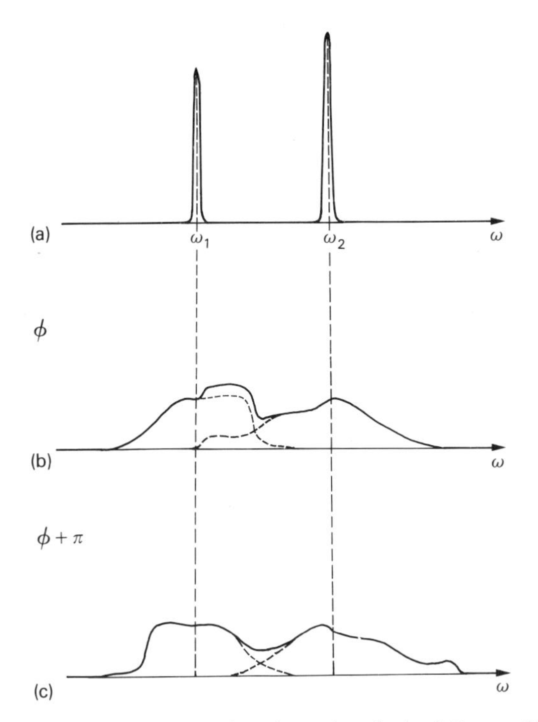

Fig. 4.17. Separation of two overlapping chemical shift profiles. (a) High-resolution spectrum in absence of field gradient. (b) Profile at angle ϕ made up of the two profiles shown as broken lines. (c) Profile at angle $\phi+\pi$: each of the two constituent profiles are mirror images of their counterparts in (b). If the individual and total intensities are denoted I^1, I^2, and I^{T} respectively, then $I^{\mathrm{T}}_{\phi}=I^1_{\phi}(\omega)+I^2_{\phi}(\omega)$ and $I^{\mathrm{T}}_{\phi+\pi}=I^1_{\phi+\pi}(\omega)+I^2_{\phi+\pi}(\omega)=I^1_{\phi}(2\omega_1-\omega)+I^2_{\phi}(2\omega_2-\omega)$, where ω_1 and ω_2 are the resonant frequencies of the two spin species in the absence of any applied gradient. Except in the trivial case of completely overlapping resonances these linear equations allow I^1 and I^2 to be separated.

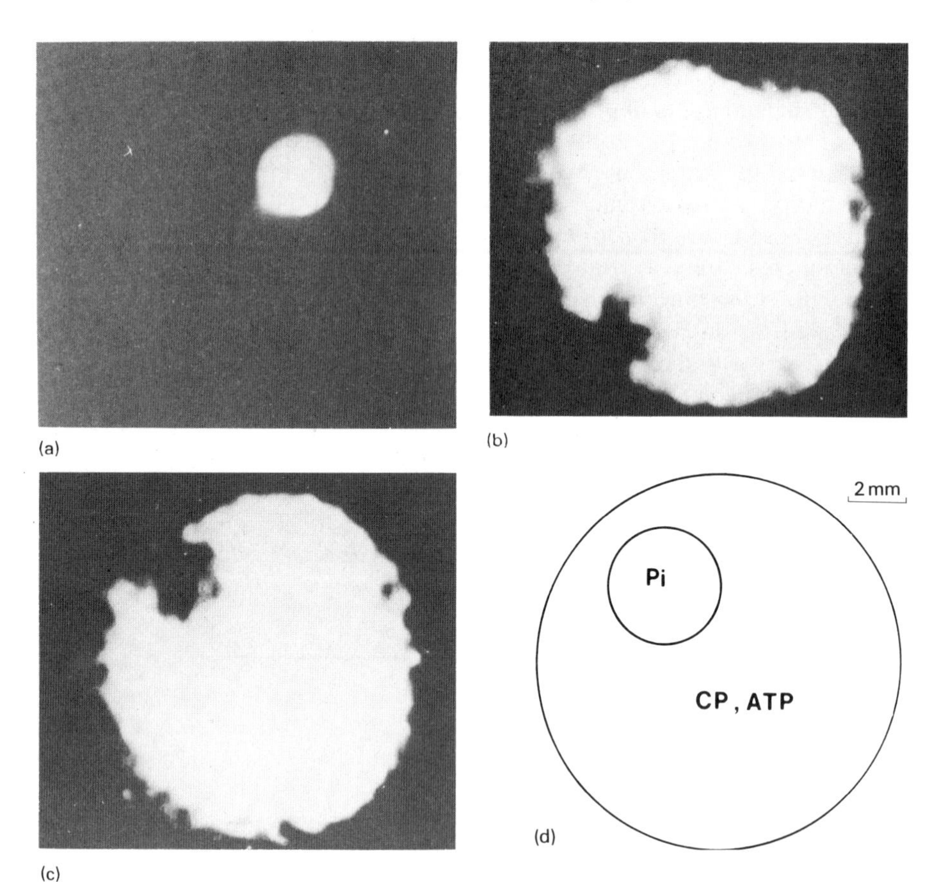

Fig. 4.18. (a) Two-dimensional projected image of the inorganic phosphate (P_i) distribution of the object shown in (d). The 65×65 point digital output of the reconstruction algorithm was smoothed by a Gaussian pointspread function and interpolated to give a 128×128 point display. (b) Reconstruction of the creatine phosphate (CP) distribution. (c) Reconstruction of the ATP distribution. (d) Horizontal cross-section of the phantom consisting of a 15 mm o.d. tube filled with 0.15 M ATP and 0.7 M PCr at pH 7.12 and an inner 5 mm o.d. tube filled with 1.6 M inorganic phosphate at pH 5.72. Note that a different intensity scale has been used for each image. (From Bendel *et al.* 1980.)

the field gradients applied are sufficiently small that the individual spectral lines are broadened to give individual profiles which do not overlap. In this simple case, reconstruction of each constituent can be carried out independently. Of course, this is only possible if the spectral lines are well separated—by at least several times their natural linewidth. In the case of two overlapping profiles, Bendel, Lai, and Lauterbur (1980) have shown that separation can be achieved by making use of the symmetry properties of n.m.r. projections, namely that the projection obtained at a field gradient oriented in the direction $\pi + \phi$ is the mirror image of that obtained at angle ϕ. The chemical shifts themselves remain unaffected by the field gradient reversal so that in many instances it is possible to separate the two components. This is illustrated in Fig. 4.17. Of course, experimental imperfections and noisy data can make the separation somewhat inexact. Figure 4.18 shows the ^{31}P n.m.r. chemical-shift images of inorganic phosphate (a) phosphocreatine (b) and ATP (c) obtained from the phantom illustrated in Fig. 4.18(d) using this straightforward approach. The high-resolution spectrum and a typical gradient broadened profile are illustrated in Fig. 4.19. Sixty projections equally distributed over 360° were obtained in all, giving 30 independent projections for image reconstruction. Similar methods have been used for ^{13}C chemical-shift studies by Hall and Sukumar (1982). Recently, the same authors (1948*a*) have developed an iterative method for separating overlapping resonances in projection reconstruction chemical shift experiments.

Lauterbur has demonstrated that, with minor modification, standard projection reconstruction methods can be used to perform chemical shift

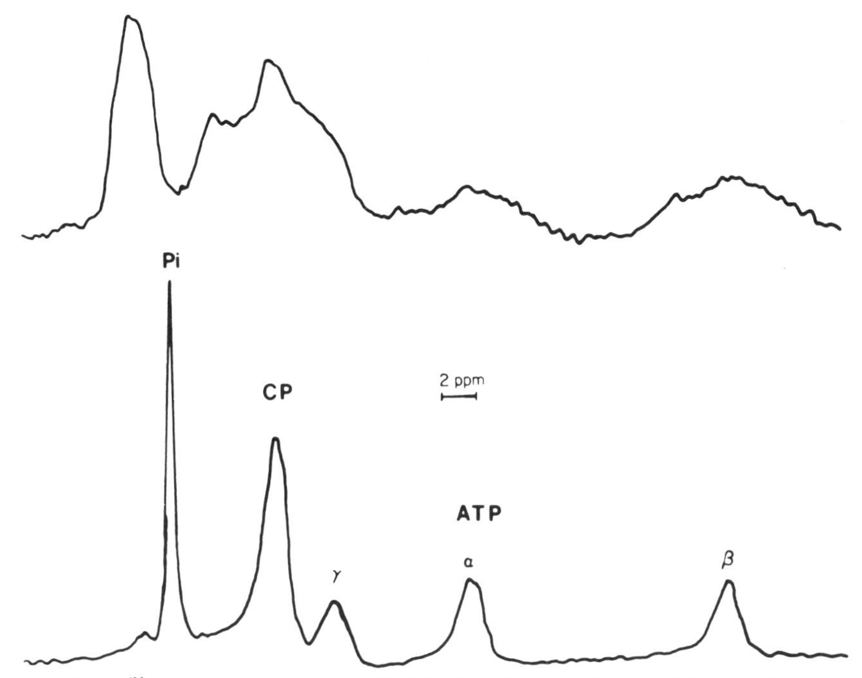

Fig. 4.19. ^{31}P n.m.r. spectra at 145.7 MHz of the object of Fig. 4.18(d) in a homogeneous magnetic field (lower trace) and broadened by a horizontal field gradient of 0.37 G/cm (upper trace). The α, β, and γ phosphate groups of ATP are labelled. (From Bendel *et al.* 1980.)

imaging. In a conventional two-dimensional experiment projections are recorded as the gradient vector executes a circular path. The chemical shift interaction produces a change in the local magnetic field of σB which can be thought of as resulting from a fixed gradient in the σ direction. To achieve a rotating 'effective gradient' vector the amplitude of the spatial gradient is varied. The projections so obtained are then scaled to correspond to a fixed magnitude effective gradient vector. The direction which lies at 90° to the chemical shift axis is not achievable with a finite gradient and the missing projections are filled in using limited angle projection reconstruction algorithms. Alternatively the problem may be avoided by choosing the angular increment such that the 90° projection is not encountered. The method has the advantage that it avoids the eddy current problems associated with switched gradient methods such as Fourier imaging. However, the non-uniform noise distribution arising from the scaling of projections will cause considerable difficulties which have not yet been fully analysed.

4.2. Fourier methods

4.2.1. Introduction

We have already introduced the concept of two- and three-dimensional Fourier transforms in connection with the projection reconstruction methods. Thus, taking a two-dimensional example, the spin density $\rho(x, y)$ can be expressed in terms of its Fourier components $F(k_x, k_y)$ as:

$$\rho(x, y) = \int_{-\infty}^{\infty} \int_{-\infty}^{\infty} F(k_x, k_y) \exp\{i(k_x x + k_y y)\} \, dk_x \, dk_y. \tag{4.35}$$

The Fourier components in turn are given by the inverse Fourier transform:

$$F(k_x, k_y) = \int_{-\infty}^{\infty} \int_{-\infty}^{\infty} \rho(x, y) \exp\{-i(k_x x + k_y y)\} \, dx \, dy. \tag{4.36}$$

Now, if we consider the response (see Fig. 4.20(a)) to a $\pi/2$ pulse or f.i.d. in the presence of a linear field gradient, G_x say, and the effects of spin–spin relaxation are (temporarily) neglected, we can express it as:

$$S(t_x) = \int_{-\infty}^{\infty} \int_{-\infty}^{\infty} \rho(x, y) \exp(i\gamma G_x x t_x) \, dx \, dy. \tag{4.37}$$

Note the similarity between this integral and that of eqn (4.36). If we now introduce a second interval t_y immediately following the $\pi/2$ pulse and during which a linear field gradient G_y is applied (see Fig. 4.20(b)), the

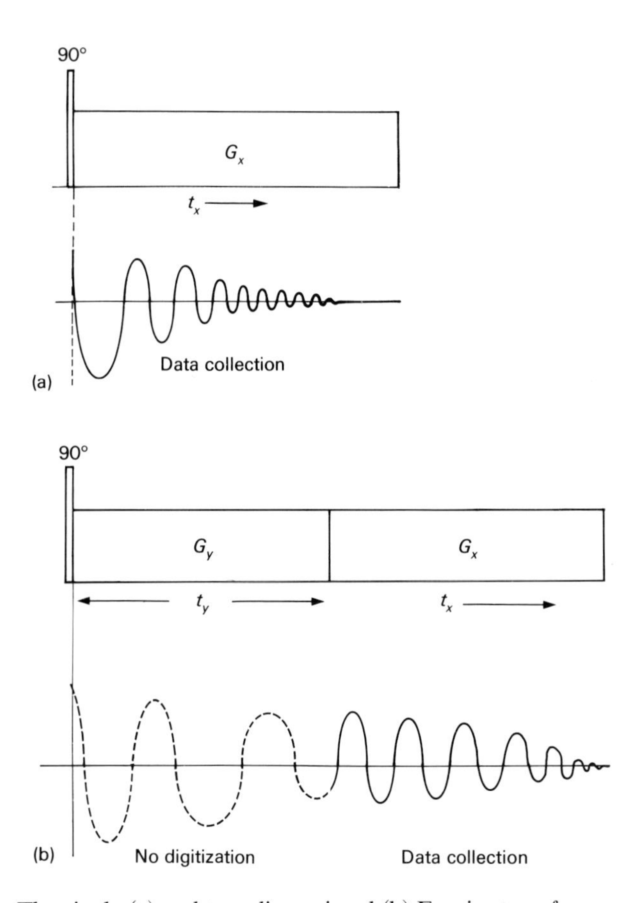

Fig. 4.20. The single (a) and two-dimensional (b) Fourier transform experiments.

corresponding signal will be:

$$S(t_y, t_x) = \int_{-\infty}^{\infty} \int_{-\infty}^{\infty} \rho(x, y) \exp\{i\gamma(G_x x t_x + G_y y t_y)\} \, dx \, dy \qquad (4.38)$$

and it is evident the values of $S(t_y, t_x)$ are, when appropriately scaled, the two-dimensional Fourier coefficients of the spin density, the Fourier conjugate variables being $\omega_x = \gamma G_x x$, t_x and $\omega_y = \gamma G_y y$, t_y.

4.2.2. Fourier zeugmatography

The simple experiment described above is the basis of the Fourier zeugmatographic method of n.m.r. imaging first proposed by Kumar, Welti, and Ernst (1975*a*, *b*). It is an example of a broader class of two-dimensional n.m.r. experiments which have found important application in simplifying the assignment of complex organic molecules (see, for example, Aue, Bartholdi, and Ernst 1976; Freeman 1980).

We now discuss the imaging experiment in greater detail, again for the present considering a two-dimensional case in the absence of relaxation effects. During the period t_x the n.m.r. signal is sampled in the normal manner at regular time intervals τ so that in the single f.i.d. a set of Fourier coefficients $S(t_0, \tau)$, $S(t_0, 2\tau)$, $S(t_0, 3\tau) \ldots S(t_0, n\tau)$ is obtained corresponding to a particular value t_0 of t_y. The value of τ is chosen so that it conforms to the sampling theorem which requires that the highest frequency present ω_{max}, given by $\omega_{max} = \gamma G_x L_x$ where L_x is the extent of the sample in the x-direction, is sampled (at least) twice per cycle. If τ is too large, the high-frequency signal, originating from the peripheral regions of the sample, is not faithfully recorded but is aliased, appearing folded back into the low-frequency part of the spectrum which corresponds to the central sample region; this is clearly a most undesirable state of affairs. If τ is too small, on the other hand, the frequency interval is too coarse and the object will fill only a small portion of the image field. Ideally, τ is chosen such that

$$\gamma G_x L_x \tau \sim \pi. \tag{4.39}$$

The total number of sample points (Fourier coefficients) n determines the number of pixels and hence the spatial resolution in the x-direction. It is usually chosen to be a power of two in order to take advantage of fast Fourier transform techniques. In subsequent f.i.d.s the period of evolution in the y gradient t_y is incremented, giving the subsequent rows in the matrix $S(t_y, t_x)$ of Fourier coefficients. Each f.i.d. or line in this matrix can be Fourier transformed as soon as it has been acquired. (Conveniently, this can be done in the time allowed for relaxation recovery between r.f. pulses). Fourier transformation in the second dimension, i.e. down the columns of the matrix, can only be carried out when all data have been collected. Often an equal number n of zero points are added to the data (zero filling) in the y dimension to take advantage of the twofold improvement in resolution which can be achieved by this means (Bartholdi and Ernst 1973). This procedure is sometimes also applied in the first (x) dimension, though it is less essential since a larger f.i.d. can always be recorded without the penalty of any concomitant increase in total T_1 recovery time.

To understand the principle of the two-dimensional transformation consider the fate of a single spin-isochromat located at a point (x_0, y_0). In the absence of relaxation, the signal in the y-gradient will be a simple harmonic (sine/cosine) function with a frequency $\gamma G_y y_0$. Correspondingly, in the x-gradient, the signal will be simple harmonic with a frequency

$\gamma G_x x_0$. The evolution of the magnetization (in the rotating frame) is illustrated in Fig. 4.21. Note, however, that in an actual experiment, data are only recorded during the t_x interval. In the first f.i.d., for which $t_y = 0$, the signal is a pure cosine wave $\cos \gamma G_x x_0 t$ and the effect of increasing t_y in successive f.i.d.s is to introduce a progressive phase shift $\phi = \gamma G_y y_0 t_y$ such that the signal is given by:

$$S(t_x) \propto \cos\{\phi + \gamma G_x x_0 t_x\}. \tag{4.40}$$

When $t_y = \pi/2\gamma G_y y_0$, for example, the recorded signal becomes a pure sine wave. The corresponding (cosine) Fourier transforms are shown alongside their respective f.i.d.s. It will be seen that in each case the result

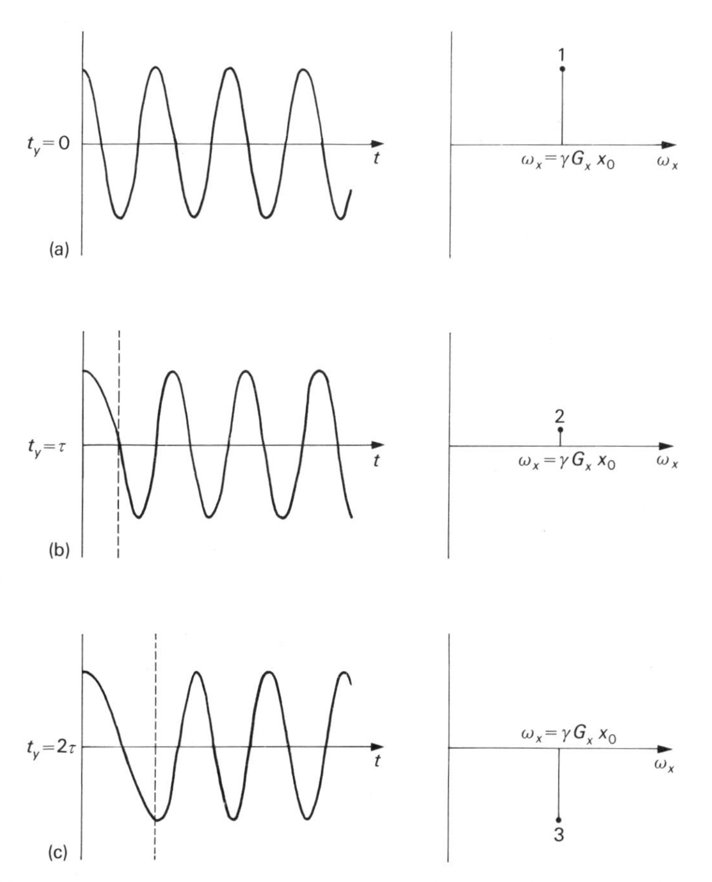

Fig. 4.21. F.i.d.s and F.t.s (in the first or x dimension) for a single-spin isochromat located at x_0, y_0 in the presence of a G_x interval and a variable G_y interval t_y corresponding to 0, τ, and 2τ for (a), (b), and (c) respectively. The F.t.s are non-zero at a single point $x = x_0$ only, giving the x-location of the isochromat.

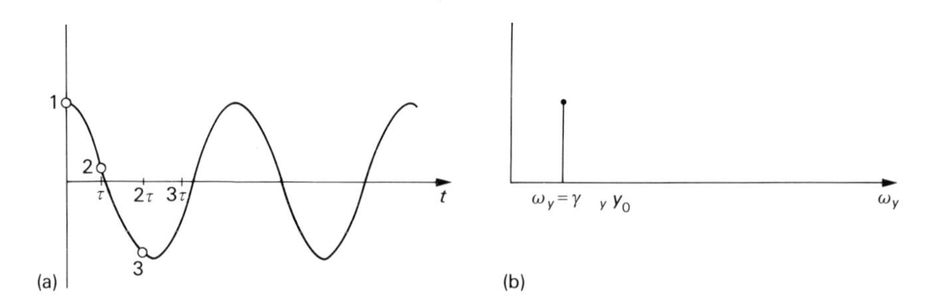

Fig. 4.22. (a) Signal obtained by taking the spectral amplitudes at $f_x = (\gamma G_x/2\pi)x_0$ of which those in Fig. 4.21 are examples, denoted 1, 2, and 3. (b) Fourier transform of (a) in the second or y-dimension showing the y-location of the isochromat.

is a δ-function at the frequency ω_x, corresponding to the position x_0. The amplitude varies in a systematic way and is in fact proportional to $\cos\{\gamma G_y y_0 t_y\}$ as illustrated in Fig. 4.22. When it comes to taking the Fourier transform in the second dimension, all columns give zero results apart from that corresponding to $\omega_x = \gamma G_x x_0$, which yields a δ-function at the position corresponding to $\omega_y = \gamma G_y y_0$ (see Fig. 4.22(b)). The two-dimensional transformed matrix therefore consists of a single point at frequency co-ordinates corresponding to (x_0, y_0) with a magnitude proportional to the spin density $\rho(x_0, y_0)$ at that point. Since Fourier transformation obeys the principle of linear superposition, viz:

$$\text{F.t.}(f(x)+g(x)) = \text{F.t.}(f(x)) + \text{F.t.}(g(x)), \tag{4.41}$$

we can apply the same reasoning to each individual spin isochromat within an extended sample with the result that we obtain a matrix which represents the full spin density distribution of the imaged object. This is illustrated in Fig. 4.23 for a simple object consisting of two parallel capillaries oriented normal to the image plane. The data show a number of imperfections arising from the non-linearity of gradients, etc., but nevertheless illustrate the principle well. They are also of historical interest being the first examples of images obtained using this technique (Kumar *et al.*, 1975*b*). Note that the capillaries are parallel to the z-axis and are separated on the x-axis. The evolution in the t_y interval should therefore be a simple cosine function, since there is no relative displacement and hence no frequency difference in the y-direction. In the period t_x, on the other hand, the f.i.d. consists of the sum of two cosine functions giving rise to a characteristic beat pattern with a beat frequency equal to $\gamma G_x \, \Delta x$ where Δx is the separation of the capillaries along the x-axis (see also the discussion of § 2.4.2.4). The corresponding Fourier transforms and the reconstructed image are also illustrated in Fig. 4.23.

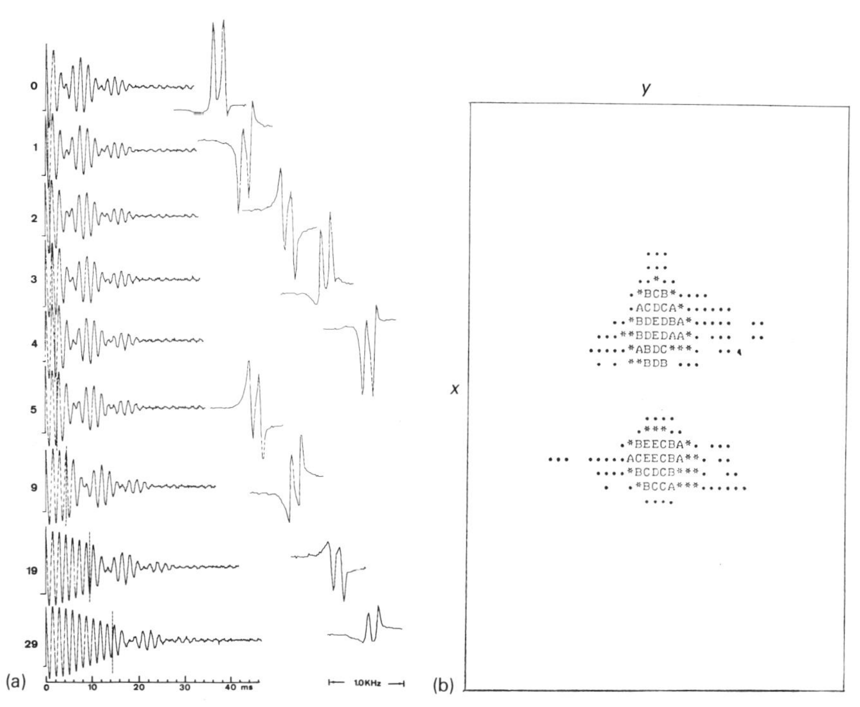

Fig. 4.23. Fourier zeugmatographic image of a phantom consisting of two capillaries with axes parallel to z and separated along x. (a) F.i.d.s showing evolution in G_y (prior to vertical broken line) and G_x gradients and the Fourier transform in the first dimension (x). (b) Reconstructed image displayed on an intensity scale consisting of, in ascending order, (blank), ·, $*$, A, B, C, D, and E. (From Kumar *et al.* 1975*a,b*.)

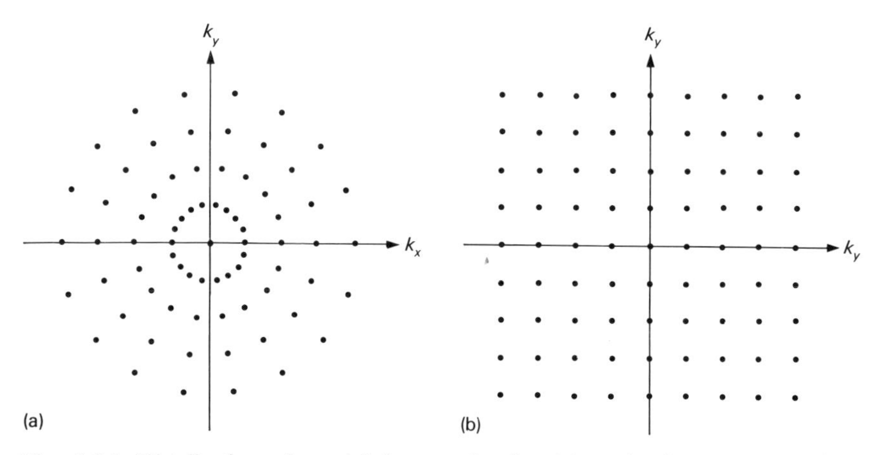

Fig. 4.24. Distribution of spatial frequencies for (a) projection reconstruction, and (b) Fourier zeugmatography imaging methods.

An interesting point arises concerning the nature of the two-dimensional array of Fourier components which, if the t_x sample interval τ is constant and the t_y increments are equal, will be rectangular. This makes interpolation prior to Fourier transformation redundant and is to be contrasted with the situation for projection reconstruction where Fourier components are equally spaced along a radius and their density therefore reduces with increasing spatial frequency $|k|$ (see discussion of § 4.1.2.4 and Fig. 4.24). This means that for Fourier imaging, high spatial frequencies are reconstructed with the same accuracy as the lower ones, resulting in an improved resolution of regions such as edges where the spin density changes rapidly.

4.2.2.1. Effects of T_2 relaxation. In the discussion thus far the effects of spin–spin relaxation have been ignored. They can, however, be easily incorporated in the analysis by damping the f.i.d. with the exponential term $\exp(-t/T_2)$ (see § 2.5.2). The equation for the spin density is:

$$\bar{\rho}(x, y) = \iint S(t_x, y_y)\exp\{-i(\gamma G_x x t_x + \gamma G_y y t_y)\}\, dt_x\, dt_y, \qquad (4.42)$$

where $\bar{\rho}(x, y)$ indicates that the spin density obtained in this way will not be the true one $\rho(x, y)$ which is given by eqn (4.35). To examine the relationship between $\bar{\rho}(x, y)$ and $\rho(x, y)$, the T_2 relaxation term is introduced and we can write:

$$S(t_x, t_y) = \int_{-\infty}^{\infty}\int_{-\infty}^{\infty} \rho(x', y')\exp\{i(\gamma G_x x' t_x + \gamma G_y y' t_y) - (t_x + t_y)/T_2\}\, dx'\, dy'. \qquad (4.43)$$

Substituting this into eqn (4.42) we obtain:

$$\begin{aligned}\bar{\rho}(x, y) &= \iiiint \rho(x', y')\exp\left\{\left(i\gamma G_x(x'-x) - \frac{1}{T_2}\right)t_x\right\}\\ &\quad \times \exp\left\{\left(i\gamma G_y(y'-y) - \frac{1}{T_2}\right)t_y\right\} dx'\, dy'\, dt_x\, dt_y\\ &= \iint \rho(x', y')\int \exp\left\{\left(i\gamma G_x(x'-x) - \frac{1}{T_2} dt_x\right)t_x\right\} dt_x\\ &\quad \times \int \exp\left\{\left(i\gamma G_y(y'-y) - \frac{1}{T_2}\right)t_y\right\} dt_y\, dx'\, dy'. \end{aligned} \qquad (4.44)$$

Neglecting the contribution from the spins below resonance, the two inner integrals of eqn (4.44) can be written as $\mathscr{L}(\gamma G_x(x'-x))$ and $\mathscr{L}(\gamma G_y(y'-y)$ where

$$\mathscr{L}(\omega) = A(\omega) + iD(\omega) \qquad (4.45)$$

is the (complex) lineshape function (see § 2.4.2.4) given, for a liquid-like sample such as the capillaries of water by the Lorentzian expression:

$$\mathrm{A}(\omega)=\frac{M_0/T_2}{(1/T_2)^2+\omega^2} \tag{4.46a}$$

and

$$\mathrm{D}(\omega)=\frac{M_0\omega}{(1/T_2)^2+\omega^2}. \tag{4.46b}$$

The measured spin density can then be written as:

$$\bar{\rho}(x, y)=\int_{-\infty}^{\infty}\int_{-\infty}^{\infty}\rho(x', y')\mathscr{L}(\gamma G_x(x-x'))\mathscr{L}(\gamma G_y(y-y')\,\mathrm{d}x'\,\mathrm{d}y', \tag{4.47}$$

and is thus the convolution of the real spin density with the natural lineshape function which, in the case of a biological sample, may well be more complex than the simple Lorentzian expression of eqn (4.46). Equation (4.47) holds, provided the ranges of the inner integrals extend for times t_x, $t_y \gg T_2$. This condition is generally satisfied for t_x but may well not be for t_y. In this event, the appropriate convolution function will be a sinc function leading (in two dimensions) to a star-like artefact familiar to workers in the optical field (see also discussion of rotating frame Fourier zeugmatography).

4.2.2.2. Phase separation. One of the problems with two (and three)-dimensional n.m.r. spectroscopy is that the absorption and dispersion terms, $\mathrm{A}(\omega)$ and $\mathrm{D}(\omega)$ respectively, become mixed together in both real and imaginary parts of the spin density. This is normally overcome by computing the absolute value $|\bar{\rho}|$. Although this procedure results in loss of spatial resolution over that for a pure absorption mode signal (the dispersion term has broader wings since it decays as $1/\omega$ rather than $1/\omega^2$ (see eqn (4.46)), provided the natural linewidth is less than the frequency separation of adjacent pixels as is generally the case, there should be no real problem. If necessary, however, it is possible to separate out the absorption from the dispersion mode by repeating the imaging experiment with appropriate gradient reversals (see Kumar *et al.* 1975*b* for details). This problem is alleviated if a spin echo experiment is performed (see § 4.2.4) and the full echo is digitized (the approximately symmetric function has a predominantly absorption mode Fourier transform). It is also possible to achieve a similar effect by imposing a pseudo echo behaviour on a decaying signal through the use of resolution enhancement techniques e.g. multiplication by a sine-bell function. The use of absolute value data does carry with it the bonus that it corrects for unwanted phase shifts arising as a consequence of r.f. skin-depth problems, for example. This may be especially important if one is attempting to obtain a proton image at an unusually high frequency, as a prelude to producing a ^{31}P chemical-shift image for example. (The high field is required to achieve reasonable sensitivity and dispersion with the low γ nucleus).

4.2.2.3. Three-dimensional imaging and other experiments. Although only two-dimensional imaging examples have been discussed, the extension to three-dimensions should be clear. This simply involves the introduction of a further variable period of evolution t_z during which a third mutually-orthogonal field gradient G_z is applied. The analysis then proceeds in similar fashion through the use of a three-dimensional Fourier transform.

Of course, the process need not stop at this point ; as well as two or three spatial dimensions it is possible to conceive further time periods during which the spins evolve under some other Hamiltonian. Provided it leads to a phase shift which accumulates linearly with time, this interaction can be separated using a multi-dimensional Fourier transform experiment. An obvious example would be the use of an 'extra dimension' to give chemical-shift resolution. Another example which has been discussed in the literature (Maudsley, Oppelt, and Ganssen 1979) uses this type of technique to measure the magnetic field inhomogeneity over the cross-section of a magnet designed for whole-body imaging. In this scheme, a sample consisting of a disc filled with water (or some other suitable proton rich substance) is placed at the cross-section of interest. A three-dimensional Fourier imaging scheme is employed, two periods of which are dedicated to producing spatial discrimination over the plane of the disc (x, y) and the third (z) to evolution in the (inhomogeneous) magnetic field alone. On performing the three-dimensional Fourier transform, the z-column of the resultant matrix, corresponding to a particular point (x_0, y_0), should have only one non-zero element in a position which corresponds to the frequency offset and hence inhomogeneity pertaining to that point. The value of this non-zero element should be constant, reflecting the uniform distribution of spin density in the disc phantom. It is therefore the location, rather than the magnitude of the matrix elements which is of importance. For display, the 'z' co-ordinate can be colour or black and white intensity coded in the usual manner (see § 5.6). Figure 4.25(a) shows an example of the use of this method. A second and more recent example, taken from Maudsley, Simon, and Hilal (1984), is shown in Fig. 4.25(b). Other planes can be investigated by moving the disc phantom through the region of interest. Alternatively, an extended phantom can be used and either a four-dimensional Fourier imaging

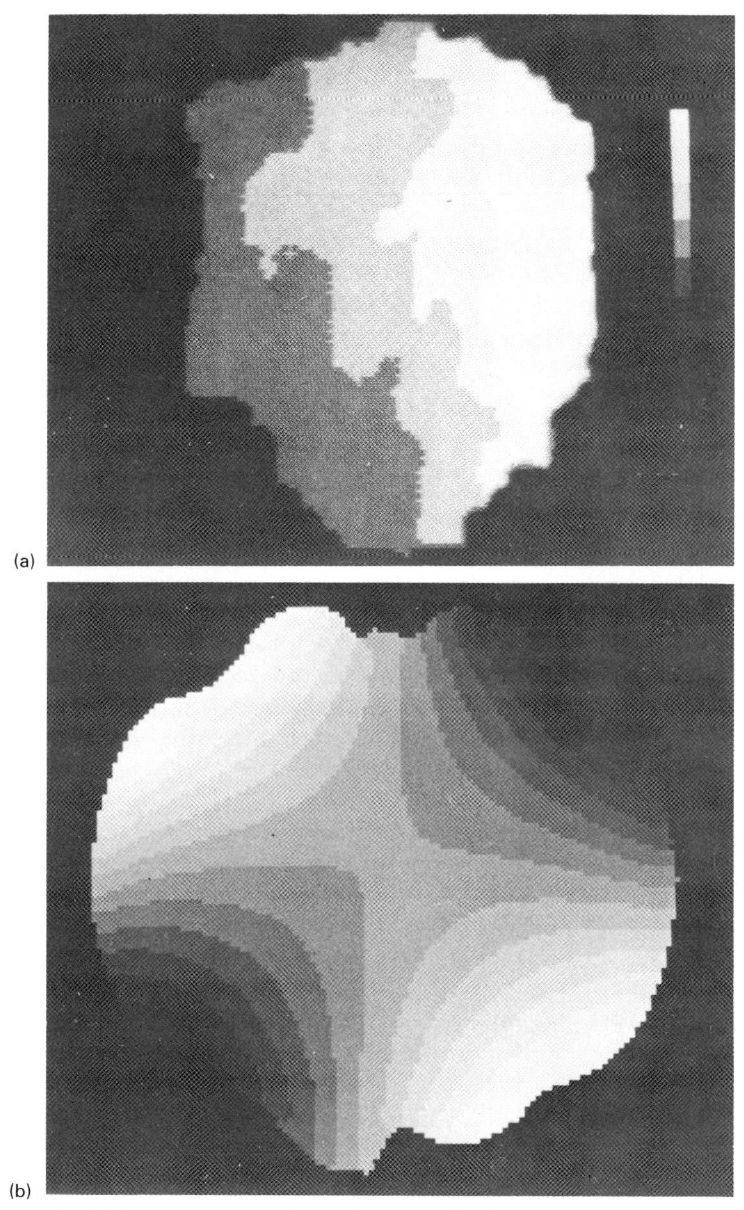

(a)

(b)

Fig. 4.25. (a) The two-dimensional field distribution over a 1.3-cm-diam. cross-sectional area. The interval between each grey level corresponds to a field change of 1.5 μT. (From Maudsley *et al.* 1979.) (*b*) Field variation in the *xy*-plane resulting from a deliberate missetting of the [XY] shims. The separation between adjacent grey levels corresponds to 2 p.p.m. (From Maudsley *et al.* 1984. Copyright Institute of Physics 1984.)

method or a three-dimensional one with selective plane excitation can be employed.

Fourier imaging in its various guises (see below) has many attractions and is a particularly efficient method for n.m.r. imaging applications. In the analysis of Brunner and Ernst (1979) it appears slightly less efficient than the projection reconstruction method since data are not acquired during the one (two-dimensional) or two (three-dimensional) phase encoding stages of the experiment. However, this slight deficiency disappears when, as is usually the case, one seeks T_1 image-dependency, which generally requires relatively long relaxation recovery periods between pulses. There is considerable interest in the Fourier method and it is perhaps the most important image option to have been made available in commercial instruments.

4.2.3. Spin-warp imaging

In 1980 the Aberdeen team proposed a new technique for n.m.r. imaging which they called the 'spin-warp' method (Edelstein *et al.* 1980). It is closely related to the two- and three-dimensional Fourier methods described above, differing in one simple but very important respect; namely, the use of constant time intervals for the phase encoding, the phase variation being achieved through the use of variable amplitude field gradients. The two-dimensional scheme is illustrated in Fig. 4.26. Concentrating on the events starting at interval three, the $\pi/2$ Gaussian-

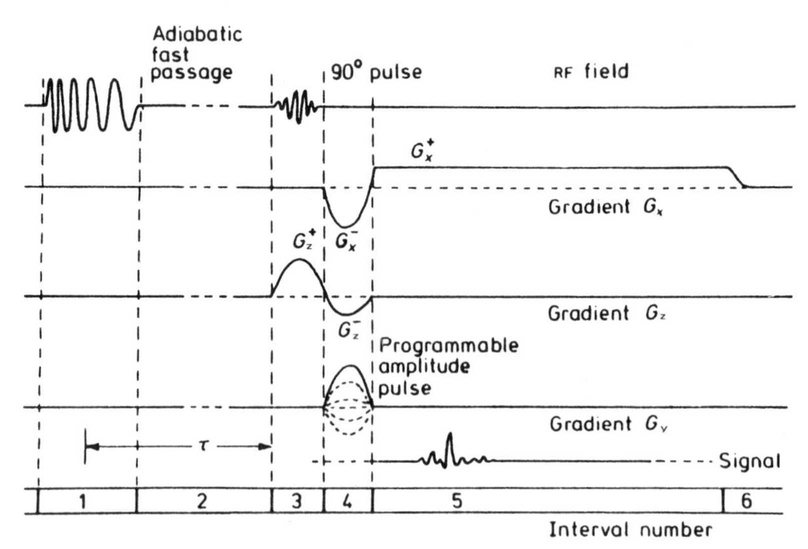

Fig. 4.26. The spin-warp imaging pulse scheme. (From Edelstein *et al.* 1980. Copyright Institute of Physics 1980.)

shaped pulse, in combination with the field gradient G_z, selects a plane of spins normal to the z-axis. In the absence of any further field gradients, the effect of the z-gradient reversal would be to refocus the spin magnetization emanating from this plane at a time corresponding to the end of interval 4 (see Sutherland and Hutchison 1978, and discussion of § 3.3.2). However, during the latter interval a reverse G_x gradient causes a shift of the echo maximum into interval 6. The rephasing and subsequent dephasing occur in the presence of the linear gradient G_x so that the Fourier transform of this echo decay will give the projection of the spin density normal to x. This process gives one dimension of spatial encoding; the other is achieved by the variable amplitude G_y gradient also applied during interval four. Its effect is to introduce a phase shift

$$\begin{aligned}\phi &= \int_0^{\tau} \gamma G_y(t) y \, dt \\ &= \gamma y \int_0^{\tau} G_y(t) \, dt,\end{aligned} \tag{4.48}$$

where τ is the length of interval four. In subsequent f.i.d.s, ϕ is varied by linear variation of $\int_0^{\tau} G_y(t)\,dt$ rather than by incrementing the time spent in a static gradient as was the case in the standard Fourier imaging experiment (see eqn (4.40)). The Aberdeen team used a half sine wave (Edelstein *et al.* 1980; Johnson, Hutchison, and Eastwood 1982) but the exact functional form of $G_y(t)$ is not critical. The name 'spin-warp' arose from a consideration of the distribution of spins along the y-axis in the rotating frame of reference (the misprinted 'spin wrap' is equally descriptive). In the absence of the other gradients, the effect of G_y applied in interval 4 is to introduce a phase shift which increases along the y-axis ($\phi \propto y$), giving the spin isochromats a spiral appearance like that of a corkscrew (see Fig. 4.27). Changing $\int_0^{\tau} G_y(t)\,dt$ alters the pitch of the spiral by causing the phase variation to occur more or less rapidly with increasing y. Reconstruction again simply involves a two-dimensional Fourier transform.

The appropriate sampling condition, corresponding to eqn (4.39) for Fourier imaging is that:

$$\gamma L_y \int_0^{\tau} G_{y\,\text{min}}(t) \, dt = \pi, \tag{4.49}$$

i.e. the phase accumulated in the smallest non-zero gradient $G_{y\,\text{min}}$ by the extremity of the sample $y = L_y$, is π.

The spin warp method has a number of advantages over straightforward Fourier imaging which derive from the fixed length of interval 4 during which the G_y gradient is applied. First, there is an absence of

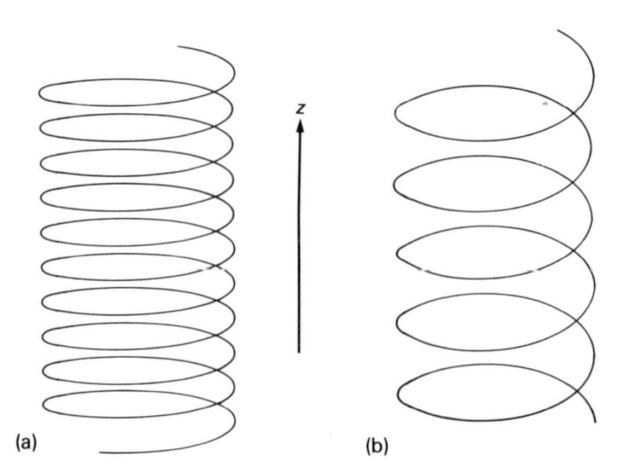

Fig. 4.27. Spin-warp imaging. Spiral phase distribution of spins with pitch corresponding to large (a) and small (b) values of G_y.

differential relaxation effects which can occur when this period is successively incremented. A second major advantage becomes apparent when the static field inhomogeneity is considered. This mimics the effect of the y-gradient, giving rise, in the case of Fourier imaging, to a spatial distortion of the image in the y-direction. Although, in the case of spin-warp imaging the inhomogeneity still leads to phase shifts during the y-encoding, they remain the same for each f.i.d. and in consequence cannot contribute to any distortion in this direction. A practical example should serve to illustrate the importance of this constraint to Fourier zeugmatography. Suppose that we arbitrarily select a minimum time increment of 0.5 ms, then, for an L_y of 0.4 m, eqn (4.39) indicates that we should use a gradient G_y of about 60 $\mu T\,m^{-1}$. If we require a modest 32-point resolution across the sample then the separation between each point will correspond to $60 \times 0.4/32$ μT or ~ 0.73 μT. A typical operating frequency for a whole-body system would be 21 MHz, corresponding to a static field of about 0.5 T. The magnet homogeneity may be 10 p.p.m. (1 part in 10^5) giving a field variation of about 5 μT, nearly seven times the impressed variation over one pixel. Unacceptable distortion would therefore result from the use of such a system. The gradient strength can be increased to try to overcome this difficulty but time intervals have to be made correspondingly shorter and one then runs into difficulties of being able to switch the gradients sufficiently rapidly. Another answer would be to switch the field gradient direction in mid-interval whilst simultaneously applying a short π pulse to the spin system. If an attempt is made to overcome the problem by improving the magnet, then the use of Fourier zeugmatography imposes severe restrictions on the field in-

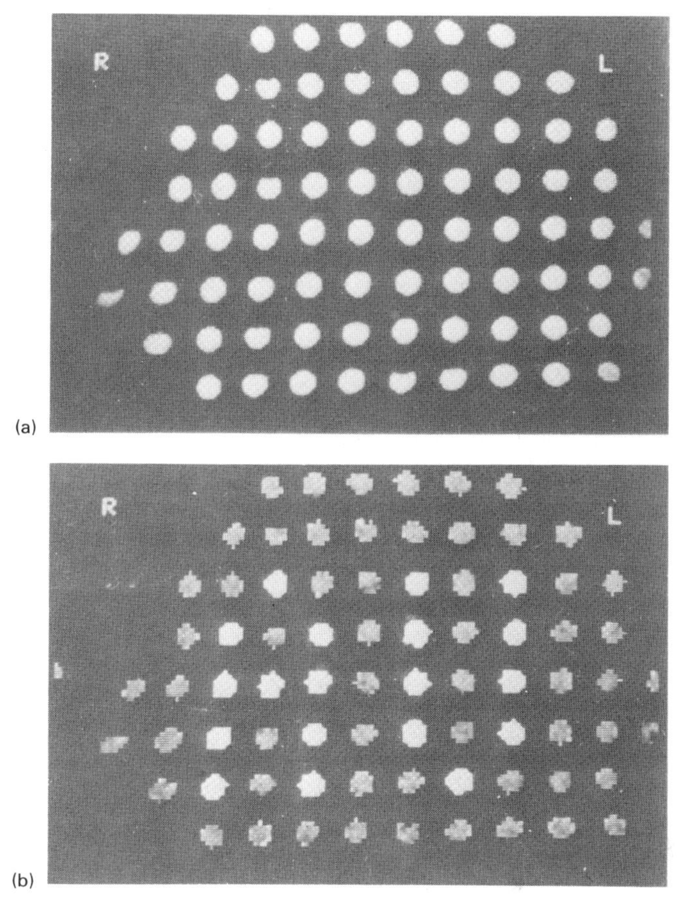

Fig. 4.28. (a) Image of a matrix of water-filled test tubes obtained by the spin-warp method, illustrating the lack of spatial distortion in the y-direction. (b) The same matrix array as in (a) but with water doped with $MnCl_2$ to give T_1 values of 200 ms (dark tubes) and 350 ms (bright tubes). (From Johnson *et al.* 1982. Copyright Institute of Physics 1982.)

homogeneity which can be tolerated. Figure 4.28, on the other hand, shows what can be achieved using the spin-warp method with a magnet having a homogeneity of 300 p.p.m. at a radius corresponding to the periphery of the human trunk; there is no visible spatial distortion in the y-direction. Curvature in the orthogonal direction is caused by magnet inhomogeneity but the G_x gradient is not subject to the same constraint as G_y (there is no field gradient switching to consider, and the sampling interval can be as short as is required) so that it can be made sufficiently large to reduce the distortion to an acceptable level. If the gradient amplitude G_y is stepped from one extreme through zero to the extreme of opposite polarity this imposes an echo behaviour in the y dimension with beneficial effect on the spatial resolution in this direction (see § 4.2.2.2).

The experiment described thus far measures the spin density distribution (provided the interpulse intervals are sufficiently long to achieve a full recovery of magnetization). The signal can be made dependent on T_1 by preceding the sequence described with a spin inversion (interval 1)). This can either take the form of a π pulse or, as in the case of the Aberdeen approach, an adiabatic fast passage. In this technique, which was commonly applied in the days of c.w. n.m.r. (see Abragam 1961, for example) the field or equivalently the irradiation frequency is smoothly swept (from +8 to −8 kHz in 10 ms in the Aberdeen case) and the magnetization, initially along the z-axis, follows the effective field in the rotating frame, ending up along the negative z-axis. The advantage of this particular technique is the achievement of good (~90 per cent) inversion over the whole sample without the need for high-power, high-homogeneity r.f. pulses which are difficult to produce in full-sized imaging systems. Following inversion of the magnetization, a period TI is allowed during which it is allowed to recover before the signal is recorded (see § 2.5.1.1) for a description of the basic inversion recovery method). If the signal obtained from such an experiment is denoted S_2 and that from the pure spin density experiment as S_1, then the relaxation time T_1 can be estimated since

$$\rho_2(x, y) = \rho_1(x, y)\{1 - 2\exp(-\mathrm{TI}/T_1)\}, \tag{4.50}$$

where $\rho_2(x, y)$, $\rho_1(x, y)$ are the effective spin densities determined from the two-dimensional Fourier transforms of S_2 and S_1 respectively. Thus:

$$T_1 = \mathrm{TI}/\log\left\{2\left(\frac{\rho_1}{\rho_1 - \rho_2}\right)\right\}. \tag{4.51}$$

This can be re-expressed as:

$$T_1 = \mathrm{TI}/\log\left\{2\frac{\mathrm{F.t.}(S_1)}{\mathrm{F.t.}(S_1) - \mathrm{F.t.}(S_2)}\right\}$$

$$T_1 = \mathrm{TI}/\log\left\{2\frac{\mathrm{F.t.}(S_1)}{\mathrm{F.t.}(S_1 - S_2)}\right\}. \tag{4.52}$$

For preference, expression (4.52) is used and the difference $S_1 - S_2$ is taken prior to Fourier transformation in order to avoid uncertainty over the sign of the signal in cases of long T_1 relaxation times. (The two-dimensional Fourier transform procedure, as we have seen in § 4.2.2.2, preserves only the magnitude of the data due to the admixture of dispersion and absorption terms). T_1 discrimination is demonstrated in the image of Fig. 4.28(b) in which the tubes contained water doped with magnesium chloride to give T_1s of about 200 ms (darker tubes) and 350 ms (lighter tubes).

The Aberdeen team have estimated (Johnson *et al.* 1982) that for a 128×128 image matrix with a slice thickness of 18 mm (equivalent rectangular width) the uncertainty in the ρ values obtained using their spin-warp method at 1.7 MHz without signal averaging is about 6 per cent. For the relaxation-time image the comparable figure is 8 per cent for T_1 values in the region of 200 ms and somewhat greater for other values. Although the T_1-determination is based on a difference method, the whole-body cross-sections, somewhat surprisingly in view of earlier problems with a similar method (Mallard, Hutchison, Edelstein, Ling, Foster, and Johnson 1980; Hutchison, Edelstein, and Johnson 1980), show very little evidence of motional artefact. Presumably the difference lies in the fact that each f.i.d. yields only a single line in the selective excitation method (§ 3.3) whereas, in the case of spin-warp imaging and Fourier methods in general, all n f.i.d.s contribute equally to each of the n image lines; since they are obtained at different (random) times the motion is effectively averaged.

To date, several hundred patients and volunteers have been examined with the Aberdeen prototype imaging system (see, for example, Edelstein *et al.* 1980; Edelstein, Hutchison, Smith, Mallard, Johnson and Redpath 1981; Smith, Mallard, Hutchison, Reid, Johnson, Redpath, and Selbie 1981; Smith, Mallard, Reid, and Hutchison 1981; Smith 1982; Hutchison and Smith 1982). No ill effects, neither during nor subsequent to scanning, have been reported. A commercial system using this principle with a higher field magnet is now available from M & D Technology (see Table 1.1).

4.2.4. *Fourier imaging: current practice*

Fourier imaging is efficient and has proved relatively straightforward to implement. It is also versatile and is fast becoming the method of choice for commercial n.m.r. imaging systems. The spatial encoding is generally similar to that used for the spin-warp method (§ 4.2.3); namely, variable amplitude gradients applied for a fixed period of time. A basic pulse scheme for a two-dimensional Fourier imaging experiment is illustrated in Fig. 4.29(a). The selective 90° pulse is applied in the presence of a refocused z gradient to isolate a slice normal to z. The variable amplitude y gradient achieves y-dimension phase encoding,and the fixed gradient G_x allows discrimination in the x direction. Notice the use of the non-selective 180° pulse to generate a spin-echo. The echo delay TE is normally short ($\lesssim 10$ ms) and allows data acquisition to be separated from

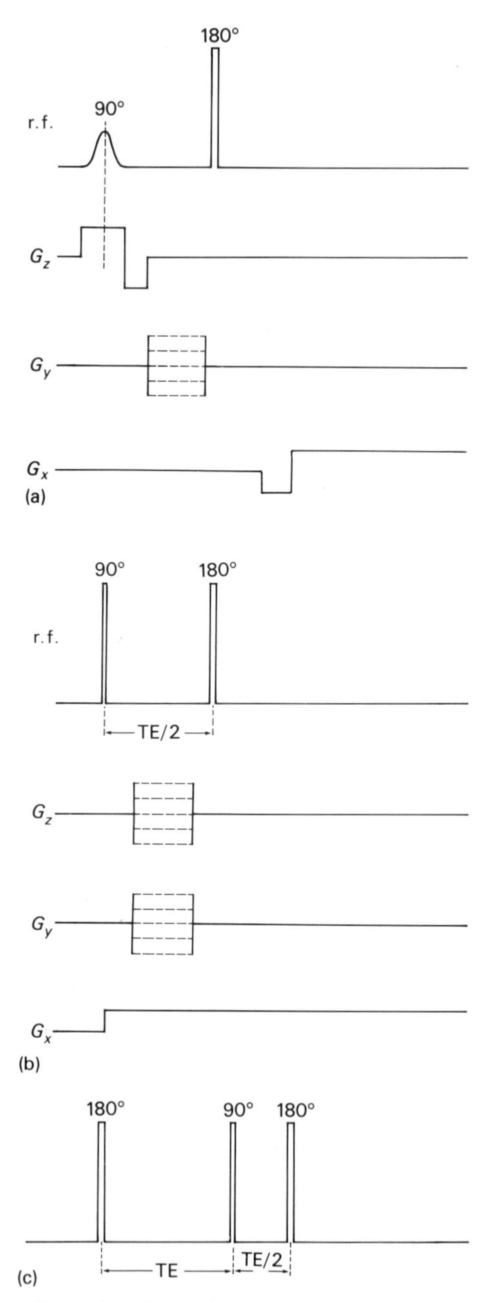

Fig. 4.29. (a) Two-dimensional Fourier imaging method. (b) Three-dimensional Fourier imaging method. (c) Inversion recovery (IR) r.f. pulse sequence for use with three-dimensional Fourier imaging methods.

the gradient switching. It also removes the effects of static field inhomogeneity during the phase-encoding period. The conceptually simpler three-dimensional experiment is shown in Fig. 4.29(b). Note that the phase encoding in z- and y-dimensions can be carried out simultaneously (although, of course, each has to be independently stepped resulting in a large number of experiments). The spatial resolution can be made isotropic by choosing equivalent gradient strengths and numbers of increments for each dimension. However, the total imaging time can then become excessive n^2 times the cycle time where n is the number of pixels across one dimension of the image field). Typical times may be several tens of minutes. It is therefore common practice to select a lower resolution in one dimension (z for example) thereby dividing the volume to be imaged into a set of contiguous slices.

If more than one slice is required the speed of the two-dimensional method can also be considerably improved by making use of the 'dead time' between the end of data acquisition and the next 90° pulse (this is required for T_1 recovery). Since only the spins in the selected slice are excited, one can use this time to excite a second slice. The process can be continued, allowing typically up to 15 slices (depending on the delay) to be simultaneously recorded. Such a scheme approaches the efficiency of the full three-dimensional method and multislice studies have become particularly popular. The method does suffer from one disadvantage however in that imperfections in the slice profile generally prevent excitation of contiguous slices. Commonly then, the slices are interleaved in two separate multislice studies to obtain full coverage of a particular volume.

If in a standard Fourier imaging experiment the cycle time (TR) is less than or of the order of T_1, spin–lattice relaxation effects will be present in the image. In fact, the repeated 90° pulse is reminiscent of the saturation recovery (SR) method of T_1 measurement (§ 2.5.1.1) and the pulse scheme is often referred to as a SR sequence. The image can be made more sensitive to T_1 effects by preceding the 90° pulse by a non-selective 180° one. This is illustrated for a three-dimensional method in Fig. 4.29(c); for simplicity only the r.f. pulses are shown. The sequence is similar to the 180°–TI–90° or inversion recovery method for T_1 measurement (§ 2.5.1.1) and the pulse scheme is generally referred to as an IR sequence. T_2 effects can also be introduced by increasing the length of the echo time, TE, so that it becomes comparable with the natural T_2 of the system being studied. For biological tissues values in the range 20–100 ms are typical. The pulse scheme is then referred to as a spin-echo (SE) sequence. SR, IR, SE and hybrid schemes are available for most imaging methods; see § 6.3.1 for a general discussion of image contrast.

It is possible to recall the echo a number of times by using successive 180° pulses in the manner of Carr and Purcell (see § 2.5.2.2). This pulse scheme is known as the multiple spin-echo or MSE sequence and is illustrated in Fig. 4.30(a). Typically 2, 4, or 8 echoes are used, though one can recall many more. It is possible using such a sequence to generate an image displaying pure T_2 information. However, careful attention must then be given to the cumulative effects of pulse imperfections (B_1 inhomogeneity, phase errors, etc.). These studies are especially popular since they generate additional data with little or no time-penalty.

The multiple spin-echo sequence can be further developed in a very

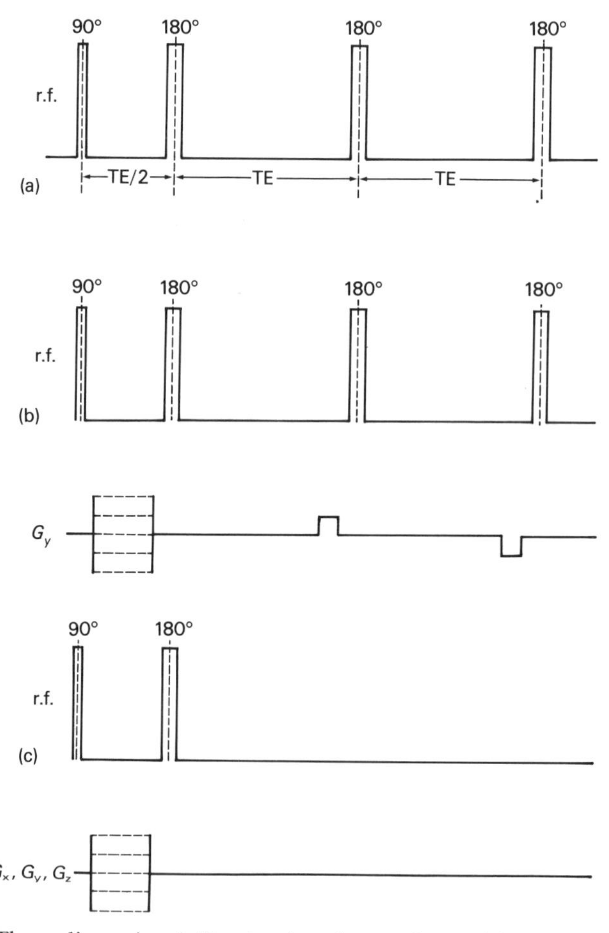

Fig. 4.30. Three-dimensional Fourier imaging variants; (a) multiple spin-echo (MSE) experiment; (b) recalled echo experiment with additional phase encoding in the y dimension; (c) chemical-shift imaging method.

interesting way. Rather than simply recall the train of echoes in the manner of Carr and Purcell, one can interpose further periods of phase encoding (see Fig. 4.30(b)) so that each echo corresponds to a separate free-induction signal in a conventional Fourier imaging experiment. Such a procedure reduces the minimum performance time for imaging by a factor equal to the number of echoes recorded. Rather than use broadband 180° pulses a similar effect can be achieved by periodically reversing the field gradient direction (Johnson, Hutchison, Redpath, and Eastwood 1983), although this has the disadvantage that it fails to refocus the effects of static field inhomogeneity and the length of the echo train is determined by T_{2e} ($1/T_{2e} = 1/T_{2n} + 1/T_{2mi}$; see § 2.5.2). Such a sequence bears more than a passing resemblance to the echo-planar technique proposed by Mansfield in 1977 (see § 4.3). In fact, 180° pulses can equally well be used with the echo-planar method as pointed out in the original papers on the subject. Such 'modified' echo-planar methods are closely related to the recalled echo sequence of Fig. 4.30(b).

It is also possible to achieve savings in processing time by paying attention to the order in which the f.i.d.s are collected (Eastwood 1983). Thus, rather than acquire f.i.d.s for monotonically increasing (or decreasing) gradients when all data has to be present before processing in the second frequency dimension can be commenced, it is possible to collect them in a 'bit-reversed' sequence allowing all but the final point calculations to proceed during signal acquisition.

Fourier imaging lends itself naturally to chemical shift studies, and is becoming increasingly popular for this purpose (see § 6.7). The basic three-dimensional imaging scheme of Fig. 4.29(b) can be used for studies of a single chemical-shift component if the 90° broadband pulse is replaced with a frequency selective one which isolates the required shift component (Hall, Sukumar, and Talagala 1984). Instead of looking at a single component, however, it is preferable to retain the multiplex advantage inherent in Fourier methods, and keep all shift components. This can be achieved by performing all phase encoding prior to the acquisition phase which thus takes place in a homogeneous field and retains chemical shift information (see Fig. 4.30(c)). Early experiments demonstrated the method for a single spatial dimension (Brown, Kincaid, and Ugurbil 1982; Haselgrove, Subramanian, Leigh, Gyulai, and Chance 1983), but these were soon extended to two and three dimensions (Maudsley, Hilal, Perman, and Simon 1983; Pykett and Rosen 1983; Hall and Sukumar 1984*b*). In the later case, data processing requires a four-dimensional Fourier transform. If high spatial and chemical shift resolution is needed this can place severe demands on the data handling system.

The advent of two-dimensional Fourier transform methods has revolutionized conventional high-resolution spectroscopy. It is too early to say whether the same will be true of Fourier imaging, but it has certainly made an impressive start.

4.2.5. *Rotating-frame Fourier zeugmatography*

The method of rotating-frame Fourier zeugmatography was proposed in 1979 by Hoult, who remains its principal proponent. The line of thought which led to the evolution of the method was one which has served n.m.r. spectroscopy well in the past; namely, whatever can be done in the laboratory frame it is also possible to do in the rotating frame. Thus, instead of the magnetic field gradient superimposed on the main magnetic field, one introduces a radiofrequency field gradient B_{1x} or B_{1y} in addition to the uniform r.f. field B_{10}. The three-dimensional version of the experiment appears as illustrated in Fig. 4.31. During time interval 1, of duration t_x, the spin system evolves under a r.f. field, $B_1 = B_{10} + B_{1x}x$, directed along the y'-axis (in the rotating frame). This causes a precession of the magnetization away from the z, z'-axis into the $z'x'$-plane. The angle of nutation is

$$\Theta_1 = \gamma(B_{10} + B_{1x}x)t_x, \tag{4.53}$$

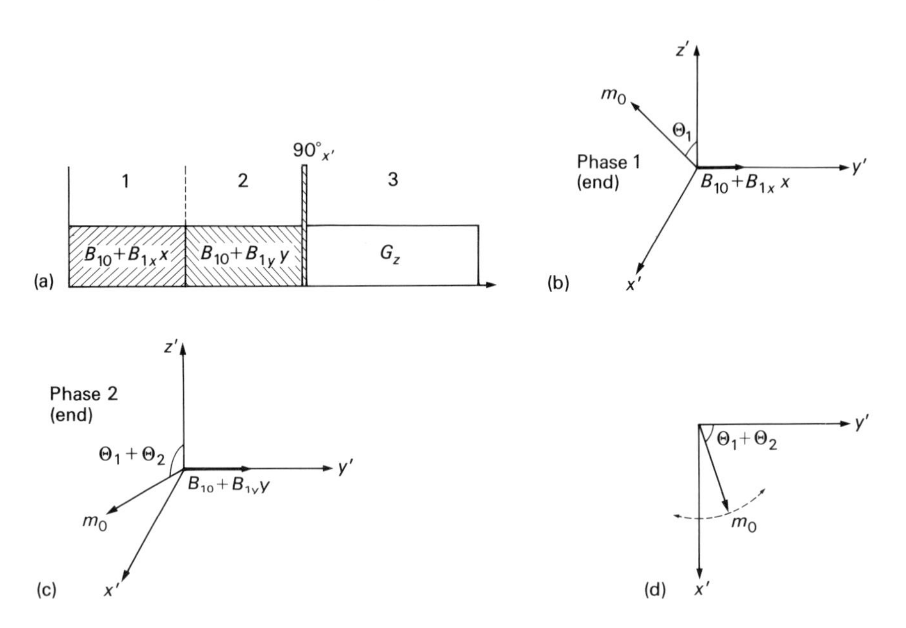

Fig. 4.31. The three-dimensional Fourier zeugmatography experiment. (a) Schematic; (b), (c) magnetization at the end of phases 1 and 2 respectively; (d) magnetization at start of observation phase (3).

and is linearly dependent on the displacement along the x-axis (laboratory frame). In the subsequent time-interval (2) a r.f. field is also applied along the y'-axis, but on this occasion its magnitude varies according to the y co-ordinate. The additional nutation is thus:

$$\Theta_2 = \gamma(B_{10} + B_{1y})t_y, \tag{4.54}$$

and the total

$$\begin{aligned}\Theta &= \Theta_1 + \Theta_2 \\ &= \gamma B_{10}(t_x + t_y) + \gamma B_{1x} x t_x + \gamma B_{1y} y t_y.\end{aligned} \tag{4.55}$$

A short (non-selective) $\pi/2$ pulse applied along the x'-axis (i.e. orthogonal to the previous pulses) nutates the magnetization into the $x'y'$-plane, where it consequently has a phase of

$$\phi = \pi/2 - \Theta. \tag{4.56}$$

In the final period (3) a static magnetic field gradient is applied to the sample and the signal is sampled in the usual manner. The imaging experiment proceeds exactly as in the case of conventional Fourier zeugmatography, with the time intervals being varied in stepwise linear fashion to give an $n \times n$ set of f.i.d.s from which the spin density can be extracted by three-dimensional Fourier transformation. The same comments made in § 4.2.2.2 with respect to admixture of absorption and dispersion terms also apply here; either an absolute value image is calculated or several sets of data are acquired from which the separation can be made (see Hoult 1979 for details).

The method of course also exists in a two-dimensional form—one can simply drop either interval 1 or 2. However, there is an alternative to this approach; rather than using the r.f. gradient to phase-encode it is made to generate an amplitude modulation. This is readily achieved by removing the $\pi/2$ pulse which separates the r.f. gradient from data acquisition in the G_z gradient. In this case, the component of magnetization in the $x'y'$-plane at the start of interval 3 has zero phase (oriented along the x'-direction) and an amplitude given by

$$m(x) = m_0 \sin\{\gamma(B_{10} + B_{1x}x)t_x\}. \tag{4.57}$$

The experiment proceeds in the usual manner with collection of f.i.d.s for different values of t_x followed by two-dimensional Fourier transformation. The advantage of the amplitude modulation technique is that absorption and dispersion terms are no longer inextricably mixed so that better spatial resolution for a given number of time increments is possible. Hoult (1979) has demonstrated this method, obtaining ^{31}P n.m.r. images in simple phantom objects. The disadvantage of the method is that, since different spatial regions are subject to different nutations, they are left with z components of magnetization which vary as

$$m_z(x) = m_0 \cos\{\gamma(B_{10} + B_{1x}x)t_x\}. \tag{4.58}$$

It is therefore essential to allow sufficient time (~5 T_1) between f.i.d.s for complete recovery of the equilibrium magnetization. This difficulty is avoided in the phase encoding method, since to a first approximation all the magnetization ends up in the $x'y'$-plane and the recovery of the z-component proceeds at a uniform rate. Plane selection for a two-dimensional experiment can be achieved in a variety of ways; for example, by using selective irradiation methods to saturate (destroy) all magnetization except that lying in the plane of interest. Hoult (1979) has advocated the use of a r.f. gradient of third or greater order for this purpose.

As an alternative to the use of a second r.f. field gradient (in the three-dimensional experiment), it is also possible to use a field gradient $G_z(t)$ oscillating at a frequency ω_1 corresponding to the rate of rotation of the magnetization in the rotating frame, i.e. $\omega_1 = \gamma B_1$. The effect, when viewed from a second rotating frame in which the B_1 precession is frozen, is exactly analogous to the effect of the B_1 field in the first rotating frame: see Hoult (1979) for details. This concept also forms the basis for a method of plane selection using 'rotating frame selective pulses' (Hoult 1980*a*). The required G_z field can be generated either through the use of the conventional field modulation coils or, more conveniently, by appropriate phase shifting of the B_1 r.f. gradient.

Radiofrequency field (B_1) gradients have also been applied to chemical-shift imaging studies. Cox and Styles (1980) used an asymmetric saddle coil to generate high-resolution spectra with one dimension of spatial localization (no static field gradient was applied in the acquisition phase). Depth-resolved chemical-shift studies have also been performed by Haase, Malloy, and Radda (1983) using a surface coil (see § 3.2.2). The B_1 field this generates is complex, but varies approximately linearly with axial distance over the range 0.2–1.4 radii from the coil centre. Chemical-shift studies are therefore most easily performed in this region.

4.3. Echo-planar imaging

The method of echo-planar imaging (Mansfield 1977; Mansfield and Pykett 1978) represents the ultimate in spatial encoding, since a complete two(or even three)-dimensional image is recoverable from a single f.i.d! The manner in which this can be achieved is perhaps most easily visualized by way of an example. Consider the matrix of test-tubes illustrated in Fig. 4.32; those tubes which are shaded are water-filled, the remainder are empty. If f.i.d.s were to be recorded in the presence of either

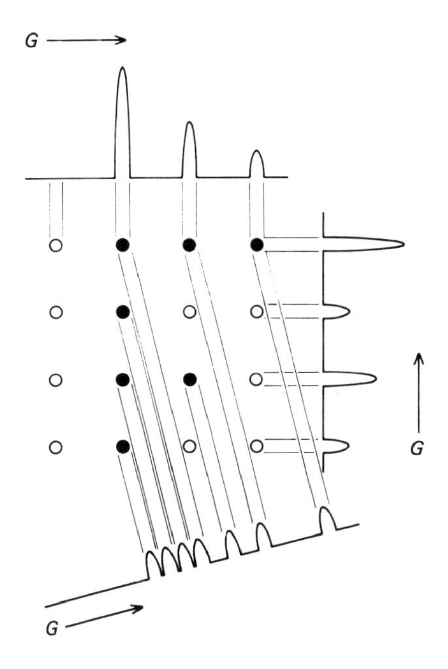

Fig. 4.32. Sketch of discrete letter F with three spin density projections corresponding to the gradient directions indicated. (From Mansfield *et al.* 1980.)

x or y field gradients and Fourier transformed, the resulting projections would be as illustrated. Neither would be unique, that is to say, given either one, it would not be possible uniquely to determine the object which gave rise to the projection. The normal solution, as we have seen, is to obtain a sequence of projections in differently oriented field gradients. However, note that in Fig. 4.32 there are special orientations of the field gradient which *do* uniquely determine the object. The requirement for this to be the case is that each point (test-tube) in the matrix appears at a unique position in the projection (see illustration). Of course, this can only be achieved for a discrete matrix—test-tubes which are physically separated. The problem of imaging a real object then is how to impose a discrete matrix on what is in reality a continuous spin distribution. The method of multiplanar imaging (Mansfield and Maudsley 1976, 1977) achieved this by selectively-exciting narrow strips of spins. However, there is a consequential decrease in sensitivity in proportion to the fraction of spins actually observed. The echo-planar method does not suffer from this drawback and achieves the discrete nature whilst still observing the entire spin distribution. The pulse sequence required to do this is shown in Fig. 4.33. We shall consider only the two-dimensional

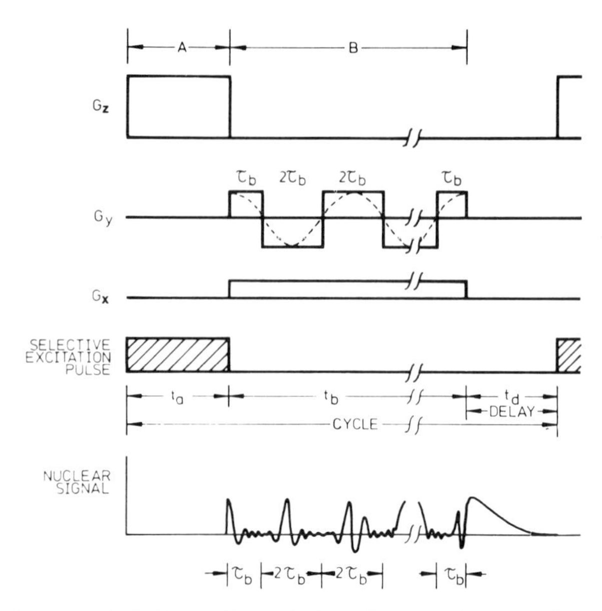

Fig. 4.33. Pulse and field gradient timing diagram for two-dimensional echo-planar imaging experiment. (From Mansfield and Pykett 1978.)

version of the experiment, so that the first phase (interval 1) is the selective excitation of the desired slice using a shaped r.f. pulse in combination with a G_z gradient and for preference using a refocusing procedure (see § 3.3.2). Ignoring the effects of the G_x line broadening gradient for the moment, we can consider the effect of the evolution in the gradient G_y. It will cause the spin isochromats to dephase giving rise to a decreasing signal envelope. In the absence of relaxation effects this decay is a coherent process which can be reversed by switching the direction of the field gradient $G_y \rightarrow -G_y$ or, alternatively, by application of a short non-selective π pulse. This gradient reversal can be periodically repeated, giving rise to a train of echoes as illustrated in Fig. 4.34. It clearly resembles the Carr–Purcell echo train, in which π pulses are used to refocus the effects of static field inhomogeneity, allowing accurate measurements of the spin–spin relaxation time (see § 2.5.2). The Fourier transform of such an echo train is easily determined with the aid of the convolution theorem (see § 4.1.2.5). Thus the echo train (see Fig. 4.35(d)) can be considered to be the convolution of a regularly-spaced set or 'comb' of δ-functions with the echo function $g(t)$ (see Fig. 4.35(a)). The convolution theorem states that the Fourier transform of the convolution of two functions is equal to the product of the Fourier transforms of the individual functions. Now the Fourier transform of a comb of

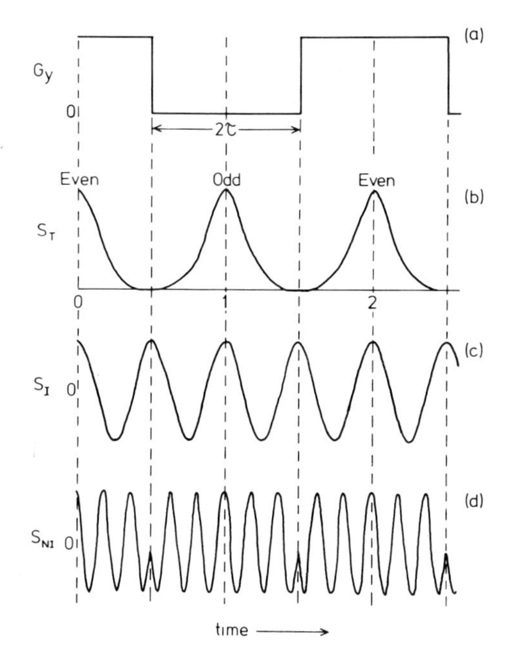

Fig. 4.34. The n.m.r. response to a switched field gradient: (a) the field gradient $G_y(t)$; (b) response envelope or spin-echo train due to switched gradient; (c) the response of a spin isochromat whose frequency is an integral multiple of the gradient repetition rate; and (d) the response of a spin isochromat whose frequency is a non-integral multiple of the repetition rate. (From Mansfield and Morris 1982.)

δ-functions is an equivalent set of δ-functions with a spacing which is inversely related to that of its Fourier conjugate, $\Delta\omega = 2\pi/\tau$ (see Fig. 4.35(b)). The echo itself represents the evolution of the spin system in the gradient G_y; its Fourier transform will therefore be the projection of the spin density orthogonal to the y-direction (see Fig. 4.35(c)). The product of the two functions is thus a series of δ-functions or 'sticks' whose amplitudes are proportional to the profile of the object normal to the y-axis (see Fig. 4.35(d)). Note, however, that the δ-functions do not simply sample the projection profile; spins lying at positions corresponding to intermediate frequencies are 'obliged' to contribute at these discrete frequencies by virtue of the temporal periodicity enforced upon them by the switched field gradient. Herein lies the reason for the great sensitivity of the echo-planar method. The spatial response function which has been determined in detail by Mansfield and Morris (1982) is illustrated in Fig. 4.36. See also Tropper (1981).

We now return to Fig. 4.33 and consider the effect of the broadening gradient G_x, which is of much smaller amplitude than G_y. Typically,

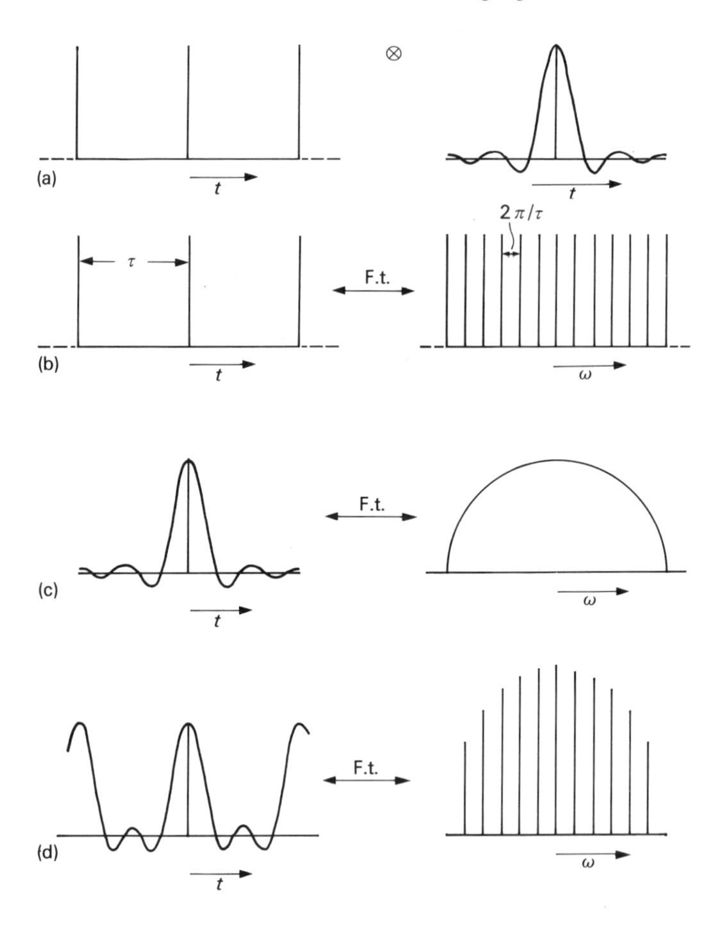

Fig. 4.35. Determination of the Fourier transform of a spin-echo train.

$G_x < G_y/n$, where n is the number of points required along an image line; this ensures that the lines do not overlap in the final spectrum. Since G_x is applied continuously and not switched, it causes a gradual evolution (decay) of the echo train envelope as described by the function $h(t)$. Its effect on the stick spectrum is to broaden each δ-function to give the full line profile along the x-axis. (This can be seen by inverse application of the convolution theorem. Thus the product of $h(t)$ with the unattenuated echo train appears after Fourier transformation as the convolution of $H(\omega)$, the Fourier conjugate of $h(t)$, with the previously discussed stick spectrum). Thus, the complete image is obtained by simple Fourier transformation of the echo train and appropriate ordering of the line profiles. Figure 4.37 illustrates the process for the case of an annulus of

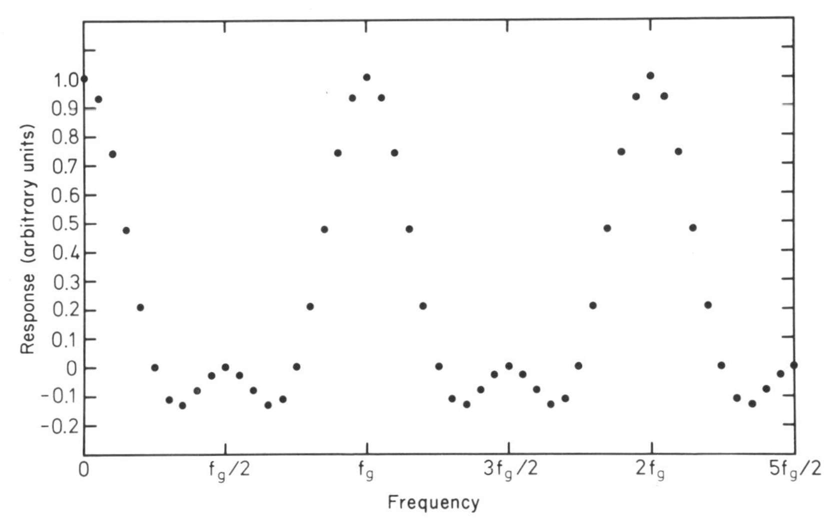

Fig. 4.36. Spatial response function of the echo-planar imaging experiment. Each point represents the total (summed) response for spins located at points along the y axis whose offset frequencies in the G_y gradient are as indicated. Spins whose offset frequency is an integral multiple of the switching frequency f_g give a maximum response. (From Mansfield and Morris 1982.)

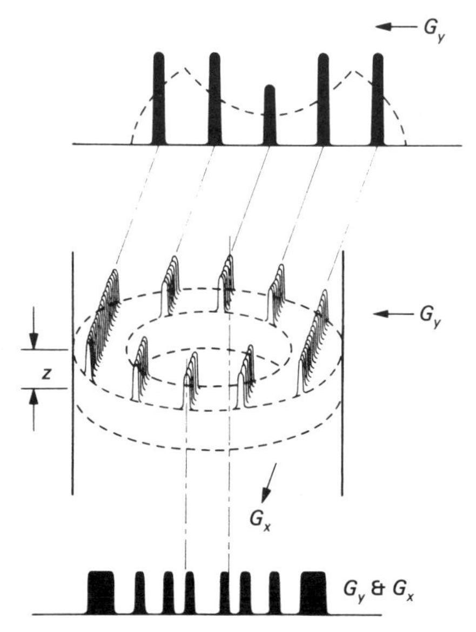

Fig. 4.37. Spin projections for a homogeneous annulus. The upper projection (broken line) is the projection in G_y alone. This turns into the discrete profile if G_y is modulated. The lower projection corresponds to the application of a modulated G_y gradient together with G_x on continuously. (From Mansfield and Pykett 1978.)

spins, both in the absence of the broadening gradient (upper part of figure) and with it switched on (lower figure). Figure 4.38 shows actual profiles obtained from a homogeneous cylindrical sample. Figure 4.38(a) illustrates the projection obtained from a single f.i.d. recorded in a static field gradient G_y whilst Fig. 4.38(b) shows the stick spectrum obtained by Fourier transformation of an echo train recorded in the absence of the broadening gradient G_x. In the latter case the G_y gradient had an amplitude of 1.26 G cm^{-1} and a repetition period of 1.28 ms (Mansfield, Morris, Ordidge, Pykett, Bangert, and Coupland 1980). Note that due to contributions from adjacent spins (see Fig. 4.36) the stick spectrum projection envelope has a much improved S/N over that diplayed in Fig. 4.38(a). The effects of the G_x-broadening are shown in Fig. 4.38(c) and the final reconstruction in Fig. 4.38(d). At this point it is worth noting that it is, in fact, necessary to separate the odd from the even echoes and Fourier transform each set separately, the reason being that the rising part of the echo is obtained in a net gradient $-G_y+G_x$, whereas the falling part is obtained in a gradient of G_y+G_x. The two images are in fact mirror images which, on reflection, can be co-added to restore the $\sqrt{2}$ loss in S/N which would otherwise ensue.

In Fig. 4.38 each echo train lasted for some 10.24 ms and could, in principle, have been used to generate a complete image. However, due to the poor S/N pertaining when this image was obtained (it was the first to be produced by the echo-planar method), some sixteen scans were co-added to give an imaging time of 8 s. Note also the presence of considerable distortion due principally to an imbalance in positive and negative excursions of the G_y gradient. Since these early experiments a one-quarter human-scale imaging system has been constructed and operates at 4 MHz with greatly improved performance. Images can be co-added to achieve a $\sqrt{N}$ improvement in image quality, and the S/N per unit time so obtained becomes comparable to that achieved by other two- and three-dimension projection reconstruction and Fourier imaging methods (Morris 1980). Or course, the real advantage of the echo-planar method will be in situations where motion is present. An obvious and extremely important example is the field of cardiology, where, provided single shot images of adequate S/N can be obtained, real-time human cardiac imaging will be possible (see Ordidge, Mansfield, Doyle, and Coupland 1982, for the details of progress in this area).

Instrumental developments are improving the quality of echo-planar images all the time. The first intermediate-sized images (Ordidge and Mansfield 1981) showing cross-sections of human arms and of small experimental animals had acceptable single-shot S/N. More recent real-time cardiac studies of rabbits, piglets, and human babies (see, for example, Rzedzian, Mansfield, Doyle, Guilfoyle, Chapman, Coupland,

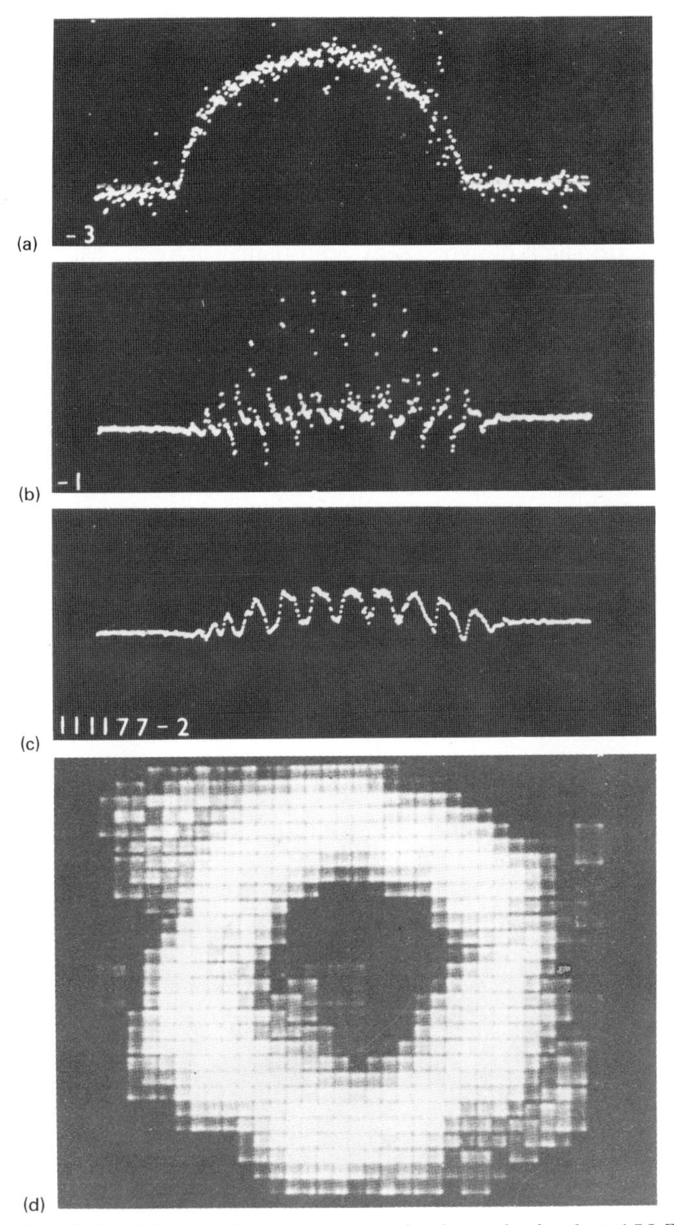

Fig. 4.38. (a), (b), (c) Experimental spin projections obtained at 15 MHz from a homogeneous cylinder of mineral oil. (a) Continuous projection with steady G_y gradient; (b) discrete projection obtained when G_y is modulated; (c) broadened discrete projection obtained as in (b) but with addition of steady G_x gradient; (d) echo-planar image of a 1.5-cm-thick selected slice through an extended cylindrical annulus of water obtained at 4 MHz in 8 s. The image field is represented by a 32×32 point array and the o.d. of the phantom was 4.8 cm. (From Mansfield *et al.* 1980.)

Chrispin, and Small 1983; Rzedzian, Doyle, Mansfield, Chapman, Guilfoyle, Coupland, Small, and Crispin 1984; Doyle, Rzedzian, Mansfield, and Coupland 1983), have achieved rather better S/N and demonstrate the truly remarkable potential of this technique. See, for example, Fig. 4.39.

One of the principal drawbacks of echo-planar imaging, aside from its increased technical requirement, is the problem of bandwidth. An example will serve to illustrate the difficulty. Suppose we require a final image consisting of 32×32 points. This can be achieved by recording an echo train consisting of 32 echoes with 32 data points recorded for each echo. Fourier transformation leads to a spectrum with 32×32 points, each one of which represents the spin density at a specific sample location. In order

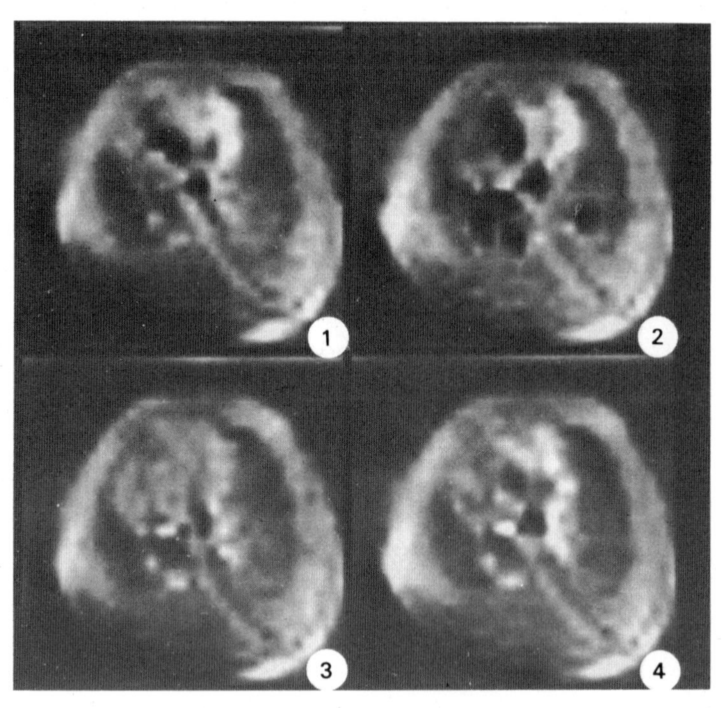

Fig. 4.39. Four e.c.g-gated echo-planar thoracic sections through an anaesthetized 21 kg piglet taken at a level near the apex of the heart. 1 corresponds to the R wave peak; 2, 3, and 4 were taken after successive delays of 150 ms. The period of the cardiac cycle was approximately 600 ms. All images clearly show the lung fields which are large at this level. The aorta with hemiazygos vein to its left and pulmonary vessels appear bright in 1, 3, 4 when the heart is at standstill, and dark in 2 where there is rapid blood flow. (From Doyle, Rzedzian, Mansfield, and Coupland 1983.)

to resolve these points, a G_x gradient of sufficient strength must be applied such that the point separation in the frequency domain, $\gamma G_x \Delta x$, where Δx is the spatial separation, is greater than the natural linewidth. Assuming a value for the latter of 25 Hz (see § 6.2.4 for a discussion of natural linewidths in biological tissue) the signal will be spread out over a bandwidth of $25 \times 32 \times 32$ Hz or ~50 kHz. If larger matrices, 64×64 or 128×128, are required then corresponding bandwidths of 200 and 800 kHz will be necessary and in order to achieve them the receiver coil Q will have to be damped (see discussion of § 5.4.1). Equally, the receiver system itself must be capable of handling such broadband signals. Note, too, that large bandwidths require fast sampling rates (see § 5.4.3) and high-speed analogue to digital converters are difficult to obtain with sufficient resolution. These factors are likely to restrict matrix sizes to 64×64 or possibly 128×128 and do not auger well for the chances of anything other than very crude three-dimensional imaging. Another problem arises when one comes to scale up the imaging system to full human dimensions. The gradient coils will of necessity be larger and have greater inductance. Switching times can then cause difficulties although these can be alleviated by making the coil part of a tuned circuit and switching at zero crossings using thyristor control (see Hoult 1980*b*, for details). Fast gradient switching can, of course, induce e.m.f.s and hence currents in the subjects being imaged. Whilst the author can personally testify to the fact that no sensation is felt in limb studies involving rates of field change as high as 250 $\mathrm{T s^{-1}}$, it is nevertheless necessary to exercise caution in this respect. Should there ever prove to be a problem an alternative does exist—instead of using switched gradients, these can be varied harmonically. It is then necessary to sample in a non-linear fashion, i.e. vary the sampling interval $t_{r+1} - t_r$, where t_r are the sampling times, such that $\int_{t_r}^{t_r+1} G(t)\,dt$ is constant. Alternatively, these can be sampled uniformly and the data weighted appropriately (Tropper 1981). Static magnetic field homogeneity can also present a major problem in the standard echo-planar experiment, since it determines the decay envelope of the echo train and hence the number of echoes which can be recalled. The problem is alleviated if 180° broadband r.f. pulses are used in place of switched gradients since dephasing due to static field inhomogeneity is then refocused.

The reader by this stage will have been left in little doubt as to the difficulties of the method. Nevertheless, its unique ability to perform ultra-high-speed imaging makes its pursuit worthwhile. Mansfield and colleagues have already demonstrated that there are solutions to all the major difficulties. In common with other n.m.r. imaging methods it too can be extended to perform chemical shift imaging studies (Mansfield 1983).

References

Abragam, A. (1961). *Principles of nuclear magnetism*, Chapt. 3. Oxford University Press.

Aue, W. P., Bartholdi, E., and Ernst, R. R. (1976). *J. chem. Phys.* **64,** 2229.

Batholdi, E., and Ernst, R. R. (1973). *J. magn. Reson.* **11,** 9.

Bendel, P., Lai, C-M., and Lauterbur, P. C. (1980). *J. magn. Reson.* **38,** 343.

Bracewell, R. N. (1956). *Aust. J. Phys.* **9,** 198.

Brigham, E. O. (1974). *The fast Fourier transform.* Prentice-Hall, Englewood Cliffs, New Jersey.

Brooker, H. R., and Hinshaw, W. S. (1978). *J. magn. Reson.* **30,** 129.

Brooks,, R. A., and Di Chiro, G. (1975). *Radiology* **117,** 561.

———, and ——— (1976). *Phys. med. Biol.* **21,** 689.

Brown, T. R., Kincaid, B. M., and Ugurbil, K. (1982). *Proc. natn. Acad. Sci. U.S.A.* **79,** 3523.

Brunner, P., and Ernst, R. R. (1979). *J. magn. Reson.* **33,** 83.

Budinger, T. F., and Gullberg, G. T. (1974). *Phys. med. Biol.* **19,** 387.

Cormack, A. M. (1963). *J. appl. Phys.* **34,** 2722.

——— (1964). *J. appl. Phys.* **35,** 2908.

Cox, S., and Styles, P. (1980). *J. magn. Reson.* **40,** 209.

Doyle, M., Rzedzian, R., Mansfield, P., and Coupland, R. E. (1983). *Br. J. Radiol.* **56,** 925.

Eastwood, L. M. (1983). Society of Magnetic Resonance in Medicine, August meeting.

Edelstein, W. A., Hutchison, J. M. S., Johnson, G., and Redpath, T. (1980). *Phys. med. Biol.* **25,** 751.

———, ———, Smith, F. W., Mallard, J. R., Johnson, G., and Redpath, T. W. (1981). *Br. J. Radiol.* **54,** 149.

Freeman, R. (1980). *Proc. R. Soc.* A **373,** 149.

Gordon, R., Herman, G. T., and Johnson, S. A. (1975). *Scient. Am.* **233,** 56.

Haase, A., Malloy, C., and Radda, G. K. (1983). *J. magn. Reson.* **55,** 164.

Hall, L. D., and Sukumar, S. (1982). *J. magn. Reson.* **50,** 161.

———, and ——— (1984a). *J. magn. reson.* **56,** 326.

———, and ——— (1984*b*). *J. magn. Reson.* **56,** 314.

———, ———, and Talagala, S. L. (1984). *J. magn. Reson.* **56,** 275.

Haselgrove, J. C., Subramanian, V. H., Leigh, J. S., Gyulai, L., and Chance, B. (1983). *Science* **220,** 1170.

Hawkes, R. C., Holland, G. N., Moore, W. S., and Worthington; B. S. (1980). *J. comput. assist. Tomogr.* **4,** 577.

Hoult, D. I. (1979). *J. magn. Reson.* **33,** 183.

——— (1980*a*). *J. magn. Reson.* **38,** 369.

——— (1980*b*). In *Magnetic resonance in biology*, vol. 1, (ed. J. S. Cohen). John Wiley, New York.

Hounsfield, G. N. (1980). *Science* **210,** 22.

Hutchison, J. M. S., and Smith, F. W. (1982). In *NMR imaging in medicine*, p. 101 (eds. Kaufman, L., Crooks, L. E., and Margulis, A. R.). Igaku-Shoin, New York.

———, Edelstein, W. A., and Johnson, G. (1980). *J. Phys.* E: *Scient. Instrum.* **13,** 947.
———, Mallard, J. R., and Goll, C. C. (1974). *Proc. Ampere Congr. 18th,* Nottingham, p. 431. North-Holland, Amsterdam.
Johnson, G., Hutchison, J. M. S., and Eastwood, L. M. (1982). *J. Phys.* E: *Scient. Instrum.* **15,** 74.
———, ———, Redpath, T. S., and Eastwood, L. M. (1983). *J. magn. Reson.* **54,** 374.
Kumar, A., Welti, D., and Ernst, R. R. (1975*a*). *J. magn. Reson.* **18,** 69.
———, ———, and ——— (1975*b*). *Naturwissenschaften* **62,** 34.
Lai, C-M. (1981). *J. appl. Phys.* **52,** 1141.
——— (1982). *J. Phys.* E: *Scient. Instrum.* **15,** 1093.
——— (1983*a*). *J. Phys.* E: *Scient. Instrum.* **16,** 1180.
——— (1983*b*). *J. Phys.* E: *Scient. Instrum.* **16,** 34.
——— (1983*c*). *Phys. med. Biol.* **28,** 925.
———, and Lauterbur, P. C. (1980). *J. Phys.* E. *Scient. Instrum.* **13,** 747.
———, and Lauterbur, P. C. (1981*a*). *Phys. med. Biol.* **26,** 851.
———, and Lauterbur, P. C. (1981*b*). *J. Phys.* E: *Scient. Instrum.* **14,** 874.
Lauterbur, P. C. (1973). *Nature, Lond.* **242,** 190.
——— (1981). *J. comput. assist. Tomogr.* **5,** 285.
———, and Lai, C-M. (1980). *IEEE Trans.* N.S. **27,** 1227.
———, Kramer, D. M., House, W. V. Jr., and Chen, C-N. (1975). *J. Am. chem. Soc.* **97,** 6866.
Lighthill, M. J. (1958). *Introduction to Fourier analysis and generalised functions.* Cambridge University Press.
Louis, A. K. (1982). *J. comput. assist. Tomogr.* **6,** 334.
Mallard, J. R., Hutchison, J. M. S., Edelstein, W. A., Ling, C. R., Foster, M. A., and Johnson, G. (1980). *Phil. Trans. R. Soc.* B **289,** 591.
Mansfield, P. (1977). *J. Phys.* C **10,** L55.
——— (1983). *J. Phys. D: Appl. Phys.* **16,** L235.
———, and Maudsley, A. A. (1976). *J. Phys.* C **9,** L409.
———, and ——— (1977). *J. magn. Reson.* **27,** 101.
———, and Morris, P. G. (1982). *NMR imaging in biomedicine,* Suppl. 2, *Adv. Mag. Res.* (ed. J. S. Waugh), Academic Press, New York.
———, and Pykett, I. L. (1978). *J. magn. Reson.* **29,** 355.
———, Grannell, P. K., and Maudsley, A. A. (1974). *Proc. Ampere Congr. 18th,* Nottingham, p. 431. North-Holland, Amsterdam.
———, Morris, P. G. Ordidge, R. J., Pykett, I. L., Bangert, V., and Coupland, R. E. (1980). *Phil. Trans. R. Soc.* B **289,** 503.
Maudsley, A. A., Hilal, S. K., Perman, W. H., and Simon, H. E. (1983). *J. magn. Reson.* **51,** 147.
———, Oppelt, A., and Ganssen, A. (1979). *Siemens Forsch.* **8,** 326.
———, Simon, H. E., and Hilal, S. K. (1984). *J. Phys. E: Scient. Instrum.* **17,** 216.
Moore, W. S., and Holland, G. N. (1980). *Br. med. Bull.* **36,** 297.
Morris, P. G. (1980). *Phys. Bull.* **31,** 306.
Ordidge, R. J., and Mansfield, P. (1981). *Br. J. Radiol.* **54,** 850.

———, Mansfield, P., Doyle, M., and Coupland, R. E. (1982). *Proc. int. symp NMR imaging*, p. 89. Bowman Gray School Medicine, Winston-Salem.

Pykett, I. L. (1982). *Scient. Am.* **246,** 54.

———, and Rosen, B. R. (1983). *Radiology* **149,** 197.

Radon, J. (1917). *Ber. Verh. Sächs. Akad. Wiss.* **69,** 262.

Rzedzian, R., Mansfield, P., Doyle, M., Guilfoyle, D., Chapman, B., Coupland, R. E., Crispin, A., and Small, P. (1983). *Lancet* **ii,** 1281.

———, Doyle, M., Mansfield, P., Chapman, B., Guilfoyle, D., Coupland, R. E., Small, P., and Crispin, A. (1984). *Ann. Radiol.* **27,** (2–3), 182.

Shepp, L. A. (1980). *J. comput. assist. Tomogr.* **4,** 94.

———, and Logan, B. F. (1974). *IEEE Trans.* **21,** 21.

Smith, F. W. (1982). *Proc. int. symp. NMR imag.* p. 125. Bowman Gray School Medicine, Winston-Salem.

———, Mallard, J. R., Reid, A., and Hutchison, J. M. S. (1981). *Lancet* **i,** 963.

———, ———, Hutchison, J. M. S., Reid, A., Johnson, G., Redpath, T. W., and Selbie, R. D. (1981). *Lancet* **i,** 78.

Sutherland, R. J., and Hutchison, J. M. S. (1978). *J. Phys.* E: *Scient. Instrum.* **11,** 79.

Taylor, D. G., May, D., and Jones, M. C. (1981). *Phys. med. Biol.* **26,** 931.

Tropper M. M. (1981). *J. magn. Reson.* **42,** 193.

5. Instrumentation

This chapter is principally concerned with instrumental requirements. For ease of reference the description of the apparatus has been divided into sections as indicated in Fig. 5.1. This is based on the concept of a 'general' n.m.r. imaging system, flexible enough to perform most, if not all, of the imaging experiments discussed in Chapters 3 and 4. Individual techniques will, of course, place more stringent requirements on certain components and will utilize them in rather different ways. Nevertheless, the basic features seen in Fig. 5.1 are common to all n.m.r. imaging systems. Figure 5.2 illustrates one manufacturer's practical realization of a clinical n.m.r. scanner. The final section (§ 5.7) of this chapter is devoted to a discussion of the relationship between imaging time, signal-to-noise ratio and spatial resolution. Estimates are given for an 'optimum' technique and the question of operating frequency is considered.

5.1. The magnet

Perhaps the single most important component of an n.m.r. imaging system is the magnet; it is certainly the most expensive, costing, with shim set, anything between £55 000 and £550 000 depending on field strength and bore (1984 prices). For clinical spectroscopic studies which require high magnetic field strength (typically >1.5 T) the only practical system is a superconducting one. Clinical proton imaging, however, has been successfully conducted at a number of field strengths in the range 0.05–2 T. The lower fields ($\leqslant 0.3$ T) do not necessitate superconducting technology, being achievable with conventional magnet systems. Indeed, iron-cored electromagnetic systems or permanent magnets were the basis for all early n.m.r. spectrometers which operated at proton frequencies of up to about 100 MHz (2.3 T); the field homogeneity requirement being met by careful shaping and polishing of the pole faces. However, even in these 'small sample' ($\sim 1\ cm^3$) instruments, magnet weights were often in excess of two tons (e.g. the Varian XL-100 spectrometer was optionally supplied with a 15 inch electromagnet weighing about 4 tons) and 'whole-body' versions, it was assumed, would be prohibitively heavy. Nevertheless, some designs using permanent magnets have been proposed and one such system is now commercially available as part of the Fonar[1] QED β3000

[1] Fonar, Inc., 110 Marcus Drive, Melville, New York 11747, USA.

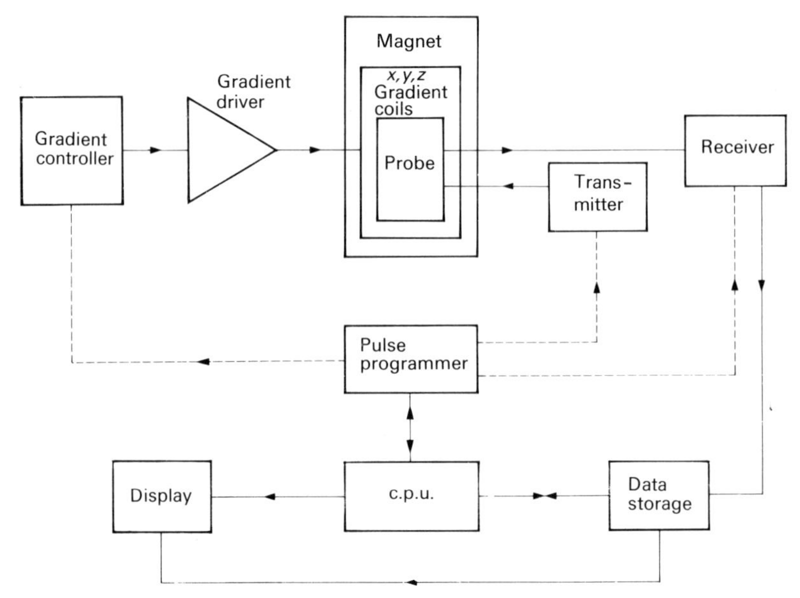

Fig. 5.1. Schematic diagram of a general n.m.r. imaging system.

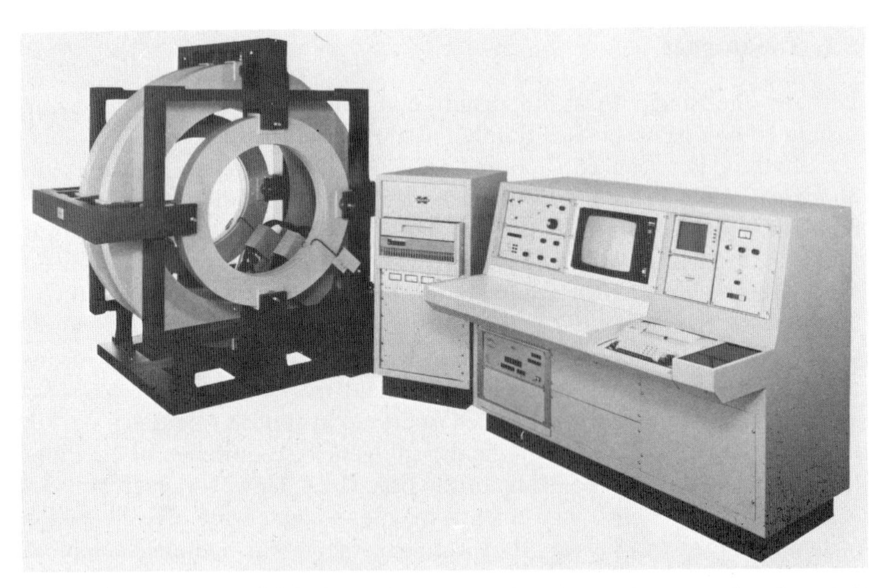

Fig. 5.2. Prototype (1981) whole-body n.m.r. imaging system showing, from left to right: 0.15 T air-cored resistive magnet; auxiliary rack containing disc drive (top) and gradient coil drivers (bottom); and main operator console with computer (bottom left), display (centre), and control monitors and radiofrequency transmitter (top right), and receiver (top left) electronics. The magnet power supply, shim, and gradient coils are not illustrated. (Courtesy of Bruker Spectrospin Ltd.)

scanner. The magnet weighs 100 tons but is assembled from magnetic 'bricks' weighing less than 100 lb each. It generates a field of 0.3 T. Iron-cored electromagnets of the Watson or 'window-frame' type (see discussion in Mansfield and Morris 1982) are particularly efficient and offer an alternative avenue of approach. The use of such a magnet in a clinical n.m.r. imaging system has been reported (Persson, Bolmsjö, and Malmgren 1983). Iron-screened electromagnets (Müller and Knüttel 1983) have similar advantages, both in the efficiency with which they generate the field and in their greatly reduced magnetic interaction with the environment. Currently, however, the two popular alternatives are air-cored electromagnets and superconducting systems. Both approaches are discussed in some depth below.

Before embarking on a detailed comparison, however, some of the basic requirements which should be met by any magnet system intended for n.m.r. imaging applications will be considered. Inevitably, many of these requirements will conflict and the optimal balance achieved between them will depend both on the particular imaging system and technique envisaged as well as on practical and financial strictures. Of course, in a commercial system, cost must be a primary factor; not just the capital cost, but also installation and running expenses. Depending on the system reliability, ease of maintenance may also be an important consideration: for example, a four-coil electromagnet is much more easily dismantled than a comparable superconducting design.

The main requirement from an imaging viewpoint is to achieve the required operating field with sufficient homogeneity over the imaging volume. As stated above, spectroscopic studies require high fields, typically $\geqslant 1.5$ T. They also need good field homogeneity; at least 1 p.p.m. to resolve the various phosphorus-containing metabolites and better than 0.1 p.p.m. if pH information is also required. However, if single point measurements only are to be made, then the volume over which high homogeneity is required is limited (for example, to a 5–10 cm diameter spherical region). In the case of whole-body proton imaging, on the other hand, operating frequencies extend over a much wider range and it is possible to work at lower field strengths. Homogeneity requirements are rather less stringent, too: high (spatial) resolution systems have successfully operated with homogeneities as low as 600 p.p.m. (Edelstein, Hutchison, Johnson, and Redpath 1980) and schemes have been published (Hutchison, Sutherland, and Mallard 1978) which allow image recovery from field distortions of up to 50 per cent! Nevertheless, the contribution of the inhomogeneity broadening should ideally be less than the natural proton linewidth, i.e. less than 20–50 Hz which is equivalent to 1–2.5 p.p.m. at an operating frequency of 21 MHz. This homogeneity must be maintained over comparatively large volumes, typically a 25-cm-

Table 5.1
Magnet parameters for n.m.r. applications

	Field strength (T)	*Homogeneity* $\Delta B/B$	*Homogeneous volume dsv*[1] *(mm)*	*Access*[2] *(mm)*
Clinical imaging				
n.m.r.	0.05–2.0	10^{-5}–10^{-4}	250–500	1000
t.m.r.[3]	1.5–4.0	10^{-7}	50–100	1000
c.s.i.[4]	1.5–4.0	10^{-7}	250–500	1000
Laboratory n.m.r.	2.3–14.1	10^{-9}–10^{-8}	5–12	50

[1] Diameter of spherical volume over which homogeneity is maintained.
[2] Diameter of magnet bore.
[3] Topical magnetic resonance.
[4] Chemical shift imaging.

diam. sphere for head studies and a somewhat larger one for whole-body work. Considerable effort is therefore required, both at the design stage, and subsequently in magnet alignment and shimming, if this optimum is to be achieved. Long and short term stability of the field is another important consideration as is the ease of access to the imaging volume. The basic requirements for both clinical spectroscopy and proton imaging studies are summarized and compared with those of conventional high-resolution spectrometers in Table 5.1.

As a final but important note, attention should be paid to the safety aspect of these large magnet systems, which typically contain some megajoules of stored electrical energy. Proper consideration should be given to ensuring that this energy is safely dissipated if, for example, a magnet supply or internal connection lead is broken, or a superconducting magnet quenches (see Mansfield and Morris 1982).

5.1.1. Magnet design: general principles

In order for a magnet to be specified realistically it is important that the prospective purchaser should have some understanding of the constraints governing magnet design and construction. In an ideal world, it would be preferable to approach the design problem by specifying the desired operating field, homogeneity, sample volume, access constraints, etc., and then calculate the configuration of current-carrying conductors required to achieve it. However, as will become apparent, the calculation of fields arising from even simple current distributions is generally non-trivial and hence extremely costly in terms of computational effort. The approach usually adopted is therefore to simplify the problem by restricting the design components to simple current elements, such as current loops, disposed in an axially symmetric manner along the magnet axis. The size, position, and relative currents of the loops are varied to optimize the homogeneity; the desired operating field can then be attained by linear scaling of all loop currents. This having been done, it must be recognized that, in practice, the currents have to be distributed over conductors of finite cross-sectional area, either to remain below the critical current density (at which point the magnet ceases to superconduct and goes 'normal') in the case of a superconducting system, or to dissipate the heat generated in a resistive system. In addition, the magnet windings will not be perfect, nor will the alignments of the constituent coils; the effects of all these imperfections will need to be considered. Even when this is done, the effects of the environment on the magnet will cause inhomogeneities to be reintroduced. It is therefore essential that provision be made for adjustment on site; we shall return to this point below.

At least in principle, it is possible to calculate the magnetic field at all points in space due to any distribution of current-carrying conductors, through the application of the Biot–Savart Law (see any standard text on electromagnetism for details, e.g. Bleaney and Bleaney (1965)), which states that the contribution **dB** to the magnetic field at a point Q arising from a current element $I\mathbf{dl}$ is

$$\mathbf{dB} = \frac{\mu_0 I \mathbf{dl} \times \mathbf{r}}{4\pi r^3}, \tag{5.1}$$

where **r** is the position vector of Q relative to **dl** and μ_0 is the permeability of free space (equal to $4\pi \times 10^{-7}\ \mathrm{Fm}^{-1}$ by definition). This is illustrated in Fig. 5.3. The total field can thus be determined by integration over the full current distribution. In the case of a single current loop of radius a, for example, the field at an arbitrary point $P(\mathbf{r}_0)$ can be written (Saint-Jalmes, Taquin, and Barjhoux 1981) as

$$\mathbf{B} = \mathbf{B}_z + \mathbf{B}_\rho, \tag{5.2}$$

where $\mathbf{B}_z$, the field component parallel to the loop axis is given by:

$$\mathbf{B}_z = \frac{\mu_0 I K}{4\pi(a\rho)^{\frac{1}{2}}} \left\{ J_1 + \frac{a^2 - \rho^2 - z^2}{(a-\rho)^2 + z^2} J_2 \right\}, \tag{5.3}$$

and $\mathbf{B}_\rho$, the radial component, which lies in the plane defined by z and $\mathbf{r}_0$ (see Fig. 5.3), is given by

$$\mathbf{B}_\rho = \frac{\mu_0 I K z}{4\pi\rho(a\rho)^{\frac{1}{2}}} \left\{ -J_1 + \frac{a^2 + \rho^2 + z^2}{(a-\rho)^2 + z^2} J_2 \right\}. \tag{5.4}$$

Here

$$K = \left[\frac{4a\rho}{(a+\rho)^2 + z^2} \right]^{\frac{1}{2}} \tag{5.5}$$

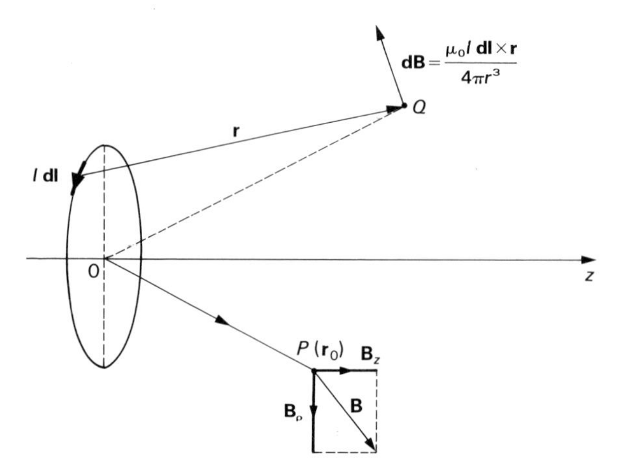

Fig. 5.3. The magnetic field **dB** generated at Q by the current element I**dl**, and the total magnetic field $\mathbf{B} = \mathbf{B}_z + \mathbf{B}_\rho$ generated at $P(\mathbf{r}_0)$ by the full Amperian loop centred at O. z and ρ are cylindrical co-ordinates with O as origin.

and J_1, J_2 are the elliptical integrals of Legendre. Thus, even for the simplest possible case of a single current loop, we have an analytical solution which gives us little insight into the manner in which elements can be systematically combined to optimize homogeneity. Since current loops are often used as surface coils (see § 3.2.2), it is, however, of interest in this context, giving the spatial variation of the B_1 field and hence also of the receiver response function (see Ackerman, Grove, Wong, Gadian, and Radda 1980, for details).

A different approach is clearly required: going back to first principles, we recall that, provided a region in which there are no conductors is considered (i.e. we calculate the field only within the magnet bore), the magnetic field **B** can be expressed as the gradient of a scalar potential V:

$$\mathbf{B} = \text{grad } V, \tag{5.6}$$

where V satisfies Laplace's equation:

$$\nabla^2 V = 0. \tag{5.7}$$

Laplace's equation occurs in many branches of physics and its solution in terms of spherical harmonics is well known (see, for example, Bleaney and Bleaney 1965). Thus:

$$V = \sum_{lm} A_{lm} r^l Y_l^m(\theta, \phi) \tag{5.8}$$

where the coefficients A_{lm} are constants, r, θ, ϕ the usual spherical co-ordinates and the spherical harmonics $Y_l^m(\theta, \phi)$ are given by:

$$Y_l^m(\theta, \phi) = (-1)^m \left[\frac{2l+1}{4\pi} \frac{(l-m)!}{(l+m)!}\right]^{\frac{1}{2}} \cdot \exp(\mathrm{im}\,\phi) P_l^m(\cos\theta), \qquad (5.9)$$

where P_l^m $(\cos\theta)$ are the associated Legendre functions. One of the important mathematical properties of spherical harmonics is that they form an orthonormal set. The practical consequence of this is that shim coils which generate spherical harmonics (Golay 1958; Hoult and Richards 1975*b*) provide field corrections which are non-interactive, a property which is particularly useful in high-resolution n.m.r. spectroscopy, where corrections up to fifth-order ($l = 5$) are commonly applied. If, as indicated above, we concentrate on finding a solution for the magnetic field which is axially symmetric, this excludes any ϕ variation, and, taking the gradient of V, we can write

$$B_z = \sum_{l=1}^{\infty} \frac{l(2l+1)}{4\pi} A_l r^{l-1} P_l(\cos\theta). \qquad (5.10)$$

It remains only to determine the coefficients A_l and this is most easily done by considering the axial field $B_z(z)$ near the centre of the magnet system ($B_z(z) = B_0$ when $z = 0$). This can be expressed in a Taylor expansion (see Jeffreys and Jeffreys 1956, for example):

$$B_z(z) = B_0 + \frac{\partial B_z(0)}{\partial z} \cdot z + \frac{1}{2!} \frac{\partial^2 B_z(0)}{\partial z^2} \cdot z^2 + \ldots. \qquad (5.11)$$

By setting $\theta = 0$ in eqn (5.10), we see that:

$$A_l = \frac{2\pi}{(2l+1)(l-1)!} \frac{\partial^{l-1} B_z(0)}{\partial z^{l-1}}, \qquad (5.12)$$

and the final field solution is:

$$B_z(r, \theta) = \sum_{l=0}^{\infty} \frac{r^l}{l!} \frac{\partial^l B_z(0)}{\partial z^l} P_l(\cos\theta). \qquad (5.13)$$

The important point to note is that we can derive the field at any point within the magnet volume provided we can determine the derivatives of the axial field at the origin. In the case of our simple current loop these are easily evaluated (symmetrical considerations eliminate the odd order contributions):

$$B_z(0) = \frac{\mu_0 I}{2a}, \qquad (5.14a)$$

$$\frac{\partial B_z(0)}{\partial z} = 0, \qquad (5.14b)$$

$$\frac{\partial^2 B_z(0)}{\partial z^2} = -\frac{3\mu_0 I}{2a^3}, \qquad (5.14c)$$

$$\frac{\partial^3 B_z(0)}{\partial z^3} = 0. \qquad (5.14d)$$

The spatial dependence $r^l P_l(\cos\theta)$ of the successive terms in the field expansion (eqn (5.13)) is illustrated in Figs 5.4(a), (b) and (c), which correspond respectively to $l = 2$, 4, and 6. Note that these field patterns apply to *any* coil system so long as axial symmetry is preserved. Since, for $r < a$, the terms in the expansion become progressively smaller the dominant field dependence is given by the first non-zero term (other than $l = 0$); hence, for example, the expression 'a sixth-order magnet'. In the case of our simple Amperian loop the field is second-order as depicted in Fig. 5.4(a) (see eqn (5.14)). By arranging elements such as current loops in such a way that higher and higher order terms cancel we thus have a systematic way of improving overall field homogeneity. The basic method for line currents was clearly discussed by Garrett (1951). This reference includes a useful tabulation of the properties of spherical harmonic functions. It is also the basis of Saint-Jalmes *et al.*'s (1981) method of 'pocket calculator magnet design'.

It has been appreciated for some considerable time that compensation to *all* orders can be achieved by a spherical surface current density distribution proportional to $\cos\theta$, and, to some extent, real magnet designs based on the use of increasing numbers of coil pairs can be considered successive approximations to this ideal. Some groups have actually constructed spherical magnets; however, in order to permit access to the central field it is necessary to be able to separate the two hemispheres (Hoult 1981) or else the spherical windings are left incomplete, allowing access from the poles (Mansfield and Morris 1982).

The first improvements we can make to the field generated by our simple loop is to add a second identical element. To maintain axial symmetry we place these loops parallel, so that the field expansion about the centre of this coil pair will contain no odd order terms. We can then adjust our one variable, the coil separation d, to eliminate the second-order term. The strength of the contribution to this term from each coil will be proportional to the second derivative of the axial field at the centre of the coil system (see eqn (5.13)) which, for a coil separation d, is

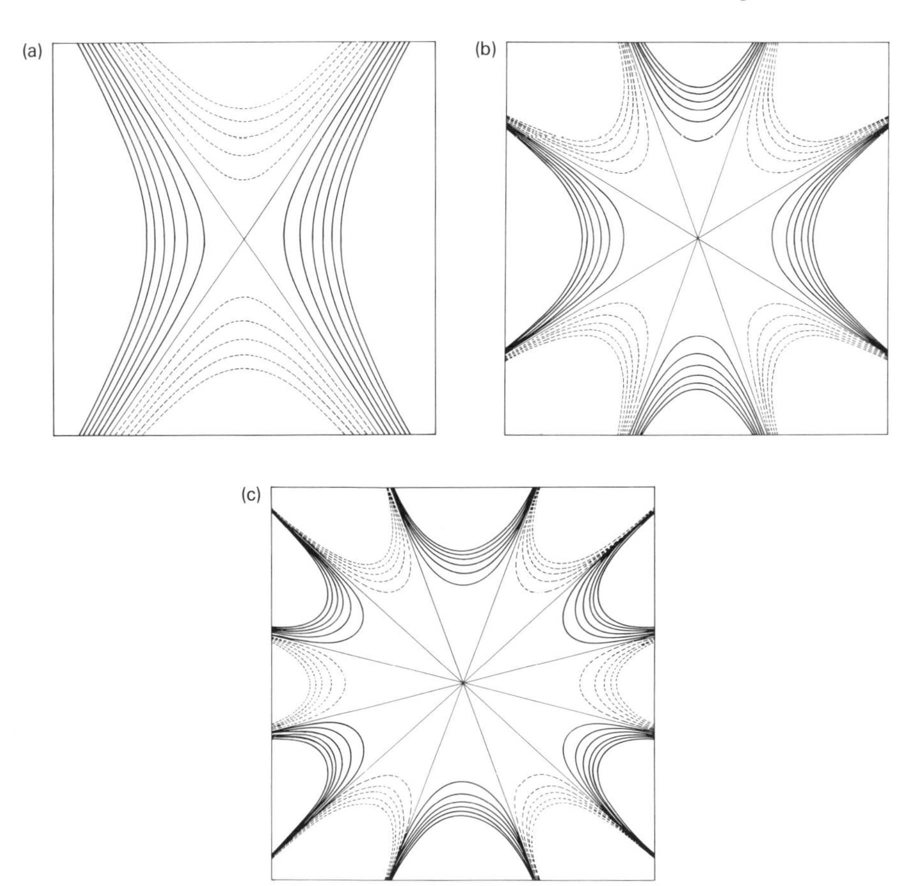

Fig. 5.4. Spatial dependence of $r^l P_l(\cos\theta)$, shown as contour plots in the r, θ-plane. $\theta = 0$ corresponds to the magnet axis and is shown horizontal. (a) Second-order ($l = 2$) contours; (b) fourth-order ($l = 4$) contours; (c) sixth-order ($l = 6$) contours.

given by (Saint-Jalmes *et al.* 1981):

$$\frac{3\mu_0 I}{2a^3}(4\xi^2 - 1)(1 + \xi^2)^{-\frac{7}{2}}, \tag{5.15}$$

where $\xi = d/2a$. This vanishes when $(4\xi^2 - 1) = 0$, i.e. when $\xi = \frac{1}{2}$ or $d = a$. Thus our two-coil system can be made fourth-order if the coils are separated by a distance equal to the loop radius; this is the well-known Helmholtz pair configuration illustrated in Fig. 5.5.

By adding a further pair of coils it is theoretically possible to achieve correction up to eighth-order and this is the design on which four-coil

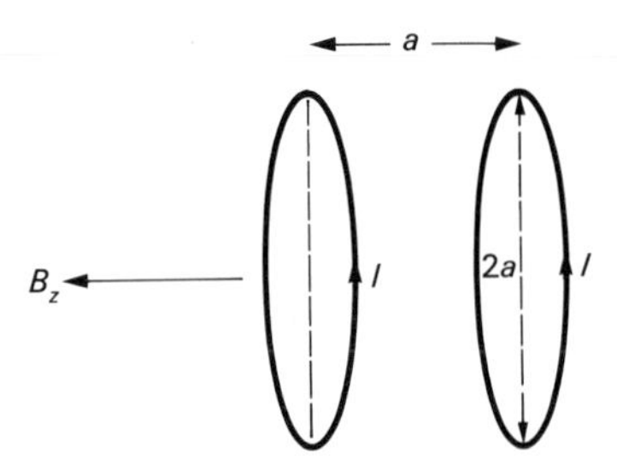

Fig. 5.5. The Helmholtz coil configuration: two Amperian loops separated by a distance equal to the loop radius.

imaging magnets are based. (For those interested in pursuing the higher-order theory, Saint-Jalmes *et al.* (1981) have derived the axial field derivatives for Amperian loops up to sixteenth-order!). When corrections for the necessarily finite cross-section of the coils are introduced, the theoretical performance reduces to sixth-order (Franzen 1962) which is what one normally achieves in practice. Saint-Jalmes *et al.* (1981) have, however, introduced a method for achieving eighth-order performance in such a system. They were also able to design a six-coil finite cross-section magnet which was compensated up to fourteenth-order! In practice, some of the coil parameters which could be varied to optimize homogeneity are adjusted to minimize power consumption or to achieve adequate access to the homogeneous volume, resulting in a degradation from the theoretical performance. Another important design consideration is the force between individual coils; this determines the strength of the framework needed to support them!

As mentioned above, it will generally be necessary to adjust the magnet to compensate for defects due to constructional tolerances; any departure from axial symmetry for example will lead to an introduction of the axial spherical harmonics $P_l^m(\cos\theta)e^{im\phi}$, $m \neq 0$ (see eqn (5.9)). In addition, the environment in which the magnet is to be placed (presence of iron girders in the building structure, for example) may also introduce inhomogeneities which need to be 'shimmed out' on site; these latter errors are usually predominantly linear gradients. Both types can be removed either by realigning the magnet (in the case of a resistive system) or through the use of a set of shim or 'Golay coils' (§ 5.2 discusses the generation of linear field gradients ($l = 1$) which are also used for spatial encoding in most imaging schemes). A further possibility is the use of judiciously-positioned ferromagnetic shims (see also § 5.1.4).

Oxford Instruments have developed autotracking systems to move a small n.m.r. probe over the magnet volume in order to construct the field plots which enable the required adjustments to be determined by computer analysis. Alternatively, an n.m.r. imaging method (Maudsley, Oppelt, and Ganssen 1979) can be used to map out the field distribution over a planar or multiplanar region. It is based on an extension of the Fourier imaging method and is discussed in § 4.2.2.3.

In the case of high-resolution n.m.r., the introduction of a sample disturbs the homogeneity to such an extent that reshimming is usually required. Fortunately, this is generally unnecessary for head or whole-body imaging studies.

5.1.2. *Resistive magnets*

The design of resistive magnets follows the prescription outlined in § 5.1.1 fairly closely. Both four- and six-coil systems are currently available, often with additional compensating coils to obviate movement of the main coils which can weigh in excess of one ton. Detailed descriptions of both resistive and superconducting systems have been given by Hanley (1982, 1984).

The field generated by a resistive system can be expressed generally (Hanley 1982) as:

$$B = \mu_0 G \sqrt{\frac{W\lambda}{a\rho}}, \tag{5.16}$$

where the Fabry factor G depends on the coil geometry, W is the power dissipated in the windings, λ is a space factor giving the proportion of the coil cross-section actually occupied by current-carrying conductor, a the inner winding radius of the coil, and ρ the resistivity of its windings. It will be noticed that the field strength depends on the square root of the power and on the inverse square root of the bore size. To minimize the power needed to achieve a specified field the magnet should therefore be small. However, this is not usually compatible with the requirements of sample access or homogeneity. Great effort has therefore been expended to achieve a good space factor λ.

The main concern when the current carrying cross-section is confined, is the dissipation of the heat generated in the resistance of the windings. This precludes the use of insulated copper wire, since heat conduction across the insulation is poor. One solution which has been adopted by Bruker Spectrospin (see Bruker Report 1/1981) is to use a hollow conductor through which coolant is circulated. This may either be purified water, in which case the inter-turn voltage must be kept low to avoid electrolysis, or else a non-conductive coolant can be buffered to a water system via a heat exchanger. With this efficient internal cooling method relatively high magnetic fields can be achieved ($\sim$0.3 T). The approach

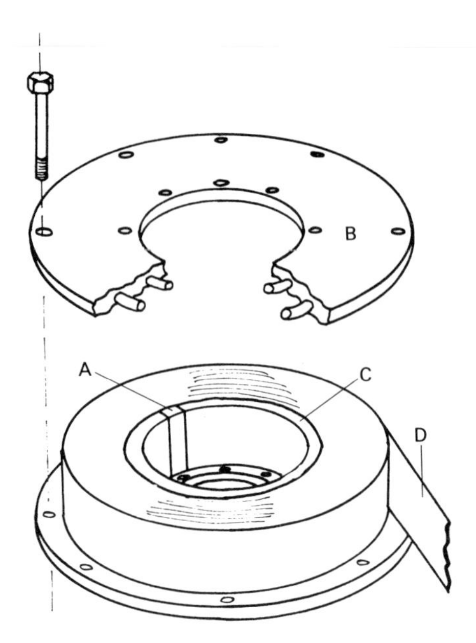

Fig. 5.6. Resistive magnet coil wound from edge-cooled aluminium strip. Key: A terminal; B water-cooled cheek; C coil former; D anodized aluminium strip. (From Hanley 1982.)

taken by Oxford Magnet Technology[1] which gives a much better space factor, although inferior heat dissipation, is to use anodized aluminium strip (rather than copper) as the conductor. Edge cooling can then be employed since thermal conduction over the aluminium strip is good (see Fig. 5.6). For air-cored resistive systems the field strength generally offered for whole-body applications is 0.15 T. This can, at the expense of extra power consumption, be increased to 0.2 T. Greater efficiency can be achieved through the use of iron-screened resistive magnets, which are currently available at a field strength of 0.23 T.

In the case of resistive magnets, the question of sample orientation arises. Although logistically the obvious configuration is one in which the patient lies along the magnet axis (Fig. 5.7(a), (b)) this requires a saddle-type receiver coil which gives a signal-to-noise ratio which is some $\sqrt{3}$ lower than is available from an equivalent solenoidal system (see § 5.4.1). Consequently, there is some merit in adopting the more awkward configuration in which the patient is rotated through 90° to enable the latter type of receiver coil to be employed (Fig. 5.7(c)). The Aberdeen

[1] Oxford Magnet Technology Ltd., is part of the Oxford Instruments Group of companies, Osney Mead, Oxford OX2 0DX, UK.

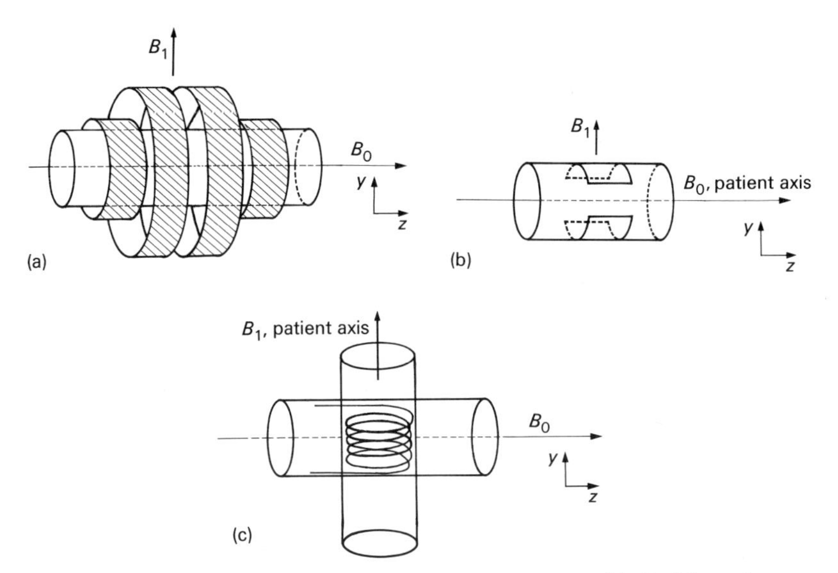

Fig. 5.7. (a) Schematic of a four-coil resistive magnet. (b) Saddle-coil system illustrating axial orientation of patient. (c) Solenoidal system illustrating transverse orientation of patient.

group have adopted this approach (Hutchison, Edelstein, and Johnson 1980; see also Fig. 5.8) and the early line scan images of Mansfield and co-workers (Mansfield, Pykett, Morris, and Coupland 1978; Morris, Mansfield, Pykett, Ordidge, and Coupland 1979) also followed this principle. However, particularly for the higher-field magnets, there is very little room to spare between the two central coils and access is, to say the least, awkward. Such a configuration is not, of course, possible with the conventional type of superconducting system, although split pair designs which would allow this approach have been proposed. Permanent and iron-cored electromagnets generate their field over the relatively narrow gap between pole faces—they are therefore ideally suited to take advantage of this transverse geometry.

One problem with resistive systems arises from the need to maintain short- and long-term stability of the field. Since imaging studies normally require the use of time varying gradients which are superimposed on the main field, it is difficult to employ the standard n.m.r. solutions to this problem, namely the feedback from a Hall probe device or a field frequency lock (essentially a simple n.m.r. spectrometer where the absorption frequency, usually of a second nucleus such as fluorine or deuterium, provides a very accurate measure of the field to frequency ratio). The Aberdeen team have successfully employed the latter method

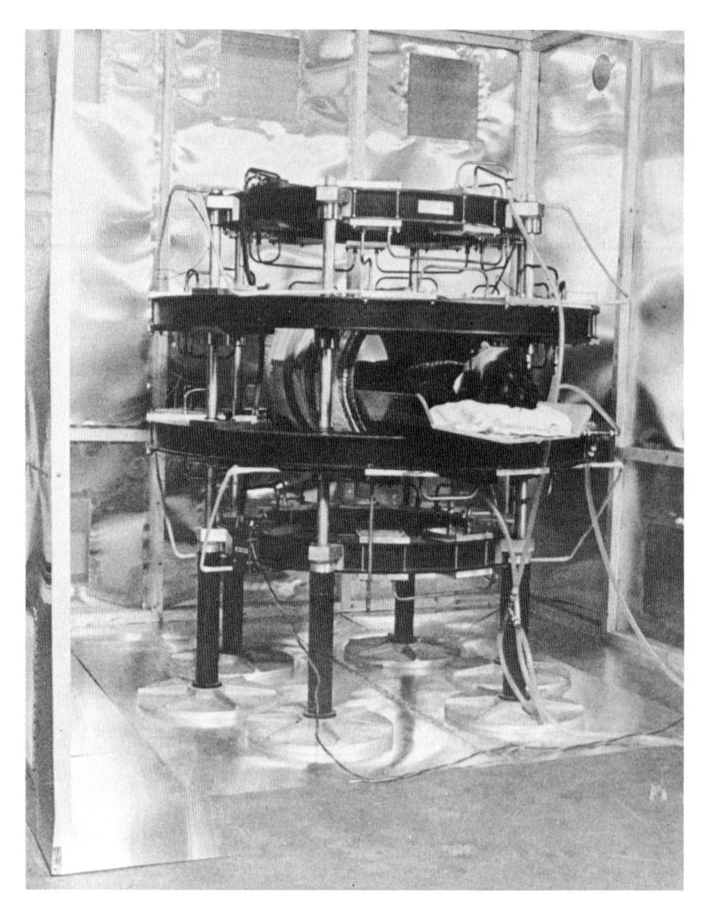

Fig. 5.8. The first Aberdeen n.m.r. imaging apparatus (Jan. 1979). The magnet was the first whole-body system to be supplied by Oxford Instruments and generates a field of 0.04 T, corresponding to a proton frequency of 1.7 MHz, in the vertical direction. (From Hutchison, Edelstein, and Johnson 1980. Copyright Institute of Physics 1980.)

by making use of gating techniques (Johnson, Hutchison, and Eastwood 1982). The solution generally adopted, however, is to achieve the requisite current stability in the magnet power supply—usually ~1 p.p.m. Unfortunately, this is not necessarily a guarantee of field stability if environmental factors, such as the temperature, change or, particularly when the magnet is first run up to field, when intercoil forces and the internal temperature of the windings cause subtle but important changes in magnet dimensions.

This section is concluded with specifications for four (Table 5.2) and six

Table 5.2
Four-coil resistive magnet for whole-body imaging

Field strength	up to 0.15 T
Power requirement	43 kW at 0.15 T
Field/current	5.3 Gauss/amp
Homogeneity	35 p.p.m. over 24 cm dsv[1] 90 p.p.m. over 40 cm dsv 90 p.p.m. over an ellipse of diameter 40 cm transverse by 28 cm axial
Clear bore	800 mm
Bore within central field for mounting gradient and RF coils	981 mm
Height of coil centre	1060 or 1165 mm options available
Magnet cooling	Water
Flow rate	50 l/min—magnet 15 l/min—power supply
Maximum height	2165 mm
Overall length	1510 mm
Overall width	2250 mm
Weight	3000 kg
Maximum power demand	60 kW; 3 phase

[1] Diameter of spherical volume
Courtesy of Oxford Magnet Technology; Feb. 1984

(Table 5.3) coil resistive magnets designed for n.m.r. imaging applications and kindly supplied by Oxford Magnet Technology. The four-coil system is illustrated in Fig. 5.9.

5.1.3. *Superconducting magnets*

The remarkable phenomenon of superconductivity in which electrical resistance drops abruptly to zero was discovered by Kamerlingh Onnes in 1911, first in mercury and subsequently in a number of other pure metals. Unfortunately, the critical magnetic field, above which superconductivity is quenched, was very low for these Type I materials, making them useless for magnet design. This latter application had to await the discovery in the 1950s of Type II superconductors. The material now used for most n.m.r. magnets, including those destined for imaging or t.m.r. applications, is an alloy of niobium and titanium in the range $Nb_{44\%}Ti$ to $Nb_{50\%}Ti$. It has a critical field H_c of about 10 T, a critical temperature T_c of 9 °K and a critical current density of 3×10^3 A mm^{-2} at 5 T. It has the great advantage of being mechanically strong and is

Table 5.3
Six-coil resistive magnet for whole-body imaging

Field strength	up to 0.15 T
Power requirement	72 kW at 0.15 T
Field/current	5.4 Gauss/amp
Homogeneity	10 p.p.m. over 30 cm dsv 30 p.p.m. over 40 cm dsv 80 p.p.m. over 50 cm dsv
Clear bore	800 mm
Bore within central field for mounting gradient and RF coils	1120 mm
Height of coil centre	1060 or 1165 mm options are available
Magnet cooling	Water
Flow rate	80 l/min—magnet 15 l/min—power supply
Maximum height	1990 mm
Length	2210 mm
Overall width	2250 mm
Weight	4470 kg
Maximum power demand	85 kW; 3 phase

Courtesy of Oxford Magnet Technology; Feb 1984.

utilized as an array of filaments embedded within a copper matrix with which it can be co-drawn. In conventional n.m.r. spectroscopy for which fields in excess of 10 T may be desirable, the magnet core uses niobium tin, an intermetallic compound. This is a comparatively brittle material which is formed *in situ* by annealing the magnet after it has been wound.

The very first superconducting magnet for n.m.r. imaging was designed and built by Damadian and co-workers (Minkoff, Damadian, Thomas, Hu, Goldsmith, Koutcher, and Stanford 1977; Minkoff, Damadian, Stanford, and Lipkowitz 1977) and produced the first ever whole-body n.m.r. image in 1977 (Damadian, Goldsmith, and Minkoff 1977). Originally envisaged as one of a pair of Helmholtz coils, their single-coil magnet proved to have sufficient homogeneity and field strength for use with their f.o.n.a.r. point imaging method.

Although some designs have been based on the resistive four coil systems discussed in § 5.1.2, the intercoil forces then necessitate sturdy supports. These in turn make it difficult to reduce heat conduction and achieve the efficient dewar design necessary to minimize boil-off of the cryogens needed to keep these magnets below their critical temperature. Designs are therefore based on the solenoid which, if infinitely long, gives a field which is uniform over its cross-section. Correction coils are added to

Fig. 5.9. Oxford Instruments 0.15 T four-coil aluminium strip resistive magnet for whole-body imaging. The clear access bore is 0.8 m. See Table 5.2 for further specifications. (Courtesy of Oxford Magnet Technology Ltd.)

compensate for finite length; in the simplest case, by adding an extra coil at each end. Once such a magnet has been wound and mounted in its dewar ('canned' in the trade parlance) it is not possible to adjust the internal configuration in the same way as for a resistive system. A superconducting shim coil assembly is therefore often included and the shim currents are set to give good homogeneity at the time the field is run up. If required, the final adjustment can then be made with a room temperature shim set which at the time of writing may contain up to eighteen individual shim coils. This allows the field to be re-optimized when a new sample is introduced—normally essential with high-resolution n.m.r. spectrometers including t.m.r. systems. Fortunately, as indicated above, the low homogeneity requirements of proton imaging systems normally enable one to dispense with this procedure which, if it became necessary, could drastically decrease patient throughput.

Whereas the design of the solenoid is the key to achieving good homogeneity, the dewar design is what determines the running costs. In order for the magnet to remain in the superconducting state it must be maintained at a temperature below T_c, and this is achieved by immersing the windings in liquid helium (total immersion is not essential since the cryostat design ensures isothermal conditions are maintained). Since liquid helium is expensive (£3.50 per litre (1984) in the UK, rather cheaper in the US, one of very few countries to count helium as a natural resource) and has a boiling point at atmospheric pressure of only 4.3 K, it is normally contained in a sophisticated dewar of the type illustrated in Fig. 5.10. The succession of super-insulation filled vacua, gas-cooled shields, and the liquid nitrogen chambers surrounding the inner liquid helium vessel containing the magnet, are directed towards minimizing heat introduced by conduction, convection, and radiation. Any supports, fill, or boil-off tubes are constructed from materials of poor thermal conductivity in an effort to reduce helium loss. The power leads are, of course, removed when the magnet has been run up and is operating in the persistent mode. The helium refill time can be considerably extended, up to a period of approximately one year, by using radiation shields maintained at low temperature (20 K, for example) by mechanical refrigeration with helium gas as refrigerant. This removes the

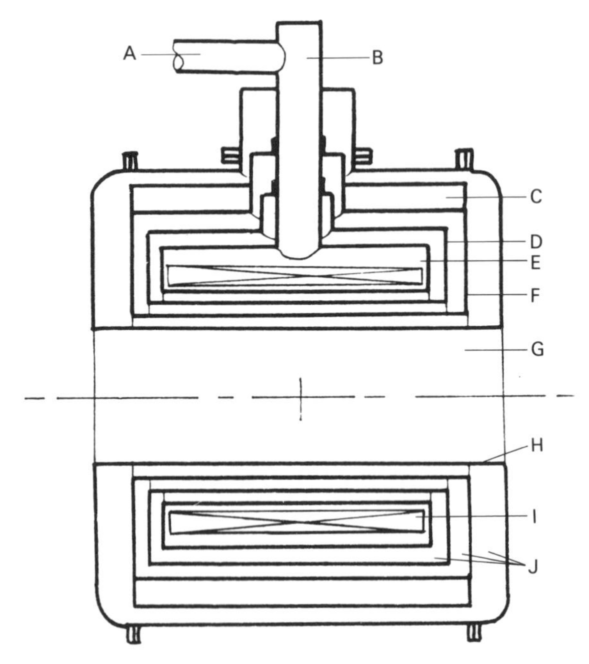

Fig. 5.10. Schematic of cryostatic construction for a whole-body superconducting magnet. Key: A helium gas vent; B neck tube; C liquid nitrogen; D gas-cooled shield; E liquid helium; F liquid-nitrogen-cooled shield; G room temperature bore; H outer vacuum case; I magnet coils; J vacuum. (From Hanley 1982.)

need for liquid nitrogen. Unfortunately, the cost of such closed cycle refrigeration is high; in the region of £25 000 at the time of writing. Since improvements in dewar efficiency have reduced the boil-off to 0.25–0.3 l/h such a capital investment will take many years to recover. This option is therefore only likely to be favoured in more remote areas where regular helium supply is difficult. Superconducting joints are so good

Table 5.4
Superconducting magnets for whole-body imaging

	Magnet			
	0.5 *T*	1.0 *T*	1.5 *T*	2.0 *T*
Magnet specification				
Field strength	up to 0.5 T	up to 1.0 T	up to 1.5 T	up to 2.0 T
Homogeneity				
basic: 50 cm d.s.v.	750 p.p.m.	750 p.p.m.	750 p.p.m.	750 p.p.m.
basic: 30 cm d.s.v.	200 p.p.m.	200 p.p.m.	200 p.p.m.	200 p.p.m.
with shim set: 50 cm d.s.v.	25 p.p.m.	40 p.p.m.	40 p.p.m.	40 p.p.m.
with shim set: 30 cm d.s.v.	5 p.p.m.	10 p.p.m.	10 p.p.m.	10 p.p.m.
Field stability	0.1 p.p.m./h	0.1 p.p.m./h	0/1 p.p.m./h	0.1 p.p.m./h
Time to energize (approx.)	4 hr	8 hr	8 hr	8 hr
Cryostat specification				
Room temperature clear bore	1050 mm	1050 mm	1050 mm	1050 mm
Patient access bore with shim set	920 mm	920 mm	920 mm	920 mm
Bore tube material	Glass reinforced plastic			
Overall diameter	1955 mm	2100 mm	2100 mm	2100 mm
Overall length: magnet only	2290 mm	2290 mm	2290 mm	2290 mm
with shim set	2350 mm	2350 mm	2350 mm	2350 mm
with shim and gradient set	2368 mm	2368 mm	2368 mm	2368 mm
Overall height	2425 mm	2400 mm	2400 mm	2400 mm
Minimum ceiling height	2740 mm	2740 mm	2740 mm	2740 mm
Height at central field	1060 mm	1070 mm	1070 mm	1070 mm
Weight: magnet	4300 kg	5715 kg	6250 kg	7250 kg
with cryogens	4800 kg	6315 kg	6835 kg	7800 kg
Cryogenic performance				
Liquid nitrogen boil-off	1.0 l/h	1.0 l/h	1.0 l/h	1.0 l/h
Time between refill	16 days	20 days	20 days	20 days
Liquid helium boil-off	0.4 l/h	0.4 l/h	0.4 l/h	0.4 l/h
Time between refill	33 days	49 days	41 days	33 days
Estimated annual helium consumption (with 0.3 G/cm gradient set, excluding 2000 l initial charge on installation)	5000 l			
Estimated annual nitrogen consumption (with 0.3 G/cm gradient set, excluding 2000 l pre-cool and initial charge	10000 l			

Courtesy of Oxford Magnet Technology, Feb. 1984

these days that there should be negligible downward drift of the field—a typical half-life for decay would be several hundred years!. The most common need for magnet service is to repump the dewars, necessary on average about once every five to ten years.

Superconducting systems are characterized by very high capital, installation, and running costs due to the nature of the constructional materials and coolants required. The principal advantages are high field and remarkable reliability and stability; field drift is in general less than 0.1 p.p.m per day. Oxford Magnet Technology, the leading magnet supplier for imaging applications, currently offers whole-body systems with a 1 m bore in four standard field strengths: 0.5, 1.0, 1.5 and 2.0 T. Specifications are given in Table 5.4, and a 1.5 T system is illustrated in Fig. 5.11. A wide range of other systems, particularly smaller bore, higher field, are also available in either horizontal or vertical geometries.

Fig. 5.11. Oxford Instruments 1.5 T superconducting magnet for whole-body imaging and spectroscopic studies. The clear access bore is 1 m. See Table 5.4 for further specifications. (Courtesy of Oxford Magnet Technology Ltd.)

5.1.4. Installation

In general, the installation of n.m.r. imaging systems is much more problematical than for other imaging modalities, particularly if high field strengths are required. First, there are the more obvious difficulties associated with site access and maintenance; the installation of a large superconducting system may require the removal and replacement of existing walls to enable entry of the magnet dewar, and provision must be made for routine access of a 500-1 transport dewar for liquid helium refills on a monthly or six-weekly basis. Then there are the environmental problems which are of a bi-directional nature. Thus, the magnetic field can affect the operation of slow electron beam devices such as the photomultiplier tubes of X-ray CT systems, and endanger personnel fitted with cardiac pacemakers (see § 6.8); equally, nearby structural ironwork will adversely affect the magnet homogeneity.

Viewed from a sufficient distance, the magnetic field of air-cored electromagnets and superconducting systems is approximately dipolar, falling off as the inverse cube of the distance from the magnet centre. Figure 5.12 shows the 5 G contours for a number of field strengths

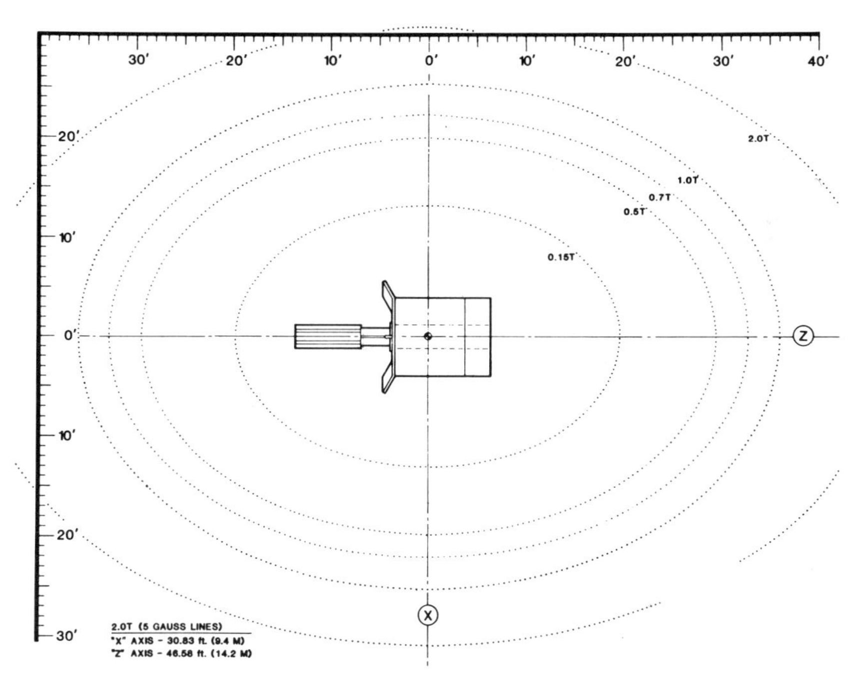

Fig. 5.12. The 0.5 mT (5 G) peripheral field lines for 0.15, 0.5, 0.7, 1.0, and 2.0 T magnets (Courtesy of Picker International Ltd.)

Table 5.5
Approximate ferrous proximities

	Field strength limits	*Magnet size*				
		0.15 T	0.5 T	0.7 T	1.0 T	2.0 T
Pacemakers, ferrous objects, mag tapes, oxygen cylinders, etc.	5 G	19′8″ (6 m)	29′6″ (9 m)	32′10″ (10 m)	37′1″ (11.3 m)	46′7″ (14.2 m)
Cars, elevators, etc.	2.5 G	24′10″ (7.56 m)	37′0″ (11.29 m)	41′4″ (12.6 m)	46′9″ (14.24 m)	58′8″ (17.89 m)
Image intensifiers, gamma camera photomultipliers, etc.	0.5 G	42′4″ (12.93 m)	63′4″ (19.31 m)	70′8″ (21.54 m)	79′11″ (24.35 m)	100′4″ (30.59 m)
Electron microscope	0.01 G	156′3″ (47.63 m)	233′4″ (71.14 m)	260′4″ (79.35 m)	294′4″ (89.71 m)	369′9″ (112.7 m)

Courtesy of Picker International Ltd.

currently in use for n.m.r. imaging or clinical spectroscopy. Table 5.5 details the fringe field limits and distances of closest approach for a number of devices. At the higher field strengths it may be difficult to meet these conditions within an existing hospital site. In such circumstances one is faced with the problem of either constructing a new building for the facility (perhaps the best solution if one envisages spectroscopic studies where high-field homogeneity is a prerequisite) or of magnetically screening the existing site. This can be a costly procedure both in terms of materials and also in terms of computational effort involved in the design, particularly if multiple magnet installations are involved or the environment is particularly hostile. Magnetic shielding is an important problem and will be even more so as field strengths increase. Oxford Magnet Technology have produced a screening yolk for this purpose: it consists of two end-plates coupled together with steel strips as illustrated in Fig. 5.13. Shielding 0.5 and 1 T magnets requires 10 and 20 ton yolks respectively. The 20 ton yolk, when used with a 1.5 T system, reduces the fringe field to that of an unshielded 0.5 T magnet.

As far as the magnetic field homogeneity is concerned, a certain amount of structural ironwork can be tolerated—the simple expedient of adding ferromagnetic 'shims' to symmetrize the situation goes a long way to help. However, if there is a substantial amount of ironwork within the 10 G fringe field of the magnet it may not be possible to shim to the specified homogeneity. Of course, permanent, iron-cored, and iron-shielded systems are largely free from these problems, though they are not available in high field strengths ($\gtrsim$0.3 T) and may require additional structural support particularly if installed other than on a ground floor.

Although not directly associated with the magnet, another major problem can arise from radiofrequency interference generated, for exam-

Fig. 5.13. Iron yolk for magnetic screening (Courtesy of Oxford Magnet Technology Ltd.)

ple, by radio transmitters, fluorescent lights, and devices such as light dimmers which use silicon controlled rectifiers. A good r.f. ground plane (earthed at a single point) is essential. If the problem is particularly severe then additional screening may be required in the magnet room walls and ceiling. The typical r.f. attenuation required of such a system might be ~100 dB at a frequency of 5 MHz. Additional filtering of the mains power supplies is mandatory.

5.1.5. Choice of magnet

The choice of a particular magnet system is dictated primarily by the field strength at which one wishes to operate (this in turn may be influenced by the budget at the disposal of the purchaser). For clinical spectroscopic studies there is little option, in view of the high operating fields required (normally 1.5 to 2.0 T), other than to go for a superconducting system. For whole-body studies a relatively large room temperature bore is necessary, perhaps 1 or 1.2 m to allow adequate space for shim and gradient assemblies and r.f. coils. For animal or human limb studies a proportionately smaller bore is acceptable, 0.3 m being a popular choice. Homogeneity should be ~0.1 p.p.m. over as large a volume as possible.

For proton imaging only, a much wider range of systems is available. The optimum operating frequency is still a matter for dispute (see § 5.7). However, the recent trend has been towards higher fields with 0.3–0.6 T superconducting systems as popular choices. Those imaging systems offering a spectroscopic capability operate in the 1–2 T range. Whilst it is true that high fields generally afford a higher signal-to-noise ratio per unit time, this approach has its problems, not least in terms of stray field containment. Lower field magnets (0.15 T air-cored electromagnets, for example) present less problems in design and installation and can yield excellent image quality at perhaps half the cost of a 2 T unit. Hospitals with limited funds available would do well to consider this alternative. Resistive magnets are available from a number of manufacturers including Oxford Instruments, Bruker Specrospin,[1] and Walker Scientific.[2] Superconducting systems are largely the province of Oxford Instruments although other manufacturers, for example Intermagnetics General Corporation,[3] have supplied such magnets. Running costs for iron-core or iron-shielded systems will be somewhat less than those for comparable air-core systems due to their rather higher efficiency. Running costs for permanent magnets are, of course, zero—a big plus in their favour.

Figure 5.14 gives an indication of the variation in capital cost of resistive and superconducting magnet systems with field strength. The graph is taken from the work of Hanley in 1982, prior to many of the large-bore high-field systems becoming available. Additional points corresponding to the costs of 0.8 m resistive and 1.0 m superconducting systems at the time of writing have been added. Magnet running costs are equally important: an indication of their magnitude is given in Fig. 5.15, also from the work of Hanley.

The conversion of a systems capital, installation, running, service, and depreciation costs into a cost per scan is difficult to assess in advance. An estimate for break-even operation and a patient throughput of 2500 patients per year is US $375 for a resistive system and US $500 for a high-field superconducting one. For comparison X-ray CT break-even costs per scan are of the order of US $250.

Magnet technology for n.m.r. imaging applications has evolved rapidly and continues to do so: experimental 1-m-bore 4 T systems may soon be available for spectroscopic studies at a cost of £550 000, and we can also

[1] Bruker Spectrospin, Silberstreifen, D-7512 Rheinstetten, W. Germany.
[2] Walker Scientific, 10 Rockdale St., Worcester, Massachusetts 01606, USA.
[3] Intermagnetics General Corp., P.O. Box 566, Guilderland, New York 12084, USA.

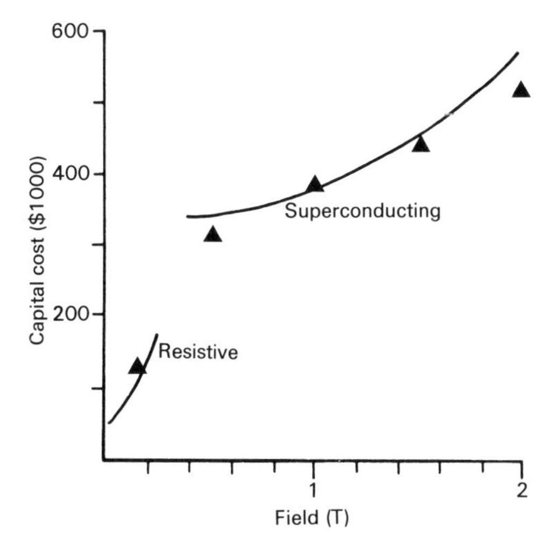

Fig. 5.14. Capital cost versus field strength for 0.8-m-bore n.m.r. imaging magnets. Additional points (▲) are approximate costs at time of writing for a 0.8 m resistive and 1 m bore 0.5, 1.0, 1.5, and 2.0 T superconducting systems. (After Hanley 1982.)

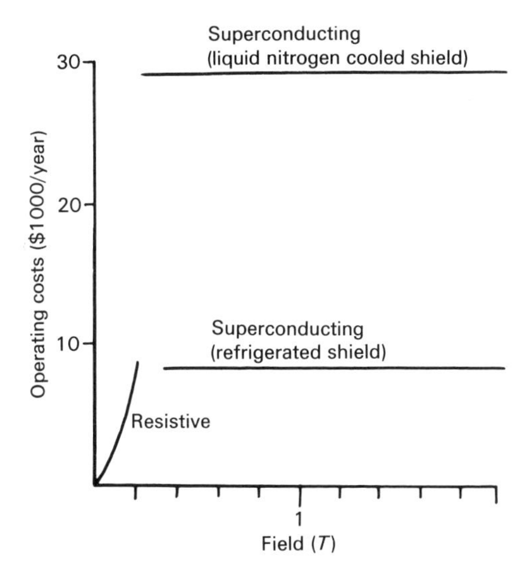

Fig. 5.15. Operating costs for 0.8-m-bore magnets versus field strength. (From Hanley 1982.)

look forward to the possibility of greatly improved homogeneity, perhaps to 5 p.p.m. over a 50 cm d.s.v., 1 p.p.m. over a 30 cm d.s.v.

5.1.6. Patient handling

The design of the patient-handling system is an important consideration. It should allow accurate, minimum-fuss access to the working region. Equally, a rapid patient removal capability is essential, since cardiac arrest and life-support systems contain ferromagnetic components and cannot therefore be kept in close proximity to the magnet. Positioning to an accuracy of the order of 1 mm can be achieved in each dimension with the aid of laser alignment methods now standard on X-ray CT units.

The magnet environment can be very claustrophobic, particularly in the case of superconducting systems, where the open ends of the dewar are sometimes sealed off with metal gauze to form a complete r.f. screen around the probe. In such circumstances it is easy for the patient to feel isolated, and it is important to maintain audio-visual contact at all times. Since consoles are normally separated from the imaging room by a glass screen, giving the option of an acoustic barrier, communication with the clinician is generally via a two-way intercom. Nursing staff, however, should always be present in the imaging room itself when a scan is in progress. The patient may be afforded the additional safeguard of a 'panic' button, ensuring his rapid exodus from the magnet.

Problems due to claustrophobia can be alleviated by appropriately shaping the patient access bore and by paying attention to lighting, ventilation, and temperature control. As the patient is normally scanned in a prone position, a mirror system is particularly beneficial in the maintenance of visual contact.

One potentially disturbing feature of some n.m.r. imaging sequences is the 'click' caused by the slight movement of the gradient coils as the current through them is switched—a problem which is particularly apparent in high field superconducting systems. Sound suppression systems have been developed to reduce this noise to an acceptable level.

5.2. Gradient coil systems

5.2.1. General considerations

We have seen in Chapter 2 that, in an n.m.r. imaging experiment, the ability to assign a signal as being derived from a particular spatial region of the sample depends on the application of a secondary magnetic field with a known spatial dependence. In the case of a f.o.n.a.r. or t.m.r.-type experiment (see Chapter 3) high-order gradients (Hanley and Gordon

1981) are used to generate a central aperture from which (in the case of t.m.r.) high-resolution spectra can be obtained. (A similar approach can also be used to define an aperture which is remote from the receiving coil, a technique christened 'inside out n.m.r.' (Cooper and Jackson 1980; Burnett and Jackson 1980; Jackson, Burnett, and Harmon 1980) which can be used as a prospecting aid, for example in the search for oil. In a conventional imaging experiment, however, the simplest and most desirable spatial dependence for the magnetic field is a linear one (see § 2.2.2). In general, three mutually orthogonal gradients $G_z = \partial B_z/\partial z$, $G_y = \partial B_z/\partial y$, $G_x = \partial B_z/\partial x$ are required, even if we require to image in only two dimensions (the third gradient is then used to select the particular slice we wish to observe). Note that the fields generated by these gradients are always in the z-direction but vary in amplitude according to the displacement along the z-, y-, and x-axis respectively. In the most flexible imaging systems the roles of these gradients can be interchanged or combined permitting software-controlled changes from axial to coronal, sagittal, or intermediate slice orientations. Some examples are given in Chapter 6.

We have discussed the choice of gradient strength required to spatially resolve all the pixels along the gradient direction in § 2.5.2.1. We can apply similar reasoning to a consideration of the linearity required of these gradients. Thus, when the image data are analysed, we assume that each discrete band of (angular) frequencies $\Delta\omega$ arises from a region of the same spatial extent, Δx say. Then

$$\Delta\omega = \gamma G_x \Delta x. \tag{5.17}$$

See Fig. 5.16(a) where the straight line corresponds to the ideal linear gradient G_x. In a real situation, however, the gradient will depart from linearity (Fig. 5.16(c)) and the widths Δx associated with each frequency interval will vary with position $\Delta x = \Delta x(x)$. In general, the non-linearity increases with distance from the origin giving rise to spatial and intensity distortions near the periphery of the image. This is illustrated in Figs. 5.16(b) and (d), which show the profiles obtained from an object of uniform cross-section extending over the range $0 < x < a$ in the gradients of Figs 5.16(a) and (c) respectively. The most stringent requirement we need to make of the gradient linearity is that the total spatial distortion over the full dimensions of the object be less than one pixel width. Thus, for a 128–point profile it should be less than 1/128, or 0.78 per cent, a tall order indeed. However, since the distortion occurs principally near the outer edges of the sample, much greater non-linearities are tolerated in practice. One pitfall which should be avoided is illustrated in Fig. 5.16(e). It involves placing a signal producing region of the object in a region where the gradient field has passed through its maximum value

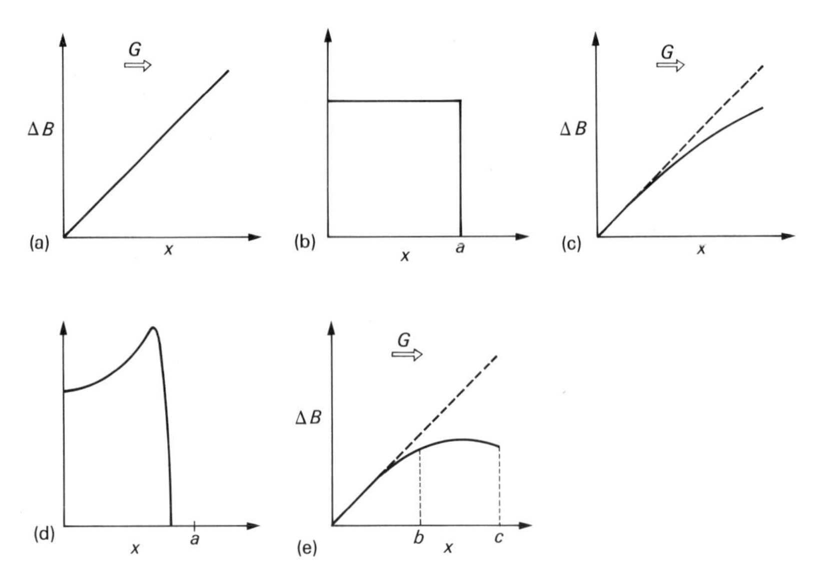

Fig. 5.16. Gradient linearity: (a) linear gradient; (b) profile generated by a 'top hat' spin distribution of width a; (c) non-linear gradient; (d) profile generated by 'top hat' spin distribution by gradient of (c); (e) gradient turnover showing non-unique aspect of field-frequency encoding.

and 'turned over' such that some of the frequency bands correspond to two distinct spatial regions, those labelled b and c for example, a situation which is clearly disastrous for image reconstruction. In practice, it is difficult to avoid such gradients, particularly along the z-direction or patient axis. It is then crucially important that the transmitter/receiver system does not excite/receive signals from parts of the sample extending into those regions.

Different techniques utilize the gradients in different ways; some requiring purely static, others oscillating, and still others rapidly-switched gradients. Strength and linearity needs also vary from method to method and may conflict with switching requirements. For example, a high gradient strength will need a coil system with many turns; good linearity over the sample volume will necessitate a large coil system, or one with many separate elements. In both cases the result is likely to be a high-inductance system with a long switching time. Note that, since substantial currents (often tens of amp) are required to energize the gradient coils, and these are located within a large uniform magnetic field, quite large forces exist on the current carrying conductors. Thus, for a static field B_z of 0.1 T and a current of 10 A, the force on a conductor lying normal to the field direction would be 1 Newton per metre length.

The coil formers should therefore be sufficiently robust to contain these forces. In addition, if the field gradient direction is switched rapidly the slight movement of the conductors to their new equilibrium position manifests itself as an audible click, which can be quite disturbing to operator and patient alike.

In what follows, there is no attempt to provide a comprehensive description of all possible coil systems, the aim is only to outline some basic design principles. The simplest of the gradients one needs to consider is the axial one, G_z. It is possible to apply similar design methods to those employed for the main magnet (§ 5.1.2), except that an antisymmetric geometry is now selected in which current directions are opposed in each member of a coil pair. The even terms, including the zeroth one, in the spherical harmonic expansion of the field, thus vanish identically. Coil currents, sizes, and separations are varied to give the desired first-order (gradient) term whilst minimizing successive higher-order ones; third, fifth, seventh, etc. For example, with a single-coil pair the most uniform gradient (third-order term zero) is achieved at a separation of $\sqrt{3}$ times the coil radius. This is the well-known Maxwell pair which has been widely used for the n.m.r. measurement of spin diffusion (Tanner 1965).

The other gradients G_x and G_y are more problematical since they do not preserve axial symmetry. However, since G_x can be derived from G_y by using the same coil assembly rotated through 90°, we need only consider one case, G_y say. Two basic methods are in popular use; those based on straight wire systems and those using saddle-type Golay coils.

5.2.2. Straight wire systems

Consider the four parallel infinite straight wires illustrated in cross-section in Fig. 5.17. The magnetic field $B_z(y, z)$ generated by a single wire at (a, b) can be expressed as (Zupančič and Pirs 1976):

$$B_z(y, z) = \frac{\mu_0 I}{2\pi r} \mathcal{R} \sum_{n=0}^{\infty} \left(\frac{\xi}{r}\right)^n e^{-i(n+1)\phi}, \tag{5.18}$$

where $\xi = y + iz$, $a + ib = re^{i\phi}$, a, b, r, ϕ, are as defined in Fig. 5.17 and I is the current. Similar expressions apply for the wires at $\pi - \phi$, $\pi + \phi$ and $-\phi$, so that when the four contributions are summed, the even-order terms can be rejected on symmetry grounds and we are left with

$$B_z(y, z) = \frac{2\mu_0 I}{\pi r}\left\{\frac{y}{r} \cdot \cos 2\phi + \frac{(y^2 - z^2)}{r^2} \cos 4\phi + \ldots\right\}. \tag{5.19}$$

The first term yields a linear gradient $\partial B_z/\partial y$ in the y-direction of

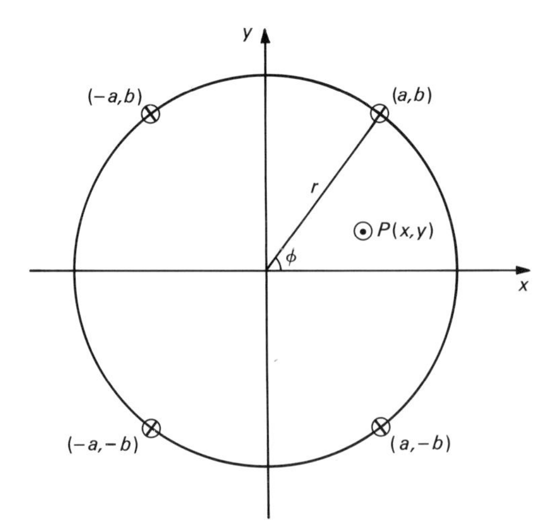

Fig. 5.17. The geometry for gradient generation by four parallel infinite straight wires.

magnitude:

$$\frac{\partial B_z}{\partial y} = \frac{2\mu_0 I \cos 2\phi}{\pi r^2}. \tag{5.20}$$

Provided $|\xi| < r$ (i.e. we consider points lying within the circle shown as a full line in Fig. 5.17), the higher order terms become progressively smaller and can be considered as corrections to this linear gradient. The first of these correction terms can be made to vanish if ϕ is chosen such that $4\phi = m\pi + \pi/2$ or $\phi = (m + \frac{1}{2})\pi/4$, m integral. Thus, for $\phi = 22.5°$ ($m = 0$) or 67.5° ($m = 1$) the gradient is linear up to fifth-order and has a magnitude of $\sqrt{2}\,\mu_0 I/\pi r^2$ (there is a reversal of gradient direction for the two angles, however). Of course, in a practical system, the line currents cannot be of infinite extent, nor can they exist in isolation; provision must be made for suitable return current paths. A number of alternatives are possible. One can, for example, make the wires sufficiently long that the return paths can be routed well away from the central region, so that the above field expressions apply. When applied in an imaging context, this solution leads to unwieldy, high-inductance coil systems, however. Another approach is to use a set of return wires in the manner of Fig. 5.18. Since these have the same angular separation, no third-order term is reintroduced, at least to a first approximation, though the gradient is somewhat diminished (Mansfield and Morris 1982). Of course, the finite length of these two sets of parallel wires and the necessity for current links between them will degrade the purity of the field gradient. Bangert

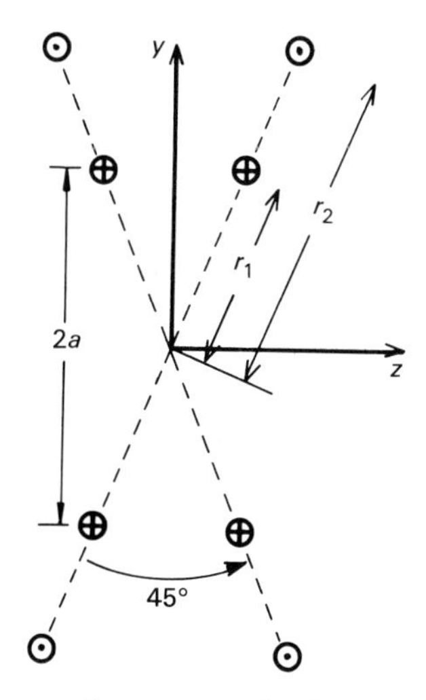

Fig. 5.18. One possible set of return paths for parallel straight wire gradient system. (From Mansfield and Morris 1982.)

and Mansfield (1982) have computer-optimized such a system with the added constraint that the entire assembly should fit within the confines of an Oxford Instruments' four-coil magnet designed for whole-body imaging. Their final configuration is illustrated in Fig. 5.19(a) and a field plot obtained from a practical realization of this design in Fig. 5.19(b). The Aberdeen team originally adopted a somewhat similar approach, although, since their gradient coil system was not required to rotate, they were less constrained by dimensional considerations (see Fig. 5.8 in § 5.1.2).

Another way in which the return paths for these four basic parallel wires can be completed is illustrated in Fig. 5.20. The attraction of this particular configuration is that the wires running along the magnet axis cannot generate a field in this (z) direction (the vector product in the Biot–Savart Law (eqn (5.1)) requires that the field produced be orthogonal to the generating current element $I\mathbf{dl}$). These wires can be run out to a considerable distance so that the final loop-completing current path produces negligible field within the sample volume. Alternatively, they can be placed such that they satisfy the condition $\phi = 22.5°$ as depicted in Fig. 5.20. They then do not generate a third-order gradient but increase the amplitude of the linear one generated by the inner wires corresponding to $\phi = 67.5°$ (Mansfield and Morris 1982). The problem with this kind of solution is again one of size; the coils tend to extend a long way down the magnet axis. They are desirable when field gradient uniformity is the primary consideration but, when switching

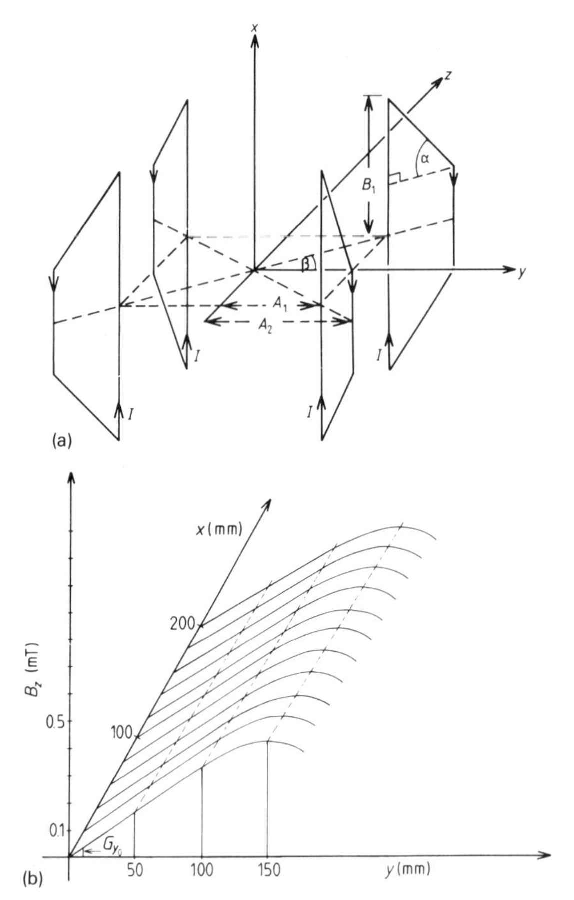

Fig. 5.19. (a) Trapezoidal gradient coil system. (b) Field generated by a 1000 amp turn gradient system with optimized parameters: $\alpha = 49.5°$, $\beta = 22.5°$, $A_1 = 278$ mm, $A_2 = 474$ mm, $B_1 = 421$ mm. The gradient is generated in the x-direction. (From Bangert and Mansfield 1982. Copyright Institute of Physics 1982.)

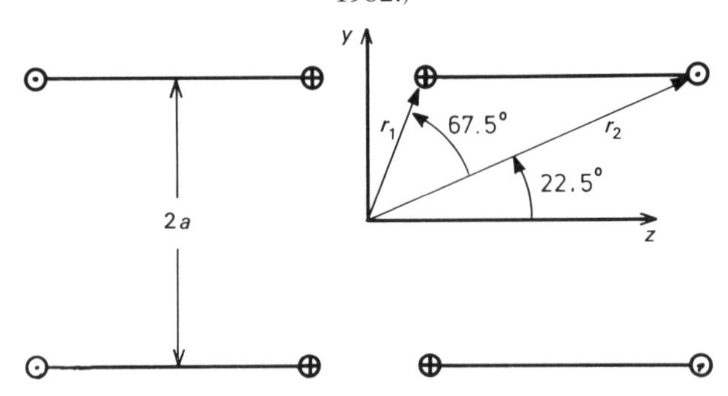

Fig. 5.20. Four straight wire gradient system with return paths parallel to the magnet (z) axis. (From Mansfield and Morris 1982.)

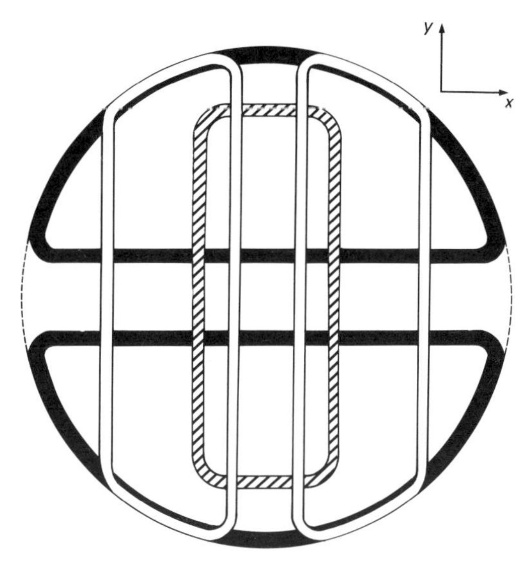

Fig. 5.21. Anderson-type flat gradient coil system. The diagram shows one of a coil pair with x (unshaded), y (full), and z (shaded) gradient coils. (From Bottomley 1978.)

speed is important, more compact versions such as that proposed by Bangert and Mansfield (1982) are preferable.

Yet another way in which return paths can be completed is to route them in long rectangular (Anderson 1961) or semi-circular paths as illustrated in Fig. 5.21. Although not yielding such uniform gradients, the advantage of such a coil system is that the x-, y-, and z-gradient coils can be mounted on two flat plates taking up a minimum of space along the magnet axis (Bottomley 1978). This is all-important if one is working with an iron-cored electromagnet. (However, this situation brings with it its own particular problems in the form of image currents induced in the adjacent pole faces.) Image currents can also be a problem in superconducting systems, giving rise to excessively long gradient rise times. For this reason the inner bores of superconducting magnets are now often made of a non-conducting material, glass reinforced plastic, for example (see Table 5.4).

5.2.3. Saddle-coil systems

Rather than use parallel current elements, which restrict sample access, it is possible to employ circular arcs as in the saddle-coil system illustrated in Fig. 5.22 (Moore and Holland 1980; Bottomley 1981*a*). This system was discussed by Hoult and Richards (1975*b*) in connection with

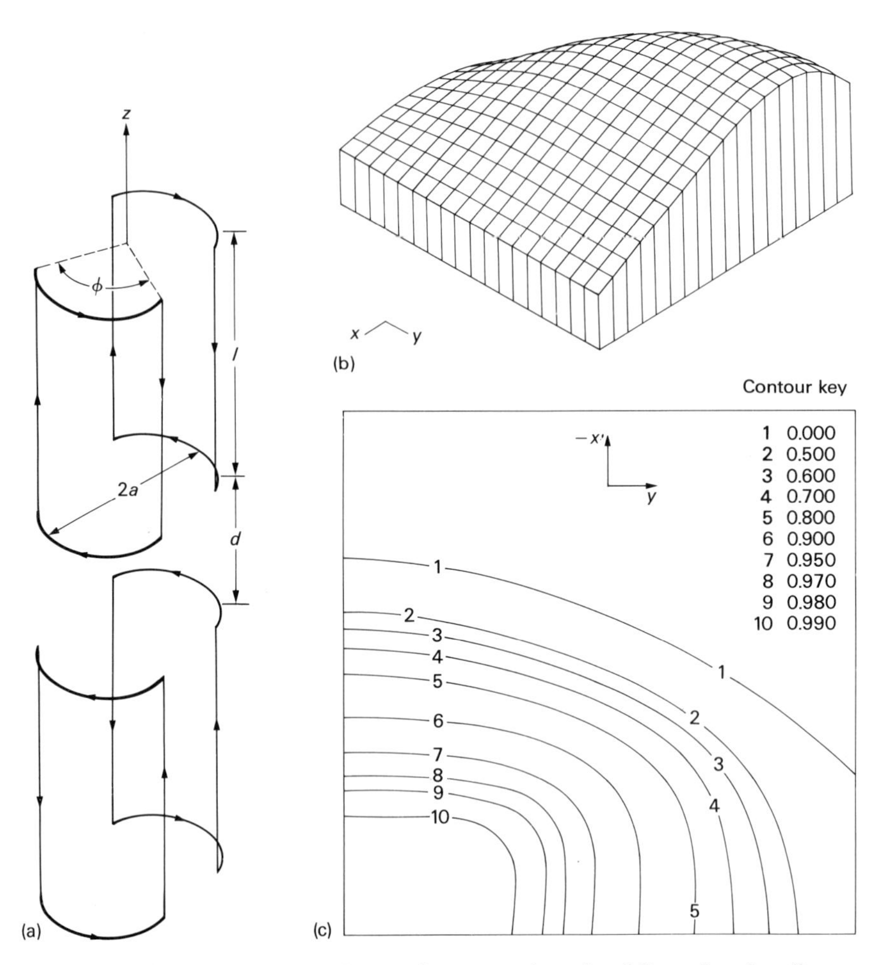

Fig. 5.22. (a) Golay-type gradient coils: two pairs of saddle coils of radius a, length l, angle ϕ and separated along the magnet axis (z) by a distance d. (b) Magnetic field (vertical axis) generated by the configuration of (a) with $d/a = 0.775$, $l/a = 3.5$ and $\phi = 120°$. One quadrant of the central ($z = 0$) field plane is shown with x- and y-axes each extending out to one coil radius. The departure from linearity is shown as a contour plot over the same quadrant in (c).

the design of shim coil sets to reduce static field inhomogeneity. Note that again the return paths are by way of axial wires which cannot contribute to the field in the z-direction. The magnetic field generated by the optimum configuration of Fig. 5.22(a) is illustrated in (b) for one quadrant of the central field plane; the departure from linearity is shown as a contour plot in (c). If multislice or volume imaging is contemplated it is

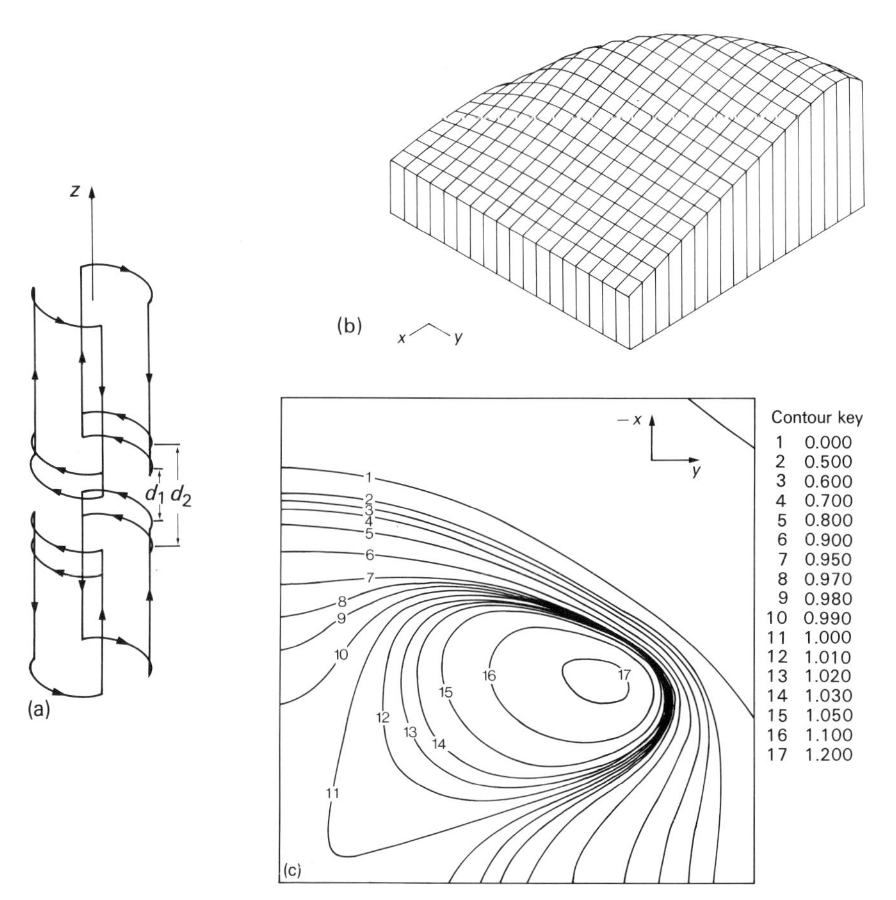

Fig. 5.23. (a) Modified Golay-type gradient coils employing four pairs of saddle coils. (b) Magnetic field generated in one quadrant of central field plane by configuration shown in (a) for $d_1/a = 0.375$, $d_2/a = 1.60$, $l/a = 3.5$, and $\phi = 120°$; the departure from linearity is shown as a contour plot in (c).

necessary to take into account the gradient outside the place of symmetry. The region in which linearity is preserved is then reduced but the error does not exceed 3 per cent over a spherical volume of radius $0.31a$.

It is possible to improve on this linearity by using a number of pairs of saddle coils with different axial separations. Figure 5.23(a) indicates the optimum arrangement for two such pairs. The field and linearity are illustrated in (b) and (c) respectively. In this case the radius of the spherical volume over which linearity is better than 3 per cent increases to $0.36a$.

5.2.4. Gradient control units

Although central to all n.m.r. imaging schemes, the field gradients G_x, G_y, and G_z are utilized by the various methods in a diversity of ways. Control over them can be exercised directly from the central processor unit (c.p.u.), which has overall responsibility for the imaging experiment, by outputting the appropriate driving voltages to the gradient amplifiers via digital to analogue converters (d.a.c.s). Alternatively, a signal pulse from the c.p.u. can be used to trigger hard-wired logic to step through a sequence of pre-determined gradient states, or to gate on an appropriate waveform generator. However, perhaps the best scheme is to use a microprocessor-based system slaved to the c.p.u. In order to avoid timing errors it is a good policy to derive the microprocessor clock signal from the instrument's master oscillator. This approach has been employed by the Aberdeen group, who use a system based on a Z80 microprocessor for gradient control (Johnson, Hutchison, and Eastwood 1982).

Bangert and Mansfield (1982) have described a system based on a National Semiconductor SC/MP microprocessor, which activates a stepper motor to rotate mechanically the whole gradient coil assembly through a predetermined angle. Although this scheme has the merit of simplicity, it throws away the advantages of purely electronic gradient rotation which can be achieved by combining x- and y-gradients in appropriate proportion. To obtain a gradient of magnitude G in the direction ϕ using this method it is necessary to simultaneously apply an x gradient of magnitude $G \cos \phi$ and a y-gradient of magnitude $G \sin \phi$ (see Fig. 5.24(a)). In the three-dimensional case of a gradient of magnitude G in an orientation θ, ϕ (see Fig. 5.24(b)) the gradients required

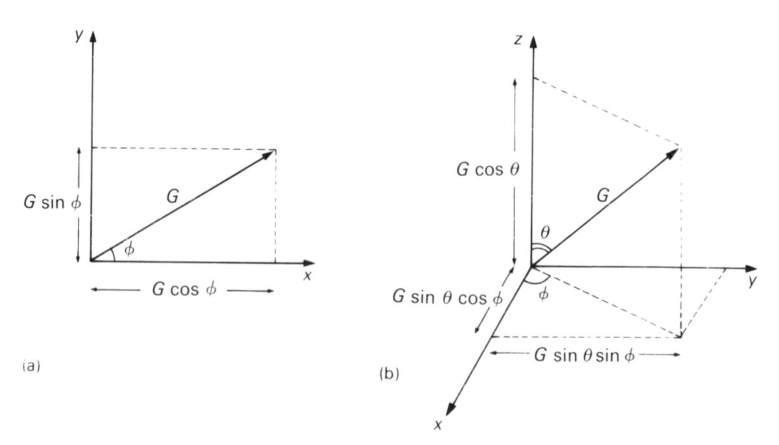

Fig. 5.24. Two- (a) and three-dimensional (b) gradient orientations.

along x-, y-, and z-axes are respectively $G \sin\theta \cos\phi$, $G \sin\theta \sin\phi$, and $G \cos\theta$. Rotation is thus simply a matter of feeding the appropriate currents to the x, y, and z coil systems to generate these gradients. Lai and Lauterbur (1980) have described a versatile three-dimensional gradient controller which uses a Motorola 6800 microprocessor. The drive voltages are generated from a sine look-up table using software multiplication to give the G_x and G_y gradients. The output from the device is to three operational amplifiers via 10-bit digital to analogue converters (AD 1979). The same principle was used by Lai, Shook, and Lauterbur (1979) in an earlier two-dimensional controller based on a SC/MP microprocessor.

Another approach recently described (Bottomley 1981*a*, *b*), uses an Analog Devices[1] AD 7524 multiplying digital to analogue converter to control the amplitude, and DTM1717 vector generators to control the orientation of the gradient. The latter device takes an 8-bit angle and generates at its two output ports the sine and cosine of the input function.

Although the power required to drive the larger gradient sets can be quite substantial, currents of 30 to 40 amp being not uncommon, commercial wideband audio-amplifiers such as the Amcron[2] M600, capable of delivering 600 W into an 8 Ω load over a bandwidth of 1 Hz to 20 kHz are well suited to the job. If system flexibility is not a primary consideration, then simpler supplies can be constructed at much lower cost. We should mention at this point that commercial packages including gradient coil sets, controller, and power supplies are available from at least two companies: Oxford Instruments and Bruker Spectrospin.

Those techniques which require rapid gradient switching pose considerable, though by no means insuperable, technical difficulties (see, for example, Hoult 1980).

5.3. The r.f. transmitter

5.3.1. Frequency source

In order to ensure coherent operation of the instrument, the r.f. source for the transmitter, the reference r.f. for the phase sensitive detectors (p.s.d.s) and, preferably, the timing for the control pulses, should all ultimately be derived from the same master oscillator. Since n.m.r. imaging systems operate in a frequency range appropriate to the construction of quartz crystal oscillators, these devices may be used directly. The operating frequency can be specified and the crystal cut

[1] Analogue Devices, One Technology Way, P.O. Box 280, Norwood, Massachusetts 02062, USA.

[2] Manufactured by Crown, 1718W Mishawaka Road, Elkhart, Indiana 46514, USA.

accordingly. Long-term stability is generally ~1 p.p.m. or 0.1 p.p.m. if the temperature is controlled (ovened) and the output is typically 1 V peak to peak into 50 Ω. There is generally no need to vary the operating frequency in an imaging system, though if such a facility were required —for example, for the imaging of different nuclei—then a frequency synthesizer source (e.g. Programmed Test Sources PT6 160)[1] would be appropriate.

5.3.2. *Gating and modulation*

Since all imaging spectrometers operate in the pulse Fourier transform mode (see Chapter 2), it is necessary to gate the r.f. source. The interval for which it is switched on, or pulse length, depends on the particular imaging scheme: for steady state free precession methods a short (~ few μs) high power pulse is essential, whereas if selective irradiation methods are being employed, much larger pulse widths (~1–50 ms) are required. In the latter case it is also necessary to shape (amplitude modulate) the r.f. envelope in order to excite spins over a narrow frequency band and hence, with suitably applied field gradients, over a small spatial region. Both gating and modulation are most easily carried out at low signal levels, i.e. before power amplification.

5.3.2.1. Gating. Transmitter gating serves two purposes; it provides a means for generating the pulses required to excite the spin system and, when in the off state, isolates both the spin system and the sensitive receiver from the transmitter r.f. In the case of high-power transmitter pulses the latter role can be quite a challenging one, since the receiver may well be required to detect induced e.m.f.s of the order of microvolts immediately following a transmitter pulse of some kilovolts. The overall voltage isolation required on the transmitter channel in this case would be in excess of 10^{10} or 100 dB. Whilst this is normally achieved progressively at several points in the transmitter system, the low level gate is the first such control point and does the lion's share of the work. During the r.f. pulse itself it is also necessary to provide some protection for the receiver. This normally takes the form of both passive and active gating—see § 5.4.

Gating to the required standard can be achieved in a number of ways, though some care is usually necessary in the design to achieve optimum results. Field effect transistors (f.e.t.s) have particularly good isolation properties and a number of designs based on these devices have been published. Hoult and Richards (1975*b*) and Baines and Mansfield (1973), for example, have described f.e.t. gates giving 80 and 90 dB of isolation respectively.

[1] Programmed Test Sources Inc., Acton, Massachusetts 01720, USA.

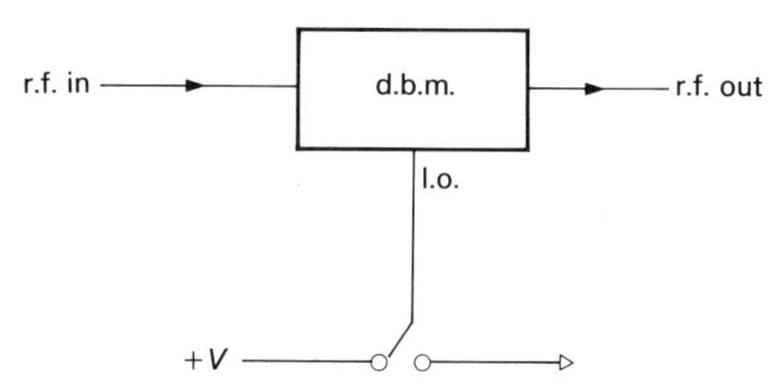

Fig. 5.25. The double balanced mixer as an r.f. gate. When the gate (LO) is grounded there is approximately 30 dB of isolation between input and output. When the gate is at +V the r.f. is allowed to pass through the device.

If a lower degree of isolation can be tolerated, the ubiquitous double balanced (d.b.m.) mixer can be used for the purpose. This acts just like a switch: when a d.c. voltage is applied to the gate (see Fig. 5.25) the r.f. power passes through the device; when the gate is grounded, input and output ports are isolated. Typically, the single stage on/off ratio is 30 to 40 dB, though the devices can be connected in series to improve this figure.

To attain the highest possible isolation, it is necessary between the pulses, to eliminate even 'low level' r.f. at the observation frequency, since even this can 'leak' into the receiver channel. One way to achieve this is to use a frequency multiplier (doubler, tripler, etc.) at the final stage of power amplification. Alternatively, the required observation frequency can be generated by mixing two frequencies (again using a double balanced mixer) one or both of which is pulsed. It is also generally desirable to detect at an intermediate frequency to avoid leakage from this source (this also enables the receiver to be optimized for a single frequency in a multinuclear instrument). Such methods are frequently employed in state-of-the-art conventional n.m.r. spectrometers.

5.3.2.2. Modulation. Modulation as well as gating can be achieved using double balanced mixers (also known as ring modulators or double balanced modulators) available from companies such as Hewlett Packard,[1] Merrimac,[2] or Hatfield[3] Instruments. The r.f. is fed into the device via port 1 and appears at port 3 amplitude modulated by the waveform applied to port 2. Although the modulation function can be produced directly using analogue circuitry (see Hutchison, Edelstein, and Johnson

[1] Hewlett Packard Components, 350 West Trimble Road, San Jose, California 95131, USA.

[2] Merrimac Industries Incorporated, P.O. Box 986, 41 Fairfield Place, West Caldwell, N.J. 07006, USA.

[3] Hatfield Instruments Ltd, Burrington Way, Plymouth, PL5 3LZ, UK.

1980 for the generation of a Gaussian envelope, for example), the most flexible systems store this function digitally in the c.p.u. or in a buffer memory from which the discrete values can be clocked out in sequence at a rate determined by the pulse controller (this may be a separate device or the task may be handled by the c.p.u. itself). The binary words have in the past been used to control binary attenuation gates directly (see Mansfield, Maudsley, and Baines 1976). However, this gives rise to a stepped, rather than a continuous, waveform with the step size governed by the voltage level corresponding to the least significant bit. A double balanced mixer fed via a suitable digital to analogue converter gives a much smoother waveform. Figure 5.26 illustrates the basic approach. Note that the d.b.m. also acts as a r.f. gate if the control line is grounded following the pulse.

The aim of the modulation, as already stated, is to excite a specific frequency band corresponding to a particular volume within the sample, e.g. a narrow plane of spins. Unfortunately, with a single channel it is not possible to excite a single frequency band centred at a frequency of, say, $\omega_0+\Delta\omega$ (where ω_0 is the carrier); one automatically also excites the symmetric sideband at $\omega_0-\Delta\omega$. This is the same problem which we encountered in Chapter 2 when we considered single channel detection (see also § 5.4.3). Not only do we waste half of our r.f. power into the unwanted sideband, we also run the risk of exciting a second region of our sample. In order to guard against this possibility, it is common practice to work at large-frequency offsets so that the spins corresponding to the other sideband are physically well-separated and therefore lie beyond the region of receiver coil receptivity. A more satisfactory solution is to employ single-sideband modulation methods which require the addition of a second r.f. channel in quadrature with the first. This is

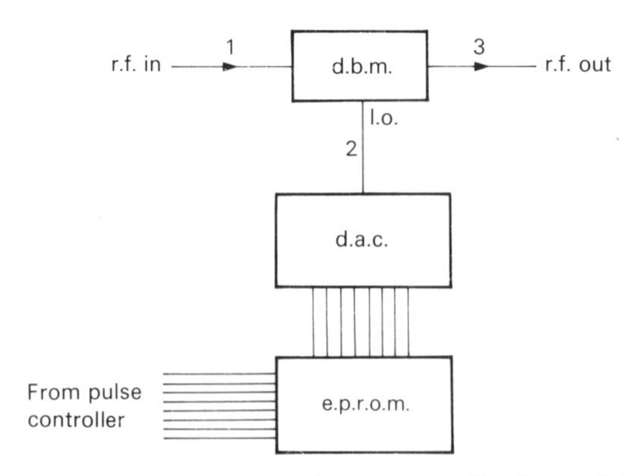

Fig. 5.26. Use of a double balanced mixer for amplitude modulation of the r.f. The e.p.r.o.m. is included to enable correction for any non-linearity of the d.b.m.

completely analogous to the use of r.f. quadrature detection methods (§§ 2.4.2.3 and 5.4.3). The practical requirements are straightforward; the r.f. is split into two components with a relative phase shift of 90°, using, for example, a quadrature hybrid, the circuitry of Fig. 5.26 is duplicated to enable modulation of the two channels which are then recombined prior to power amplification (see Mansfield and Morris 1982).

5.3.3. Power amplification

To achieve the hard (short) 90° and 180° pulses required in human imaging systems with their necessarily large transmitter coils, substantial peak power is required (typically several kilowatts). Selective irradiation procedures have more modest power needs (the line-scanning system of Mansfield *et al.* (1978), for example, operated with a peak power of only 4 W), but it is nevertheless still necessary to amplify the modulated r.f. signal which typically has a peak amplitude of about 1 V, equivalent to a power of only some 10 mW. This amplification is normally achieved in several stages, often combining solid-state drivers with a valve end-stage. Completely solid-state units can be constructed (Baines 1974;[1] Holland and Heysmond 1979) but, since currently-available r.f. power transistors are normally only capable of delivering 20–40 W, it is necessary to construct an amplifier 'tree' ending up with perhaps 64 or more parallel amplifier modules. The outputs from these are then combined to yield the desired output power. With so many expensive semiconductor units (typically £15 or more per transistor) at risk, it is absolutely essential to provide adequate protection against an imbalance resulting from a faulty module or an open circuit output caused by the operator failing to connect the transmitter coil!

The most efficient end-stage has a tuned class *C*-configuration in which the positive and negative parts of the signal are amplified by separate components. This has the additional advantage that, since the power devices do not conduct when gated off, the amplifier generates very little noise. For this reason it is a very popular choice in conventional n.m.r. spectroscopy. Unfortunately, since it is also highly non-linear, it cannot be used for those imaging applications which require modulation of the r.f. In these cases linear amplifiers (class A or class AB) are called for, and, since these do generate noise in the 'off' state it may well prove necessary to incorporate some active gating at this stage.

Commercial units, including some specifically designed for n.m.r. applications, are readily available from companies such as Amplifier Research[2]

[1] Baines, T., University of Nottingham, Private communication, Oct. 1974.
[2] Amplifier Research, 160 School House Road, Saunderton, PA 18964, USA.

or ENI.[1] They are, however, rather expensive in the high-power ranges. A low-cost 350 W class C amplifier based on a single m.o.s.f.e.t. device has been described (Darrasse 1982). It is driven by low-level logic circuitry, operates up to 15 MHz, and can be upgraded to deliver several kW of power if required.

5.3.4. The transmitter coil circuit

As discussed in Chapter 2, we require a r.f. magnetic field to interact with the spin system, and this is most simply achieved by placing the sample within a coil or inductance through which the r.f. is passed. If this coil is made part of a tuned circuit, resonant at the Larmor frequency, an amplification of the current (and hence the r.f. magnetic field) by a factor equal to the Q (see § 2.4) of the circuit takes place. Figure 5.27 illustrates such a typical transmitter tuned circuit. To a first approximation one can consider the transmitter coil L and the variable (tuning) capacitor C_2 as forming a parallel tuned circuit which resonates at a frequency ω_0 given by

$$\omega_0^2 L C_2 \sim 1. \tag{5.21}$$

The power amplifier will generally have been designed to have an output impedance of 50 Ω; that is, it delivers maximum power into a 50 Ω load. Viewed from a point such as A, however, the network would, if the r.f. resistance r of the coil windings were small, have a very high input impedance, perhaps in the range 10–100 kΩ and, if connected directly to the power amplifier, would result in a gross mismatch with most of the power being reflected back. To avoid this mismatch the variable (matching) capacitor C_1 is included: it transforms the impedance of the tuned circuit to 50 Ω. Provided that C_1 is a high-quality (high Q) device, this transformation is accomplished with virtually no power loss. Both C_1 and C_2 should be capable of withstanding moderately high voltages (perhaps up to a kilovolt). Typically, C_2 lies in the range 20–200 pF; it should certainly be greater than 10 pF since the stray capacitance will be of this order of magnitude. C_1 is usually somewhat smaller, a typical figure being about 15 pF.

Diodes have the property that they will only conduct when the forward bias voltage exceeds a certain level (about 0.6 V in the case of silicon devices). The crossed diodes D of Fig. 5.27 therefore provide a threshold barrier, removing low-level noise and cutting off the falling edge of the transmitter pulse. They should be of the high-frequency variety (low capacitance) and be capable of withstanding high-peak currents. If necessary sets of crossed diodes can be connected in parallel and/or in series.

[1] ENI, 3000 Winton Road South, Rochester, New York, NY 14623, USA.

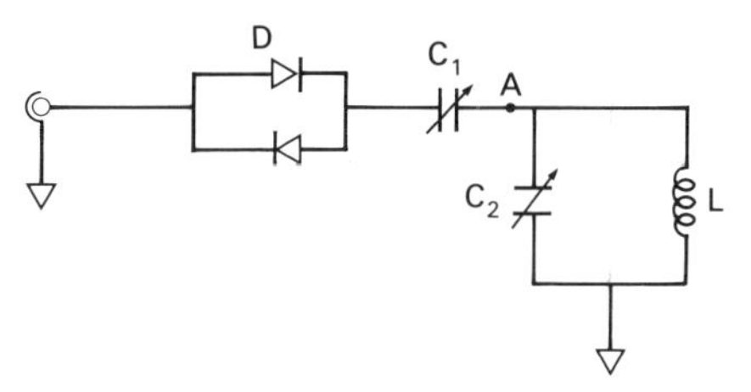

Fig. 5.27. Transmitter coil circuit. L, the transmitter coil, and C_2 a variable tuning capacitor form a parallel-tuned circuit which is matched to 50 Ω by the variable capacitor C_1. The crossed-diodes D curtail the r.f. pulse when the amplitude falls below 0.6 V.

The most crucial element in the circuit is the transmitter coil (L) itself, and numerous design possibilities have been considered for it. We will not involve ourselves in a detailed discussion here, but review some of its basic requirements. First, it should have a reasonable Q, though not too large a one since the pulse ringdown time then becomes excessive. (To take a mechanical analogy, if we set a mechanical oscillation in motion and then remove the driving force, the time for which the oscillation persists depends on the degree of friction or damping present: in electrical terms this friction is provided by the resistance of the coil which is inversely proportional to Q (see eqn (2.33))). Secondly and most importantly, the coil should generate a uniform r.f. field over the region of the sample to be excited. This is a particularly difficult condition to fulfil and normally requires a coil considerably larger than the sample. See, for example, Fig. 5.28 which shows the field profile for the popular saddle coil arrangement, wound according to Ginsberg and Melchner's (1970) 'optimum' design. Major distortions are evident at radial distances from the centre in excess of $0.7a$, particularly in the vicinity of the wires. Since we need to produce a relatively large B_1 field in our chosen volume, a third requirement is that the coil assembly should not be too large. This is also desirable from the tuning point of view: the inductance of a coil increases in proportion to its linear dimension, and we must ensure that it does not become so large that the self-resonant frequency (the frequency at which the coil resonates with its own interturn capacitance) becomes comparable with the operating frequency. We thus have two conflicting requirements; fortunately a good compromise solution is generally possible.

The saddle-coil configuration of Fig. 5.28 is a popular choice of transmitter coil, though usually the length to radius ratio (l/a) has to be reduced to between one and two in order to comply with our third requirement.

Substantially better r.f. homogeneity can be achieved by using a pair of

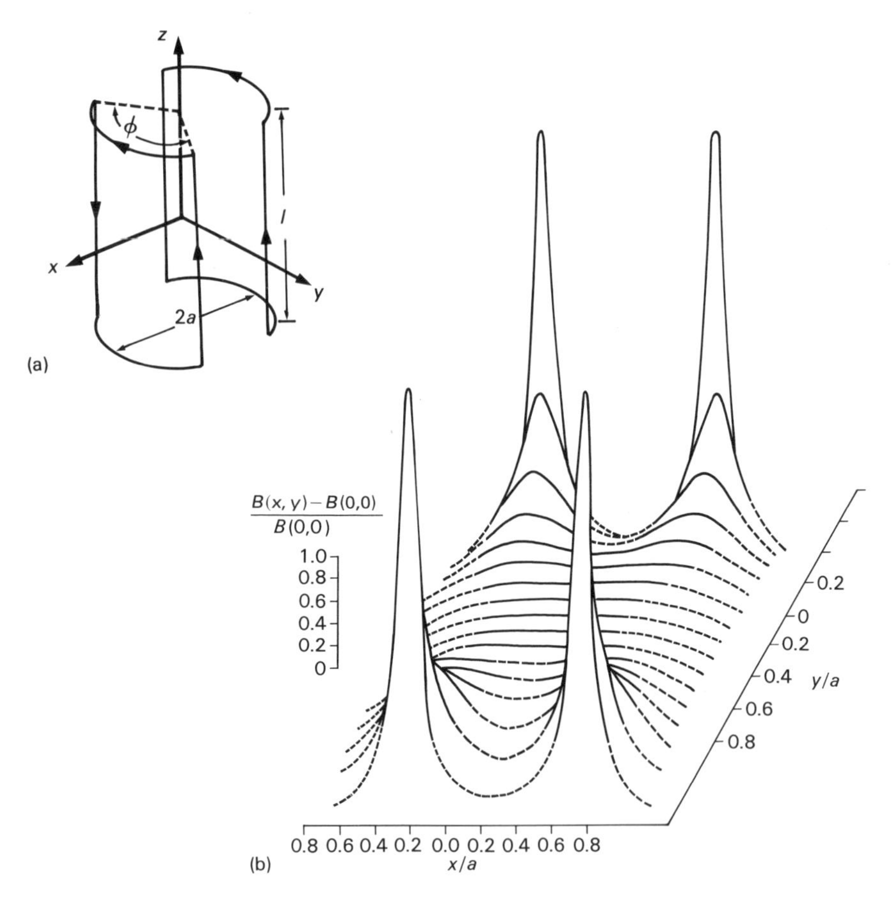

Fig. 5.28. (a) Saddle-type transmitter coil, wound to the Ginsberg–Melchner optimum configuration ($\phi = 120°$, $l/a = 4$). (b) Field profile for configuration shown in (a); field values are normalized to unity at the origin.

parallel rectangular coils as illustrated in Fig. 5.29. Such a system has been used in Mansfield's imaging laboratory for a number of different imaging schemes. It has the disadvantage that it does not fit so conveniently into the (usually) cylindrical magnet bore; a factor which has precluded its more widespread use in conventional n.m.r. spectroscopy.

Many other designs have been proposed, some of which are described in Mansfield and Morris (1982), Chapter 8. Note that for some experiments, namely rotating-frame zeugmatography (Hoult 1979*a*; see also §4.2.2), r.f. field gradients are required. These are readily achieved by having a different number of turns on the two halves of a coil pair.

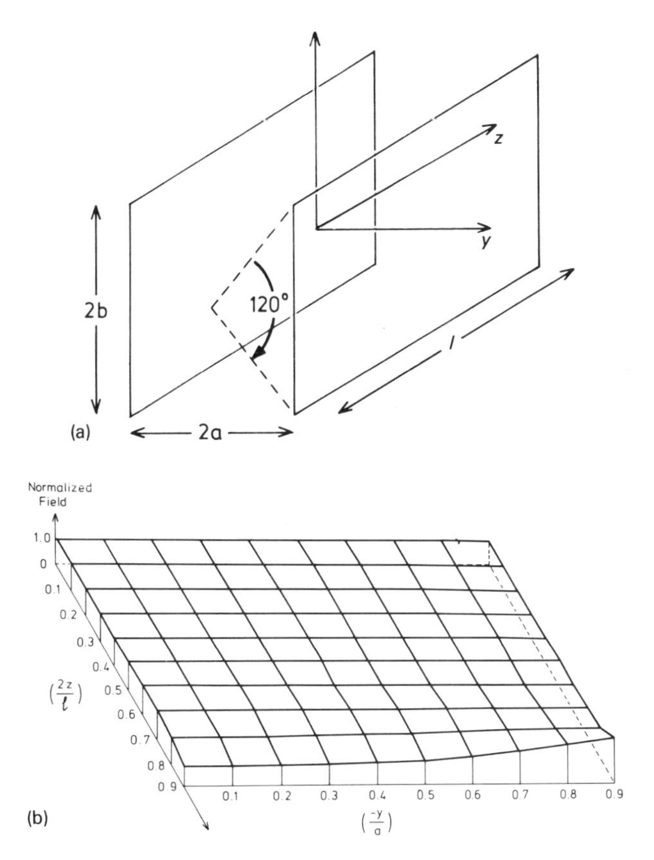

Fig. 5.29. (a) Rectangular transmitter coil assembly. (b) Normalized field distribution over one quadrant of central field plane for configuration shown in (a) with $l/a = 2\sqrt{3}$. The field is constant to better than 5 per cent over an elliptic region of major axis $0.9l$ and minor axis $1.8a$. (From Mansfield and Morris 1982.)

5.4. The receiver system

5.4.1. The receiver coil and matching network

Whereas the transmitter design leaves some margin for error, it is crucially important to have a good receiver since it directly determines the final image quality. The block diagram of Fig. 5.30 illustrates the general features of a typical receiver system. The first element in the chain, the receiver coil, is outwardly very similar to the transmitter coil; indeed, in some cases a single coil serves both functions. However, as we shall see below, the qualities required of our receiver coil conflict rather

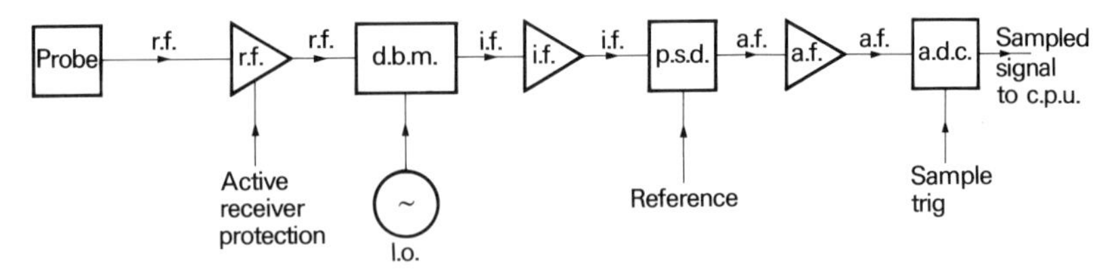

Fig. 5.30. Block diagram of the receiver system. Abbreviations: r.f. radiofrequency; intermediate frequency; a.f. audio frequency; p.s.d. phase sensitive detector; a.d.c. analogue to digital converter; c.p.u. central processor unit.

strongly with those of the transmitter coil. First, we require a coil with the highest possible Q, since in general this will give us the best signal-to-noise ratio. For the same reason we would prefer a small coil tightly coupled to the sample—this picks up signal efficiently and, since the resistance will depend on the length of conductor making up the coil, will generate less thermal noise.

Whereas a coil with a good quality factor gives a high S/N it does so at the expense of the bandwidth (the frequency separation between the points at which the signal is 3 dB down on the maximum is given by $\Delta\omega_0 = \omega_0/Q$: see Fig. 5.31). To understand the implications of this, consider an imaging scheme such as the echo-planar one, which is capable of resolving all image points in a single f.i.d. If we assume that the bandwidth allocated to each of these points is equal to the natural proton linewidth, i.e. ~25 Hz, then the total bandwidth required for a 128×128 image matrix is about 400 kHz. At an operating frequency of 21 MHz, this requires the receiver coil Q to be limited to $\leqslant 50$. This is to be compared with Q values of two or three hundred, which can be achieved

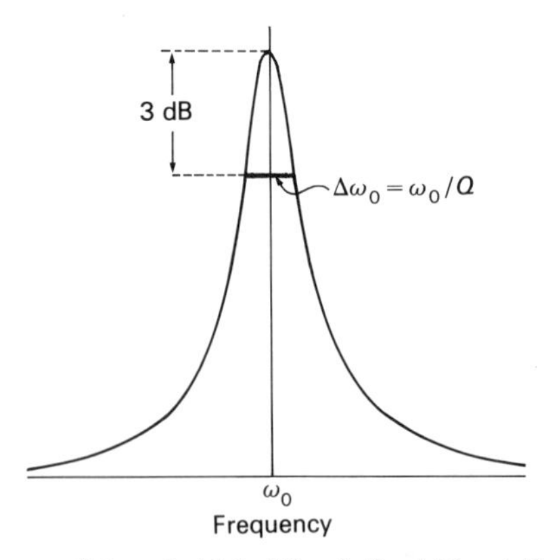

Fig. 5.31. The receiver coil bandwidth. The full width at the 3 dB point is given by ω_0/Q.

in practice and which are desirable from the S/N point of view. As the matrix size increases the situation gets worse, and this is one of the more fundamental problems of n.m.r. imaging. Obviously, it is preferable not to reduce the Q through the simple expedient of adding a series resistor, since this will generate noise. However, if the resistor is immersed in liquid nitrogen, or better still liquid helium, this problem is alleviated. A more elegant solution has been proposed by Hoult (1979*b*) which involves the use of a pre-amplifier with negative feedback to dampen the coil whilst retaining the S/N of the undampened circuit. Effectively, the feedback is used to synthesize electronically a resistor at 0 °K. Of course, the extent to which this bandwidth restriction is a problem depends on the information content of the f.i.d. Those techniques which are most efficient in this respect suffer the greatest: it is the principal factor which limits the spatial resolution (number of pixels) in an echo-planar imaging study.

We must also consider the uniformity of signal response which is analagous to the need for good r.f. homogeneity in the transmitter case. In fact, these two requirements are one and the same; they are related through the 'principle of reciprocity' (see Hoult and Richards 1976). Thus the e.m.f. ξ induced by an oscillating dipole $\mathbf{m}$ is given by:

$$\xi = -\frac{\partial}{\partial t}(\mathbf{B}_1 \cdot \mathbf{m}), \qquad (5.22)$$

and is directly proportional to the r.f. field $\mathbf{B}_1$ generated at $\mathbf{m}$ by unit current passing through the coil.

In summary, the primary consideration for receiver coil design is for a high S/N with secondary consideration given to response uniformity. This should be contrasted with the transmitter coil case where r.f. homogeneity is of prime importance. In conventional spectroscopy, r.f. homogeneity requirements are generally less stringent (unless one is using a pulse sequence which requires accurate 90° or 180° pulse settings, for example an inversion recovery T_1 determination or a polarization transfer experiment), and uniformity of receiver response is far from essential. It is not surprising then that the same coil, designed for high S/N, should serve both purposes satisfactorily. (Usually a simple saddle arrangement is used for superconducting spectrometers at frequencies $\leqslant$400 MHz.) Single coils have been widely used in n.m.r. imaging systems, too, though the sacrifice of B_1 homogeneity required to achieve good S/N is then much more questionable. In such cases it is desirable to make use of a 'Q switch', a device which allows a low Q during a transmitter pulse and reverts to high Q for signal reception (see, for example, Grannell, Orchard, Mansfield, Garroway, and Stalker 1973).

Some thought should be given to the proper isolation of the receiver

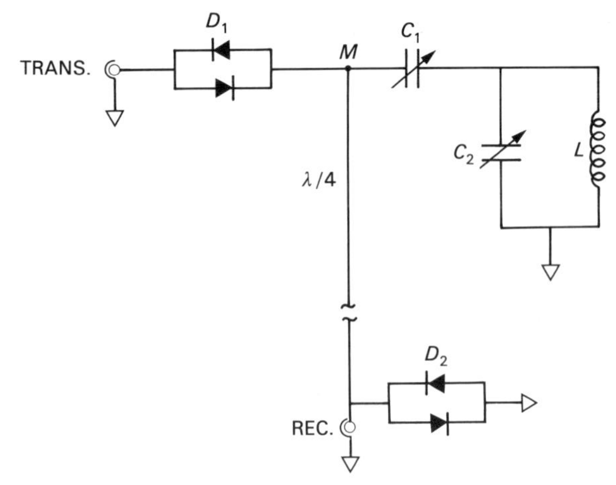

Fig. 5.32. Passive receiver protection using crossed-diodes and a quarter wavelength ($\lambda/4$) coaxial line.

during the transmitter pulse. In addition to active receiver blanking, the passive circuit of Fig. 5.32 can be used to alleviate this problem. During the transmission phase, both sets of crossed diodes conduct, those at the end of the quarter wave line effectively shorting out the receiver. This short circuit 'looks' like an open circuit when viewed from point *M*, so virtually all the transmitter power passes through to the tuned circuit. In the receiver phase, the induced e.m.f. is far too small to bias the diodes on, so the transmitter is effectively isolated and the short at the receiver input is removed.

From the above discussion it may be correctly assumed that the author's preference is for a two-coil system, since this allows the transmitter and receiver to be optimized independently. Both coils must generate or pick up r.f. magnetic fields which are orthogonal to the static field B_0 and it is usual to ensure that these r.f. fields are also mutually orthogonal. In this crossed-coil configuration, cross talk is minimized, giving some additional receiver protection.

The design of the receiver coil depends on the magnet geometry and on the patient orientation. The most efficient coil geometries from the S/N point of view are solenoidal (or ellipsoidal) ones: however, they can only be used if the patient is positioned transverse to the magnet axis (see § 5.1.2). In the more usually preferred configuration, in which the patient lies along the magnet axis, the most usual choice is a saddle-coil arrangement (either one with conventional circular or else with square end connections). The S/N obtained from this type of system is a factor of about $\sqrt{3}$ times worse than that from a comparable solenoid and was the principal reason why the first high-field superconducting spectrometers failed to achieve their promised improvement in S/N (Hoult and Richards 1976). It is possible to retrieve a factor of $\sqrt{2}$ in S/N by combining the signals from two orthogonal receiver coils of the saddle type (Chen, Hoult, and Sank 1983). This coil arrangement, when used as a transmitter, enables a rotating r.f. field to be generated, thus avoiding the loss of half of the incident r.f. power into the counter-rotating component which occurs when a linearly polarized r.f. field is employed (see § 2.4.2.1). This is of particular significance for high field imaging applications where r.f. powers are approaching the US safety limits (see § 6.8.3). Problems arising from skin depth effects are also alleviated by this approach. Surface coils have good S/N properties (see § 3.2.2) and are often used as receiver coils for n.m.r. imaging studies of regions of limited spatial extent. For example, they permit submillimetre resolution to be achieved in images of the eye and ear. They do have a rather non-uniform spatial response, however. Although various other coil arrangements have been employed, for example crossed ellipse systems (Moore and Holland 1980; Mansfield and Morris 1982), a loss in sensitivity is generally the price which must be paid for experimental convenience: it is basically associated with the increased quantity of wire required in the construction of the coil leading to generation of more Johnson noise. On this note, it is important to use a heavy gauge wire or pipe (a central core is not essential since, due to the 'skin depth' effect, the current is carried in the surface layer; see also eqn (5.26)). Hoult and Richards (1976) have shown that the design philosophy should be to produce a coil which maximizes $B_1/\sqrt{r}$, where r is the coil resistance at the operating frequency. This is equivalent to maximizing Q in the case of a solenoidal system but not, in general, for other geometries. For those interested in constructing their own coils the *Radio Engineer's Handbook* (Terman 1943) is a useful source of reference; see also Hoult and Lauterbur (1979).

Since, as mentioned above, there is a danger of coils of large dimensions becoming self-resonant at too low a frequency, it may sometimes be necessary to parallel rather than series-connect the members of coil pairs. In this way coils with undamped Qs in excess of 200 are fairly easily constructed (see Hutchison *et al.* 1980 for details of a coil with an unloaded Q of 590).

At the higher imaging frequencies currently being used, cavity arrangements, of the type used in electron-spin resonance spectroscopy, have their attractions. Split ring or slotted tube resonators have been used for n.m.r. spectroscopy; the break(s) in the conductor providing the necessary tuning capacitance (Schneider and Dullenkopf 1977*a*,*b*; Hardy and Whitehead 1981; Froncisz and Hyde 1982). More sophisticated arrangements have been proposed (McGurrin and Gully 1983) and a hybrid system is available in which additional discrete tuning capacitors are added to the basic cavity arrangement (Alderman and Grant 1979). The latter system was originally designed for high-powered ^{1}H decoupling experiments in conventional n.m.r. spectroscopy, its attraction being the localization of the electric field in a region remote from the sample—a property which also makes it an attractive design for imaging applications (see below).

The need to obviate leakage between transmitter and receiver systems has been stressed. It is equally important to protect the receiver from other sources of noise which may be radiated, for example, from a nearby ring main or from control lines. Microprocessor or computer clock pulses are also a frequent source of interference, as are radio stations. (Typically the n.m.r. imaging frequency for low field human applications will lie in the middle of the maritime frequency band.) One effective solution is to shield the receiver coil either by placing the entire magnet system in a screened enclosure and carefully filtering all power and control lines entering it, or, in the case of a superconducting magnet, by sealing off its open ends. These relatively expensive solutions, though likely to succeed, are not always essential—it depends very much on the particular machine environment and, somewhat annoyingly, on atmospheric and other 'intangible' factors!

The introduction of a patient into the receiver coil has a number of consequences. First, it detunes the resonant circuit so that it has to be retuned and matched. Fortunately, provided successive patients can be located with reasonable consistency, no subsequent retuning is usually required. Secondly, the loaded receiver Q drops dramatically as a consequence of the body's finite resistivity. This in turn means a lower S/N than would be expected from a non-conducting sample. Indeed, at the upper end of the operating frequency range, the dominant source of noise will be the patient himself. This noise originates from two sources: electric and magnetic. The electric component can be screened out using a simple Faraday shield (a useful precaution from the patient's point of view in any case), but the e.m.f.s, and hence currents induced by the r.f. magnetic field component, are unavoidable. Such considerations mean that more sophisticated receiver coils in which the resistance of the windings is reduced by cooling with liquid nitrogen or helium may offer at best only marginal improvements in sensitivity.

Considerable improvements with regard to detuning and S/N performance can be achieved by using a balanced matching system, rather than the usual arrangement in which one end of the receiver coil is grounded (Fig. 5.32). Details can be found in Murphy-Boesch and Koretsky (1983); see also Hutchison *et al.* (1980).

5.4.2. *The receiver amplifier*

The preamplifier is the most important link in the receiver amplifier chain since any noise which it adds to the signal will be amplified by the subsequent units. Noise performance is normally measured in terms of the noise figure *F*, which is defined as the ratio of the noise power at the output to that at the input, expressed in decibels. Thus:

$$F = 10 \log \frac{N_A^2 + N_S^2}{N_S^2}, \qquad (5.23)$$

where N_A and N_S are the r.m.s. voltages of the amplifier and signal noise. A noise figure of 3 dB indicates that the noise power is doubled by the amplifier and the S/N thereby reduced by a factor of $\sqrt{2}$. It is therefore important to ensure that the preamplifier selected has a noise figure which is less than 3 dB. This is normally achieved using field effect transistors which are inherently low-noise devices. Commercial preamplifiers with noise figures of about 2 dB are available (for example, from Bruker Spectrospin) and Hoult (1979*b*) has described a suitable low-frequency amplifier which uses gallium arsenide f.e.t.s (GaAs f.e.t.s: Plessey,[1] type GAT1) to achieve a noise figure of 1.3 dB (see also Hoult and Richards 1975*a*, for a high-frequency design with a noise figure of 0.3 dB). GaAs f.e.t.s have the merit that they can be operated cryogenically, when noise figures of less that 1 dB can be achieved. Since the breakdown voltage of f.e.t.s in general is rather low (~12 V) it is essential to provide adequate protection during the transmitter pulse. At minimum, this could be a pair of crossed diodes, but for preference, active gating should be used (see, for example, Hoult 1979*b*). Much less stringent requirements apply to the remainder of the amplifier chain and suitable modular devices are readily available (e.g. Avantek[2] GPD series). The total receiver gain needs to be about 10^4 but should be variable so that the analogue-to-digital converter can be filled but not saturated.

5.4.3. *Signal detection and digitization*

Having amplified the nuclear signal to a level of perhaps a few volts, the next stage is detection, i.e. removal of the r.f. component (see § 2.4.2.3). For single-phase detection this can be done directly (see Fig. 5.33(a)) but for quadrature detection, now almost universally employed, two channels are necessary. The reference r.f. is split into two components which are

[1] Plessey Optoelectronics and Microwave Ltd., Wood Burcote Way, Towcester, Northamptonshire NN12 TJN, UK.
[2] Avantek, 3175 Bowers Avenue, Santa Clara, California 95051, USA.

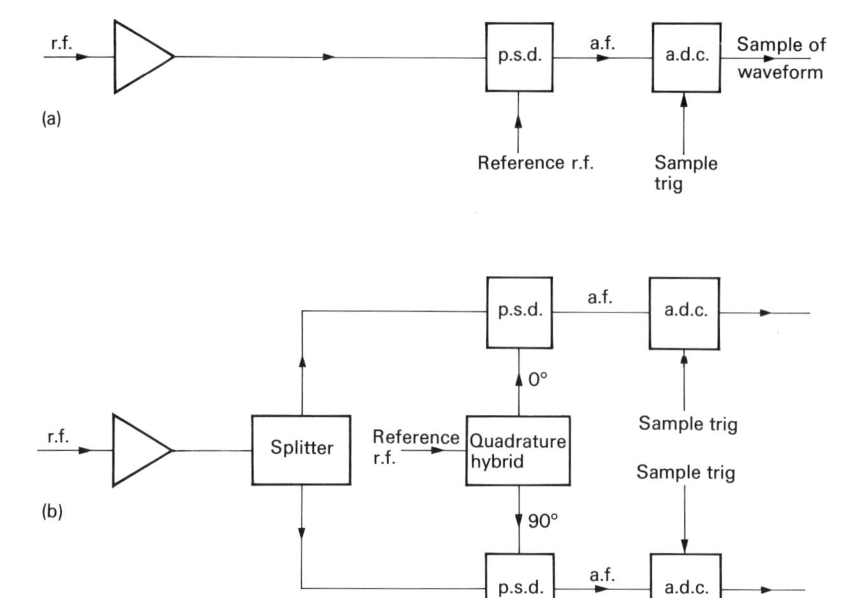

Fig. 5.33. Single (a) and quadrature (b) phase-sensitive detection schemes.

phase-shifted with respect to each other by 90° using a quadrature hybrid, for example. These reference signals are used to detect the n.m.r. signal which is also split into two channels (Fig. 5.33(b); compare with the discussion of § 5.3.2.2 on single side band irradiation).

The advantage of using quadrature detection lies in its ability to discriminate between positive and negative frequency offsets. This allows the reference (detection) frequency to be placed in the centre of the receiver frequency band (Fig. 5.34(a)), rather than to one side of it (Fig. 5.34(b)). Following single-phase detection noise components in the range $\omega_0 - \Delta\omega_0$ to ω_0 are 'folded over' into the ω_0 to $\omega_0 + \Delta\omega_0$ range, doubling the noise *power* and thereby reducing the S/N by a factor of $\sqrt{2}$. Before quadrature methods came into general use, some improvement was obtained by removing the low (or high, depending on the reference position) frequency noise with a high Q crystal filter.

Ultimately, the aim is to digitize the nuclear signal so that it can be processed in the system's central computer (central processor unit or c.p.u.). However, the maximum frequency which can be unambiguously assigned in a digitized signal is $f_{\max} = 1/2\tau$ where τ is the sampling interval; higher frequencies suffering aliasing (see Fig. 5.35). It is thus important to choose τ such that the required signal bandwidth is adequately covered: again, this requirement is most severe in the case of

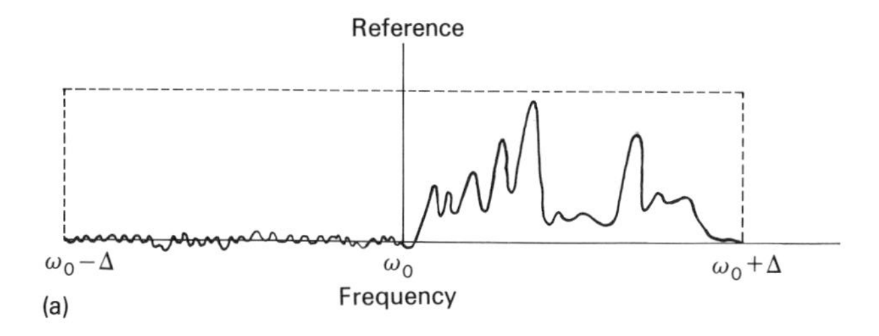

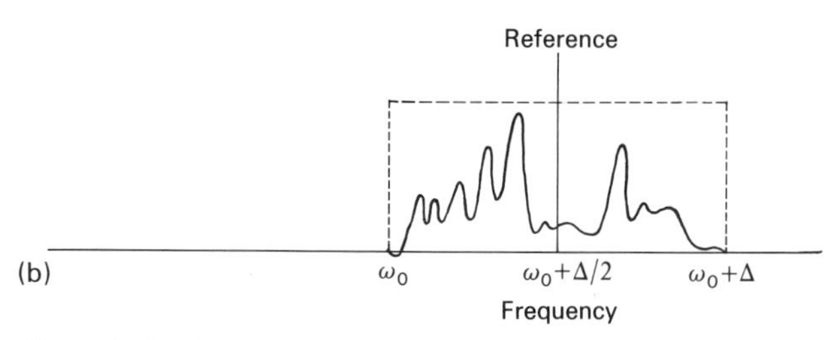

Fig. 5.34. Signal behaviour during single (a) and quadrature (b) phase-sensitive detection. In (a) the reference frequency must be placed at one extreme of the spectrum of width $\Delta\omega$ and, since positive and negative frequencies are not distinguished, noise in the range $\omega_0-\Delta\omega$ to ω_0 is folded onto the spectrum. The problem is avoided in (b) by locating the reference frequency at the centre of the spectrum, allowing a filter of half the width to be employed. (Filter functions are shown as broken lines in (a) and (b)).

echo-planar imaging. For the example discussed in § 5.4.1, a digitization frequency, $1/\tau$, of about 800 kHz is required. If the matrix size or the bandwidth per point is increased the corresponding figure will exceed 1 MHz.

It is also necessary to precede the digitization with suitable audio filters (e.g. of the Butterworth type) to avoid aliasing of the noise lying outside the digitized bandwidth (broken line of Fig. 5.34(b)). These filters themselves cause problems by introducing phase shifts into the signal. Fortunately, they are predictable (two pole Butterworth filters, for example, give rise to a phase shift which increases linearly with frequency over the pass band) and can be corrected at a later stage, following Fourier transformation.

Figure 5.36 illustrates one way in which digitization can be implemented; that is, by using two independent analogue-to-digital converters (a.d.c.s) triggered simultaneously by the pulse programmer or c.p.u. Alternatively, since such devices can be expensive (particularly if both high resolution and a fast digitization rate are required), two triggered track and hold circuits can be used and their outputs multiplexed to a

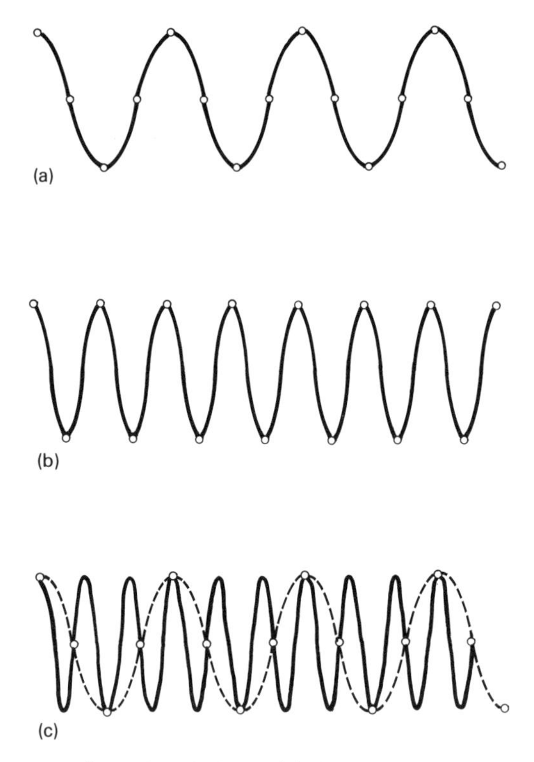

Fig. 5.35. Demonstration of 'aliasing'. (a) Sine wave of frequency $f_N/2$ over sampled (4 times per cycle). (b) Sine wave of frequency f_N sampled at the Nyquist frequency (twice per cycle). (c) Sine wave of frequency $3f_N/2$ under sampled (4/3 times per cycle). Note that in this case the sampled waveform appears to have a frequency of $f_N/2$, shown as a broken line (compare with (a)).

single a.d.c. This latter approach also has the merit of requiring only one data channel but, of course, reduces the maximum sampling rate by a factor of two. For most imaging applications (with the notable exception of echo-planar imaging) this presents no difficulty and such a method has been used by the Aberdeen team to implement their spin-warp imaging method (see Fig. 5.36(b)).

A question arises over the problem of dynamic range or, equivalently, the resolution of the a.d.c.—should it be a 16-bit device, for example, or would an 8-bit device suffice? This clearly will depend on the nature of the imaging scheme; the greater the volume of sample contributing to the signal, the greater the resolution required. If the receiver gain is adjusted such that the root mean square (r.m.s.) noise just registers in the least significant bit of the a.d.c., then the digitizer resolution should be chosen

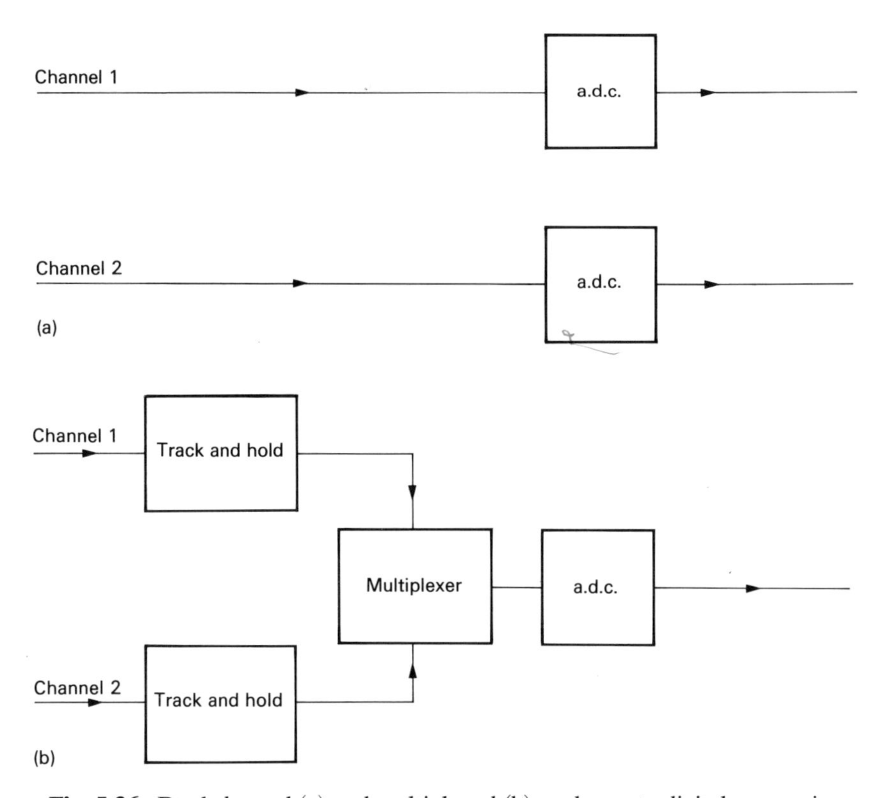

Fig. 5.36. Dual channel (a) and multiplexed (b) analogue to digital conversion.

such that the maximum signal does not overflow the device. As an example, let us assume a S/N of 250 : 1 per ml of tissue contributing to the signal (see § 5.7). Consider then a 1-cm slice through the head. Taking its cross-sectional area to be 400 cm^2 this gives us a total S/N of $250 \times 400 =$ 100 000 : 1 requiring, ideally, a 16-bit a.d.c. ($2^{16} = 65\,536$). If full three-dimensional imaging were required then ideally the a.d.c. should be a 20-bit device! Should it become necessary to compress the signal by reducing the receiver gain then 'digitization noise' will start to degenerate the image. One way around this difficulty is to compress the data in a non-linear fashion using, for example, logarithmic amplifiers.

The digitized signal may be transferred directly to the c.p.u. for subsequent processing, preferably on a direct memory access channel which leaves the c.p.u. free to perform other tasks. Alternatively, if a direct memory access channel is not available, or is too slow, a buffer memory can be interposed.

5.5. The pulse controller and central processor unit

It should be clear by now that the n.m.r. imaging experiment requires a succession of different operations (pulse generation, gradient switching, receiver gating, analogue to digital conversion, etc.), some sequential, some simultaneous, which must be performed according to a strict timing schedule (some of the 'control points' are indicated in the block diagram of Fig. 5.1). In the past these tasks have often been performed with the systems minicomputer. However, increasingly it is delegated to a specialist pulse controller which derives its timing from the master oscillator. In its simplest form this pulse controller has a set of output lines (typically 16 or 24) whose logic levels change according to the bit patterns of the words which are sequentially loaded into the output register from the controller's memory (for this reason the pulse controller is also often known as the pulse programmer). If the controller is handling the data sampling, its cycle time may need to be short (1 μs or less). The memory contents can be entered manually, permanently etched onto an electronically-programmable read-only memory, but, most conveniently in the case of a research instrument, are loaded via the c.p.u. Other controllers are based on a set of intercommunicating microprocessors, giving a very powerful and flexible system which can easily respond to the needs of the most exacting imaging method. Such systems also permit future modifications or even a complete change of imaging scheme to be achieved with the minimum of hardware rearrangement and obviously represent a good design philosophy for research applications.

The task of the c.p.u. is system co-ordination and management. It provides the interface between operator and machine, allowing selection of a particular imaging sequence and acquisition parameter set. This is often achieved through the use of software keys such as sense switches or touch-sensitive plasma screens, an approach which lends itself to a simple menu-driven operation, minimizing difficulty for the inexperienced operator whilst retaining versatility for the experienced research worker. It also allows simple system updating via changes in the support software: in a well-designed system there would be no need for a service visit, the new software could be sent through the post or down a telephone data link.

Once selected, the imaging sequence and parameters would be used to instruct the appropriate pulse and gradient controllers and the scan initiated. Data from the receiver would be returned to the c.p.u. for processing. Depending on the particular imaging scheme in use it is often possible to commence the calculation whilst further data are being acquired; in some cases it is actually possible to keep pace. The reconstructed image should in any event be available as soon after completion of the scan as possible. Since large data sets, particularly three-dimensional image arrays, may be handled, it is imperative to ensure that reconstruction does not become the rate-limiting step. A number of options are available, the fastest being to 'hard-wire' the whole reconstruction process. However, this solution lacks flexibility and the alternative one of adding an array processor to the system is usually considered preferable.

The c.p.u. should have a memory which is sufficiently large to handle the maximum image size of interest. Typically, an image might be $256 \times 256 \times 10$ bits, requiring a storage capacity of 80 kbytes. For higher spatial resolution, a $512 \times 512 \times 10$-bit image requires 320 kbytes of memory and a $128 \times 128 \times 128 \times 10$-bit volume image a full 2.5 Mbytes, though in the latter case it is not essential to have the entire data set core resident at any one time. The requirements on the data storage facility are also demanding. In order to gain rapid access to a large number of patient files, a powerful disc-based system is required. Additionally, magnetic tape can provide cost-effective archiving, and floppy discs are useful for maintaining individual patient records. Since storage and display will be among the more frequent operations required of the c.p.u. it is important to have one or more direct memory access channels so that data transfers can proceed efficiently as a background activity, leaving the c.p.u. free to perform other jobs.

In order to improve efficiency, separate viewing consoles are often provided to allow images to be assessed at leisure by the clinicians concerned. These satellite stations may be autonomous but generally are supported by the c.p.u. on a time-sharing basis. Images can be recalled from the main system store or, more conveniently, from floppy disc if a separate drive is provided with the station.

In the early years of the subject's development, the c.p.u. was usually a moderately-powerful 16-bit minicomputer such as a Data General[1] Nova, or Eclipse. In some cases it also played the role of pulse controller and display driver. However, the computational requirements of a clinical instrument can be demanding, more so than was the case for X-ray CT systems, particularly if volume imaging is intended. The modern tendency therefore has been to opt for powerful 32-bit machines which give good multi-tasking capability and greatly improve speed and accuracy. Generally, between 0.5 and 2 Mbytes of core are provided, backed by a disc system with upwards of 50 Mbyte storage capacity. Often two such disc drives are provided to enable storage of operating system and machine software on one, leaving the second free for image storage.

[1] Data General Corporation, Southboro, Massachusetts, USA.

5.6. Image display

Let us assume that we have stored in the c.p.u. memory a matrix containing our image data. Each element of this array contains values of the quantity which we wish to display, for example proton density or spin lattice relaxation time. The position within the array corresponds to the location of the volume element within the sample. To display the image on a black and white monitor such as the Tektronix[1] model 603, for example, the electron beam can be stepped sequentially from one pixel to the next along each of the image lines by suitably incrementing the x and y amplifier drives. At each point the appropriate element in the image array is used to intensity or time modulate (see, for example, Baines and Mansfield 1976) the electron beam. In addition, high-frequency modulation (preferably triangular or sawtooth) may be impressed on the x and y amplifiers to spread the point focus of the beam into a small rectangle or square of uniform intensity. Provided the whole display scan is completed in a time which is short compared with the visual persistence of the human eye ($\sim$0.1 s), a clear, flicker-free, image is seen.

Alternatively, a high-resolution raster scan monitor can be used. In this case, display hardware is available commercially, for example from Matrox.[2] Essentially, this hardware consists of a fast memory in which the image matrix is stored and from which it is clocked out through a high-speed (video) digital to analogue converter. Horizontal and vertical synchronization pulses are added to generate a composite video signal which is fed to the monitor. Typically, a 256×256 matrix is displayed, though a larger format capability is useful for presentation of higher-resolution images or for simultaneous multiple image display.

The human eye is capable of distinguishing in the region of 16 grey levels so that a 4-bit ($2^4 = 16$) display would give acceptable image quality. It is, however, common practice to work to rather higher accuracy, 8-bit (256 level) systems being especially popular. Taking this particular case we have to convert our image data, which may be stored as 10- or 12-bit words, to 8 bits. This can be done in a straightforward manner by retaining only the 8 most significant bits, corresponding to a simple scaling. It is however more useful to have a 'windowing' facility available. In the case of 12-bit data the parameter we are imaging, spin density for example, can take on any integer value between 0 and 4095, with perhaps 0 corresponding to air and 4095 to pure water. Usually, however, we are interested in the soft tissues which on this scale would occupy a range of some 600 units centred at a value of about 2800. We

[1] Tektronix, P.O. Box 4828, Portland, Oregon 97208, U.S.A.
[2] Matrox, 5800 Andover Ave, TMR, (Montreal), Quebec, Canada H4T1 H4.

would thus like to adjust our grey scale to span only these data values, setting higher or lower ones to white or black respectively. This can be accomplished with the aid of window level and width parameters which could be adjusted via front panel joystick controls to enable rapid visual assessment of the resulting image. It is also possible to use double window settings to display contrast at two levels or even a non-linear display to accentuate differences at one end of the scale. In all such operations a good guideline to follow is to ensure that each level is separated from its neighbours by at least one standard deviation. If this is not the case the resulting artefactual 'fine structure' can be very misleading.

Other features commonly provided on display systems include isodensity plots—the display of pixels with a particular data value or range of values, profile cuts—pixel values along a line through the image at an arbitrary orientation, histograms—showing the distribution of pixel values throughout a region of interest defined perhaps with the aid of a light pen, area measurement—particularly useful if in the case of multi-slice images the algorithm can be extended to volume measurements, and distance and orientation determinations. Some systems have a zoom facility, permitting enlargement of selected image regions (a feature which should not be confused with the genuine improvement in spatial resolution achievable by recording images in higher field gradients). The ability to pre-select a number of images for rapid consecutive viewing or photography is useful. Advantage can also be taken of the increasing number of sophisticated computer graphics algorithms becoming available. For example, Axel, Herman, Udupa, Bottomley, and Edelstein (1983) have used a three-dimensional surface detection algorithm to generate separate images of myocardium and cardiac blood vessels which were displayed as 'animated' transparent or opaque solids. Such methods give one a much better feel for the true three-dimensional nature of the object under investigation.

It is also possible to colour code the image data (see, for example, Holland and Bottomley 1977), and many commercial systems are available for this purpose (for example, from Ramtek[1]). We will not enter too deeply into the arguments for black and white versus colour displays—both have their strong advocates. Suffice it to say that because the untrained eye is much less likely to be confused by a black and white display this is consequently to be preferred in most imaging applications. There are, however, situations in which colour can be used to advantage. By combining colour and intensity information it is possible to display two parameters in the same image, for example, proton density colour-coded and T_1 intensity-coded. Whether both sets of information can be

[1] Ramtek, Lawson Lane, Santa Clara, California 95050, U.S.A.

effectively assimilated when presented in such a manner is a different matter, however.

For 'hard copy' images one can, of course, photograph the monitor screens directly. Polaroid cameras used either with the main image monitor or more conveniently with a smaller subsidiary monitor assigned for this purpose, are a particularly convenient source of 'instant' copies. For higher quality reproductions the use of a multiformat camera such as the model MI-7 130 from Matrix[1] has become standard. Images are recorded on to photographic film which, in terms of the number of distinguishable grey levels is vastly superior to paper (it is possible using an optical densitometer to read back 8-bit data from a photographic negative, for example). They may be viewed using the 'light box' displays familiar in all radiology departments.

5.7. Signal-to-noise estimates

The n.m.r. signal originates from the e.m.f. induced in the receiver coil by the rotating nuclear magnetization (see § 2.4.2.3). From Faraday's law of induction, the amplitude of this e.m.f. is given by the rate of change of flux linking the receiver coil, which will be proportional both to the magnetization and to its rotational frequency. As discussed in Chapter 2, the magnetization precesses at the Larmor frequency and its magnitude is proportional to the applied magnetic field B_0 and hence also to the Larmor frequency. The signal therefore increases as the square of the operating frequency. The noise also increases with operating frequency, but there is considerable disagreement concerning the strength of the functional dependence.

For non-conducting samples, organic solvents for example, the problem is relatively straightforward, since the noise is thermal, originating in the resistance of the receiver coil windings. However, even in this situation different expressions exist. The classical relationship (Abragam 1961) suggests that the noise is proportional to $\omega^{\frac{1}{2}}$ giving a S/N proportional to $\omega^{\frac{3}{2}}$, whereas Hoult and Richards (1976) argue that the noise is only proportional to $\omega^{\frac{1}{4}}$, so that the S/N increases as $\omega^{\frac{7}{4}}$.

The situation in a conducting sample is considerably more complex since, in addition to the r.f. receiver coil, the sample itself becomes a source of noise. Simple models have led to expressions for the equivalent noise resistance associated with dielectric and inductive losses (see Hoult and Lauterbur 1979; Gadian and Robinson 1979). The former can be eliminated through the use of a Faraday shield interposed between coil and sample; the latter has to be lived with. In the model of Hoult and

[1] Matrix, 230-T Pegasus Ave., North Vale, New Jersey 07647, U.S.A.

Lauterbur (1979) the noise from this source is proportional to ω, giving a S/N which increases linearly with frequency.

The S/N relationship for clinical n.m.r. imaging is therefore likely to be complex, increasing as $\omega^{\frac{3}{2}}$ or $\omega^{\frac{7}{4}}$ at low frequency, switching to a linear dependence at higher frequency where the sample noise dominates. This is illustrated schematically in Fig. 5.37. The scale of the frequency axis has deliberately been omitted: however, it is likely that for n.m.r. imaging at all but the lowest frequencies, noise from the sample will dominate. The important point to note is that the signal grows with frequency faster than the noise does, suggesting that one should operate at the highest possible frequency consistent with safety requirements (see § 6.8). This has indeed been the design philosophy adopted by a number of manufacturers. However, there are a number of factors which mitigate against this approach. The most serious objection, raised in the days before such high-field experiments had been tried, arose from possible 'skin depth' problems associated with the screening of the interior of the body from the r.f. by currents circulating in its surface layers. The importance of this latter effect can be estimated by using the well known expression for the 'skin depth', derived for plane waves impinging on a flat surface. In such a situation the current density J is given by:

$$J = J_0 \exp(-x/\delta), \tag{5.24}$$

where J_0 is the surface current density, x the distance from the surface,

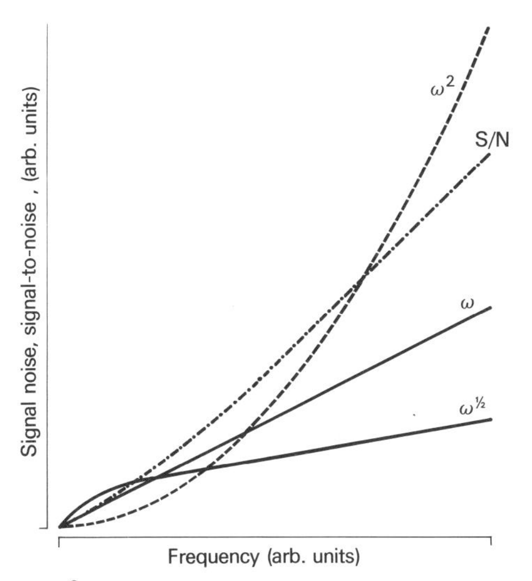

Fig. 5.37. Signal (- - - ω^2), noise (sample (—— ω) and coil (—— $\omega^{\frac{1}{2}}$)), and signal-to-noise ratio (—·—S/N) as a function of frequency. This diagram is schematic only, the frequency axis is deliberately left blank.

and δ, the 'skin depth', is given by

$$\delta = \sqrt{2\sigma/\mu_0\omega}. \tag{5.25}$$

The frequency-dependence of the conductivity σ and also the dielectric constant ε have been measured over the range of interest to n.m.r. imagers for a number of biological tissues. Useful tables of values have been collated in the *Radiofrequency Radiation Dosimetry Handbook* (Durney, Johnson, Barber, Massoudi, Iskander, Lords, Ryser, Allen, and Mitchell 1978; see also Bottomley 1978*b*, and Bottomley and Andrew 1978). By way of example, the skin depths at 5 ($\sigma = 0.385\ \Omega^{-1}\ \mathrm{m}^{-1}$) and 50 MHz ($\sigma = 0.556\ \Omega^{-1}\ \mathrm{m}^{-1}$) for liver are estimated as 14 cm and 5 cm respectively. More realistic models are available (see Mansfield and Morris 1982), but the results are qualitatively similar, suggesting that at $\sim$10 MHz the r.f. only partially penetrates the body and that at frequencies much above this gross phase and intensity distortions would result. These effects are indeed observed when one images phantoms containing physiological saline; however, they are virtually absent in whole-body images obtained at frequencies as high as 63 MHz. Presumably, this is the result of tissue interfaces providing effective barriers to current flow, thereby restricting the size of the current loops. It certainly calls into question our detailed understanding of electrical conduction in the body and one should beware of placing too much reliance on noise estimates based on simple tissue models. Although as yet unobserved, 'skin depth' effects will be present in some measure and may yet prove limiting at higher frequencies.

Another factor against high-field imaging is the increase in tissue conductivity with frequency (see tables of electrical properties of tissues in references quoted above). Since the sample noise depends on $\sigma^{\frac{1}{2}}$ the S/N will be diminished proportionately. Fortunately, with the possible exception of lung, and to a lesser extent brain, this is a small effect, likely to be relatively unimportant. Rather more difficult to assess are the problems of probe design at high-frequency where the wavelength of the radiation becomes comparable with the coil dimensions. Certainly it is a difficult art—nevertheless, a number of manufacturers have demonstrated their ability to master it.

The S/N is not the only issue in question, however—infinite S/N would be of no practical value if all tissues gave the same signal. What really needs to be maximized is the contrast-to-noise ratio which can be obtained per unit time. We therefore also need to consider the origins of tissue contrast (§ 6.2) and the details of the particular pulse sequence used to generate it (§ 6.3.1). This brings new frequency effects, which should be examined. For example, it is worth noting that the slower spin–lattice relaxation at high frequency will generally require longer interpulse intervals, which in turn will have an adverse effect on the contrast/noise/unit time (see also eqns (5.26), (5.27)).

The question of which frequency is *the* right one for n.m.r. imaging is an open one, currently surrounded by much controversy. The present trend is towards higher-field instruments, nevertheless it is certainly true that, in the right hands, low-field systems can yield excellent image quality.

Notwithstanding the difficulties discussed above it is still worthwhile to explore the relationship between imaging time, S/N, and resolution. Rather than derive expressions for the S/N per unit time afforded by each of the imaging schemes discussed, a single estimate has been made, based on an 'optimum technique'. Some indication of the extent to which each particular technique matches up to this ideal has been given in Chapters 3 and 4 (see also Brunner and Ernst 1979, and Mansfield and Morris 1982). This problem has been described by Mansfield and Morris (1982) who obtain expressions for the imaging times required to achieve a given S/N both for cubic volume elements (side Δx) and for slices of thickness

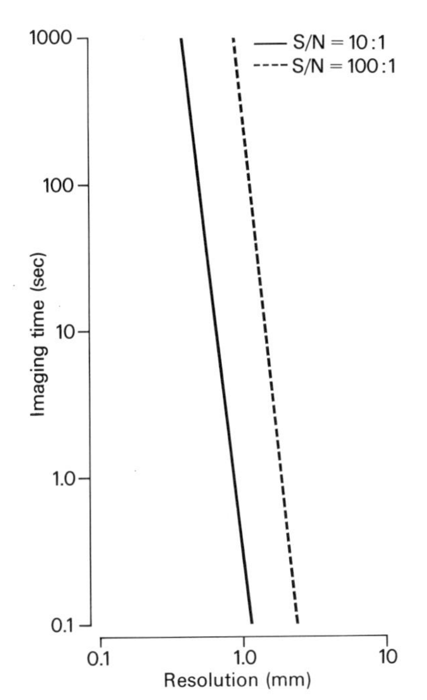

Fig. 5.38. Estimate of imaging time t_{vol} versus resolution (eqn (5.26)) for $a = 0.2$ m, $C = 0.5$, $T_1/T_2 = 8$, $\nu = 21$ MHz and S/N = 10 and 100.

Δz. Thus:

$$t_{\mathrm{vol}} \simeq \left(\frac{S}{N}\right)^2 a^2 \left(\frac{T_1}{T_2}\right) \frac{1.418 \times 10^{-15}}{\nu^{\frac{7}{2}} C} \left(\frac{1}{\Delta x}\right)^6, \tag{5.26}$$

and

$$t_{\mathrm{slice}} \simeq \left(\frac{S}{N}\right)^2 a^2 \left(\frac{T_1}{T_2}\right) \frac{1.418 \times 10^{-15}}{\nu^{\frac{7}{2}} C} \left(\frac{1}{\Delta z}\right)^2 \left(\frac{1}{\Delta x}\right)^4. \tag{5.27}$$

C is a constant which depends on the pulse repetition rate and flip angle used, but in the optimum case tends to a value of $\frac{1}{2}$, a is the receiver coil radius and ν is the frequency in MHz. Note in these expressions the expected dependence of the imaging time on the square of S/N. Note, too, the very strong dependence on resolution, inverse sixth- and inverse fourth-order respectively, indicating the extreme price to be paid for improvements in this area. Expressions (5.26) and (5.27) are plotted in Figs 5.38 and 5.39 respectively for a frequency of 21 MHz and typical values of the various constants. It is necessary to emphasize the approximate nature of these expressions, which should be used as guidelines

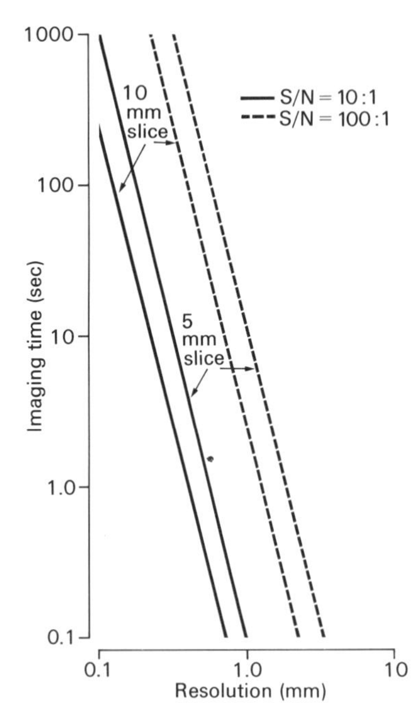

Fig. 5.39. Estimate of imaging time t_{slice} versus resolution (eqn (5.27)) for slices of thickness 5 and 10 mm. Remaining parameters as for Fig. 5.38.

only, though their functional form, with the exception of the frequency-dependence, is instructive. Also, it must be re-emphasized that such calculations apply to an optimum imaging method, so that some regions of these plots will be inaccessible to certain techniques, particularly in the minimum performance time-limit where only echo-planar imaging thus far offers the possibility of discriminating between all pixels in a single f.i.d.

References

Abragam, A. (1961). *The principles of nuclear magnetism.* Clarendon Press, Oxford.

Ackerman, J. J. H., Grove, T. H., Wong, G. G., Gadian, D. G., and Radda, G. K. (1980). *Nature, Lond.* **283,** 167.

Alderman, D. W., and Grant, D. M. (1979). *J. magn. Reson.* **36,** 447.

Anderson, W. A. (1961). *Rev. scient. Instrum.* **32,** 241.

Axel, L., Herman, G. T., Udupa, J. K., Bottomley, P. A., and Edelstein, W. A. (1983). *J. comput. assist. Tomogr.* **7,** 172.

Baines, T., and Mansfield, P. (1973). *J. Phys.* E: *Scient. Instrum.* **6,** 462.

———, and ——— (1976). *J. Phys.* E: *Scient. Instrum.* **9,** 809.

Bangert, V., and Mansfield, P. (1982). *J. Phys.* E: *Scient. Instrum.* **15,** 235.

Bleaney, B. I. and Bleaney, B. (1965). *Electricity and magnetism.* Claredon Press, Oxford.

Bottomley, P. (1978*a*). Ph.D Thesis, University of Nottingham.

——— (1978*b*). *J. Phys.* E: *Scient. Instrum.* **11,** 413.

——— (1981*a*). *J. Phys.* E: *Scient. Instrum.* **14,** 1052.

——— (1981*b*) *J. Phys.* E: *Scient. Instrum.* **14,** 1081.

———, and Andrew, E. R. (1978) *Phys. med. Biol.* **23,** 630.

———, Hinshaw, W. S., and Holland, G. N. (1978). *Phys. med. Biol.* **23,** 309.

Brunner, P., and Ernst, R. R. (1979). *J. magn. Reson.* **33,** 83.

Burnett, L. J., and Jackson, J. A. (1980). *J. magn. Reson.* **41,** 406.

Chen, C.-N., Hoult, D. I., and Sank, V. J. (1983). *J. magn. Reson.* **54,** 324.

Cooper, R. K. and Jackson, J. A. (1980). *J. magn. Reson.* **41,** 400.

Damadian, R., Goldsmith, M., and Minkoff, L. (1977). *Physiol. Chem. Phys.* **9,** 97.

Darrasse, L. (1982). *Rev. Scient. Instrum.* **53,** 1338.

Durney, C. H., Johnson, C. C., Barber, P. W., Massoudi, H., Iskander, M. F., Lords, J. L., Ryser, D. K., Allen, S. J., and Mitchell, J. C. (1978). *Radiofrequency radiation dosimetry handbook* (2nd edn). USAF School of Aerospace Medicine (RZP) Aerospace Medical Division, Brooks Air Force Base, Texas, 78235.

Edelstein, W. A., Hutchison, J. M. S., Johnson, G., and Redpath, T. (1980). *Phys. med. Biol.* **25,** 751.

Franzen, W. (1962). *Rev. Scient. Instrum.* **33,** 933.

Froncisz, W., and Hyde, J. S. (1982). *J. magn. Reson.* **47,** 515.

Gadian, D. G., and Robinson, F. N. H. (1979). *J. magn. Reson.* **34,** 449.

Garrett, M. W. (1951). *J. appl. Phys.* **22,** 1091.

Ginsberg, D. M., and Melchner, M. J. (1970). *Rev. Scient. Instrum.* **41,** 122.

Goldsmith, M., Damadian, R., Stanford, M., and Lipkowitz, M. (1977). *Physiol. Chem. Phys.* **9,** 105,

Golay, M. J. E. (1958). *Rev. scient. Instrum.* **29,** 313.

Grannell, P. K., Orchard, M. J., Mansfield, P., Garroway, A. N., and Stalker, D. C. (1973). *J. Phys.* E: *Scient. Instrum.* **6,** 1202.

Hanley, P. (1982). *Proc. int. symp. NMR imaging,* p. 41. Bowman Gray School Medicine, Winston-Salem.

——— (1984). *Br. Med. Bull.* **40** (2), 125.

Hanley, P. E., and Gordon, R. E. (1981). *J. magn. Reson.* **45,** 520.

Hardy, W. N., and Whitehead, L. A. (1981). *Rev. scient. Instrum.* **52,** 213.

Holland, G. N., and Bottomley, P. A. (1977). *J. Phys.* E: *Scient. Instrum.* **10,** 714.

———, and Heysmond, E. (1979). *J. Phys.* E: *Scient. Instrum.* **12,** 480.

Hoult, D. I. (1979*a*). *J. magn. Reson.* **33,** 183.

———, (1979*b*). *Rev. scient. Instrum.* **50,** 193.

——— (1980). *Magnetic resonance in biology,* Vol. 1. (ed. J. S. Cohen), J. Wiley, New York.

——— (1981). *J. comput. assist. Tomogr.* **5,** 291.

———, and Lauterbur, P. C. (1979). *J. magn. Reson.* **34,** 425.

———, and Richards, R. E. (1975*a*). *Electron. Lett.* **II,** 596.

———, and ——— (1975*b*). *Proc. Roy. Soc.* A **344,** 311.

———, and ——— (1976). *J. magn. Reson.* **24,** 71.

Hutchison, J. M. S., Edelstein, W. A., and Johnson, G. (1980). *J. Phys.* E: *Scient. Instrum.* **13,** 947.

———, Sutherland, R. J., and Mallard, J. R. (1978). *J. Phys.* E. *Scient. Instrum.* **11,** 217.

Jackson, J. A., Burnett, L. J., and Harmon, J. F. (1980). *J. magn. Reson.* **41,** 411.

Jeffreys, H., and Jeffreys, B. (1956). *Methods of mathematical physics.* Cambridge University Press.

Johnson, G., Hutchison, J. M. S., and Eastwood, L. M. (1982). *J. Phys.* E: *Scient. Instrum.* **15,** 74.

Lai, C-M., and Lauterbur, P. C. (1980). *J. Phys.* E: *Scient. Instrum.* **13,** 747.

———, Shook, J. W., and Lauterbur, P. C. (1979). *Chem. biomed. environ. Instr.* **9,** 1.

Mallard, J. R., Hutchison, J. M. S., Edelstein, W. A., Ling, C. R., Foster, M. A., and Johnson, G. (1980). *Phil. Trans. R. Soc.* B **289,** 519.

Mansfield, P., and Morris, P. G., (1982). 'NMR imaging in Biomedicine', in *Advances in magnetic resonance,* suppl. 2 (ed. J. S. Waugh). Academic Press, New York.

———, Maudsley, A. A., and Baines, T. (1976). *J. Phys.* E: *Scient. Instrum.* **9,** 271.

———, Pykett, I. L., Morris, P. G., and Coupland, R. E. (1978). *Br. J. Radiol.* **51,** 921.

Maudsley, A. A., Oppelt, A., and Ganssen, A. (1979). *Siemens Forsch.,* **8,** 326.

McGurrin, M., and Gully, W. J. (1983). *Rev. scient. Instrum.* **54,** 770.

6. Application of n.m.r. to biological systems

6.1. Introduction

This chapter will mainly be concerned with proton imaging, and will be heavily biased in favour of medical applications. In § 6.2, there will be discussion of some of the basic n.m.r. properties of biological tissues in order to provide some insight into the origins of tissue contrast. An elementary introduction for those unfamiliar with imaging terminology is given in § 6.3, followed by a general discussion of some important and novel features of proton n.m.r. imaging. Small-scale studies are the subject of § 6.4, and §§ 6.5 and 6.6, deal specifically with medical applications. Head scanning is considered in § 6.5, where the results of preliminary clinical trials have already demonstrated the great potential of the method, and § 6.6, deals with whole-body clinical studies which have also proved very encouraging. Comparisons with other imaging modalities, in particular X-ray CT, are made, though it should be stressed that such assessments can only be considered preliminary at this stage. Far more extensive clinical studies are necessary to establish the full potential of n.m.r. imaging. Multinuclear and chemical shift imaging studies are dealt with in § 6.7.

One of the great attractions of n.m.r. as opposed to an X-ray- or radio-isotope-based system is the lack of ionizing radiation; indeed, the technique promises to be an extremely safe one. However, as should always be the case with new imaging modalities, it is prudent to err on the side of caution. An assessment of potential hazards is given in § 6.8, quoting the National Radiation Protection Board's guidelines for n.m.r. exposure, based on the limited experience so far derived from systems in service.

6.2. Origin of proton n.m.r. signals

6.2.1. Properties of tissue water

On average, water accounts for about 55 per cent of the total human bodyweight. Of this, some 10 per cent occurs in blood plasma, 20 per cent in extracellular and 70 per cent in intracellular fluid. If we consider only the soft tissues, then water constitutes from 60 to 90 per cent of their weight, the remainder arising largely from proteins and cell membranes.

Minkoff, L., Damadian, R., Thomas, T. E., Hu, N., Goldsmith, M., Koutcher, J., and Stanford, M. (1977). *Physiol. Chem. Phys.* **9,** 101.
Moore, W. S., and Holland, G. N. (1980). *Phil. Trans. R. Soc.* B **289,** 381.
Morris, P. G., Mansfield, P., Pykett, I. L., Ordidge, R. J., and Coupland, R. E. (1979). *IEEE Trans.* **26,** 2817.
Müller, W. H.-G., and Knüttel, B. (1983). Bruker Medical Report (1) 4.
Murphy-Boesch, J., and Koretsky, A. P. (1983). *J. magn. Reson.* **54,** 526.
Onnes, H. K. (1911). *Akad. van Wetenschappen* (Amsterdam). **14,** 113.
Persson, R. B. R., Bolmsjö, M., and Malmgren, L. (1983). *Proc. Soc. magn. reson. Med.* (Aug.).
Saint-Jalmes, H., Taquin, J., and Barjhoux, Y. (1981). *Rev. scient. Instrum.* **52,** 1501.
Schneider, H. J., and Dullenkopf, P. (1977*a*). *Rev. scient. Instrum.* **48,** 68.
———, and ——— (1977*b*). *Rev. scient. Instrum.* **48,** 832.
Tanner, I. E. (1965). *Rev. scient. Instrum.* **36,** 1086.
Terman, F. E. (1943). *Radio engineer's handbook* (1st edn). McGraw-Hill, New York.
Zupančič, I., and Pirš, J. (1976). *J. Phys.* E: *Scient. Instrum.* **9,** 79.

Unless the special line-narrowing techniques of solid state n.m.r. are applied, the large structural proteins such as collagen give rise to extremely broad (several kHz) lines which, in the presence of a narrow (20–50 Hz), intense water peak, are difficult to observe by conventional n.m.r. and are totally lost to n.m.r. imaging studies. Although the membrane components (largely phospholipids) are rather more mobile, they do not contribute significantly to the proton image. The smaller proteins and metabolites present in solution do produce sharp resonances and can be studied *in vitro* by high-resolution n.m.r. (see, for example, Wüthrich 1976). However, as illustrated in § 2.6.1, they are not present in sufficiently high concentration to be observed in imaging experiments. Thus, as far as proton n.m.r. imaging is concerned, the signal is predominantly derived from water.

The only other significant sources of proton signal are the lipids which occur as adipose tissue in some regions of the body, e.g. the mammary glands. The chemical shift difference (~3 p.p.m., see § 2.6.1) is sufficient to allow resolution of the glyceride —CH_2— protons from the water resonance with the homogeneities currently available from whole-body magnets (see § 5.1). For high field studies, the signals from these two sources may be separated in order to avoid the degradation of spatial resolution which results from the two associated images appearing out of register. In general, however, the separation of chemically shifted components introduces an additional level of complexity which is unwarranted. Without this selection, proton images display the total mobile proton density, reflecting predominantly the water distribution, but also including lesser contributions from lipids. Fortunately, the latter tend to have rather different n.m.r. relaxation properties to water and, in some cases, can be distinguished on this basis. Leaving aside the possibility of signals derived from fatty tissue for the moment, we can ask what sort of tissue contrast would be available from an image based solely on water content. A glance at Fig. 6.1, indicates that, in normal soft tissues, the water content spans a range of some 15 per cent. Skeletal muscle, for example, has a particularly high value (~80 per cent) whereas that for liver is especially low (~70 per cent); we shall return to this point in connection with tissue relaxation times in § 6.2.3. Notice also the difference in water contents between white and grey matter, suggesting that it should be possible to discriminate between them on this basis alone. Together with the high water-content of the cerebrospinal fluid (c.s.f.) in the ventricles, this provided an early indication that well-resolved images of the brain should be a practical proposition; an expectation which was beautifully fulfilled in the publication of a series of n.m.r. head scans obtained at the Hammersmith Hospital with a prototype machine provided by Picker International (Doyle, Gore, Pennock, Bydder, Orr,

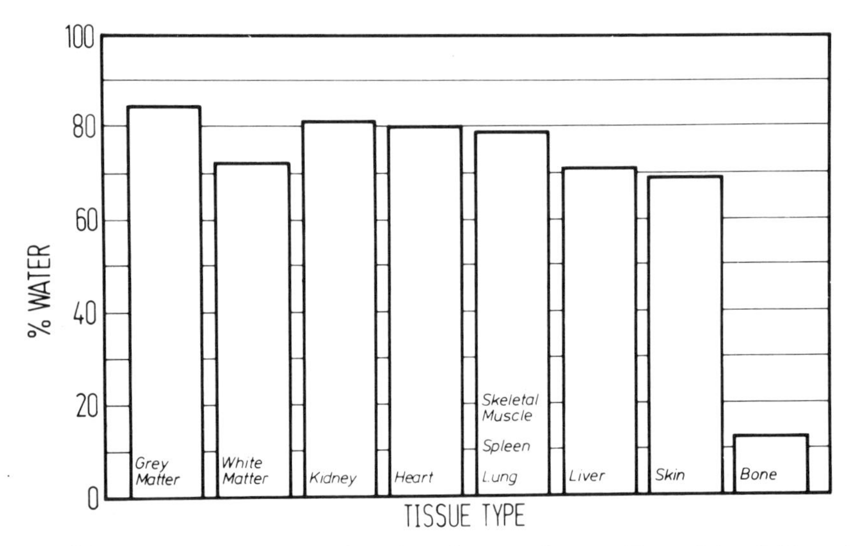

Fig. 6.1. The water content of various human tissues. (From Mansfield and Pykett, 1978.)

Steiner, Young, Burl, Clow, Gilderdale, Bailes, and Walters 1981; Young, Hall, Pallis, Bydder, Legg, and Steiner 1981; Young, Bailes, Burl, Collins, Smith, McDonnell, Orr, Banks, Bydder, Greenspan, and Steiner 1982*b*).

When it comes to pathological studies, a wide range of conditions are likely to give rise to changes in water content, some of the more obvious examples being fluid-filled cysts and oedema. Tumours also generally have increased water contents associated with them.

Although mobile proton density measurements do provide one basis for distinguishing between tissues, a number of other parameters, notably the relaxation times T_1 and T_2, are also influential. As discussed below, the best tissue contrast is generally afforded by images containing a strong element of relaxation time discrimination.

6.2.2. *Proton n.m.r. relaxation properties of tissues*

In pure water at standard n.m.r. frequencies the molecules are in rapid motion (in the sense of § 2.5.1) and the correlation time τ_c is correspondingly short ($\omega_0\tau_c \ll 1$ where ω_0 is the Larmor frequency). In this case, the spin–spin and spin–lattice relaxation times are almost equal, having a value of about 3 s. (T_2 is slightly less than T_1 due to ^{17}O *J*-coupling to the protons, which is unresolved because the frequency of proton exchange is of the same order as the coupling constant at neutral pH.) The situation with tissue water is radically different. Spin–spin relaxation often cannot be fitted to a single exponential and has to be analysed in terms of two or more exponential processes with T_2s which are typically in the range 20–100 ms. Although spin–lattice relaxation in tissues generally is exponential, the relaxation times at typical human imaging frequencies (1–20 MHz) lie in the range 100–500 ms, corresponding only to some 3–16 per cent of the pure water values. There are also differences in the diffusion properties. Thus, whereas the rotational diffusion coefficient determined by dielectric measurements is roughly comparable with that of pure water, the translational coefficient determined by n.m.r. pulsed field gradient studies is found to be about one-half of the pure water value (Chang, Hazlewood, Nichols, and Rorscharch 1972).

A number of theories have been put forward concerning the nature of tissue water, the simplest being that it does not differ in any important respect from bulk water (Dick 1971). Whilst it might be true as far as solvent properties are concerned, this naive assumption is clearly untenable from the n.m.r. viewpoint. Two theories do account, at least qualitatively, for the observed reduction in n.m.r. relaxation times. In the structured water theory (Ling 1962; Hazlewood 1976), the macromolecules and ions confer on the intracellular water a kind of semi-crystallinity, and it is this reduction in motional freedom which accounts both for the shorter relaxation times and for the lower translational diffusion coefficient. The more traditional view is that the majority of tissue water behaves just like bulk water, but that there is a small but significant fraction bound to the macromolecules. There is direct evidence for this in the observation that a certain fraction of tissue water remains unfrozen at temperatures substantially below 0 °C (Kuntz, Brassfield, Law, and Purcell 1969; Kuntz 1971). It is tempting to associate this component directly with the bound phase (Belton, Packer, and Sellwood 1973; Woodhouse, Derbyshire, and Lillford 1975) although this assignment is not universal (Foster, Resing, and Garroway 1976). The bound fraction, with severely restricted motional freedom, has a short T_1. However, it is assumed to be in fast exchange with the remaining 'free' water, which accounts for the observation of a single reduced T_1. The reduction in translational diffusion coefficient can be ascribed (tentatively) to a barrier effect of the macromolecules. In the past there has been considerable controversy between proponents of these two rival theories (Kolata 1976). For the present purpose interpretation of the observed relaxation times in terms of the two-phase fast exchange model has been decided upon, both because it is the more widely-held view and also because some of its more important predictions can be subjected to experimental test. However, it should be recognized from the outset that the complex structural nature of the cellular environment (see, for example, Fulton 1982) cannot properly be described by a two-phase, two correlation time, model, particularly when it comes to spin–spin relaxation, for which the effects of anisotropic motion may well be important. Nevertheless, such a model does offer some insight into possible relaxation processes in biological tissues.

The expressions for relaxation in a system consisting of n exchanging phases were derived by Zimmerman and Brittin in 1957. In the fast exchange limit ($\tau_e < T_i$, where τ_e is the exchange time) they obtain

$$M(t) = M_0 \exp\left[-t\Big/\Big(\sum_i P_i/T_i\Big)\right]. \tag{6.1}$$

Here P_i is the fraction of spins and T_i the (generalized) relaxation time in the ith phase. The spin system thus relaxes exponentially at a rate which is the weighted average of the rates in the individual phases. Thus

$$1/T = \sum_{i=1}^{n} P_i/T_i. \tag{6.2}$$

In the other limiting case, slow exchange ($\tau_e > T_i$), each phase relaxes independently and the relaxation recovery, though not itself exponential, can be resolved into a series of exponential components corresponding to the non- or slowly-exchanging phases. An extreme example for which eqn (6.2) applies is the spin–lattice recovery of breast tissue, which contains a high proportion of fat in addition to water. Protons from these two sources do not exchange ($\tau_e \to \infty$) and they therefore relax independently. The resultant biphasic recovery can be resolved into its two component exponentials, giving the relative amounts of fat and water (P_i) and their corresponding relaxation times (Bovée, Creyghton, Getreuer, Korbee, Lubregt, Smidt, Wind, Lindeman, Smid, and Posthuma 1980).

Many images which purport to show spin–lattice relaxation information are derived from two separate scans taken with different relaxation recovery intervals, TR (see § 2.5.1.1). T_1 values are calculated from the ratio of signal intensities for each image region, based on the assumption of a simple exponential recovery. Clearly, for tissues such as breast, this leads to anomalous T_1 values which are dependent on TR. In order to understand the origins of tissue contrast it is important that more careful T_1 studies be performed. However, from the diagnostic viewpoint, such time-consuming experiments are unlikely to be feasible on a routine basis. Rather, operating conditions should be preferred which maximize contrast for the particular clinical condition for which screening is being carried out; for example, breast cancer (Mansfield, Morris, Ordidge, Coupland, Bishop, and Blamey 1979).

6.2.3. *Relationship between water content and relaxation times*

We now examine possible correlations between water content and spin–lattice relaxation time. If we assume the two-phase fast exchange model of tissue water then, using eqn (6.2), we can write the spin–lattice relaxation rate ($1/T_1$) as

$$\frac{1}{T_1} = b\left(\frac{1}{T_1}\right)_{\mathrm{b}} + (1-b)\left(\frac{1}{T_1}\right)_{\mathrm{f}}, \tag{6.3}$$

where b is the fraction of the total tissue water in the bound phase and $(1/T_1)_{\mathrm{f}}$, $(1/T_1)_{\mathrm{b}}$ are respectively the free and bound relaxation rates. Equation (6.3) can simply be rearranged to give

$$\frac{1}{T_1} = \left(\frac{1}{T_1}\right)_{\mathrm{f}} + b\left\{\left(\frac{1}{T_1}\right)_{\mathrm{b}} - \left(\frac{1}{T_1}\right)_{\mathrm{f}}\right\}, \tag{6.4}$$

whence it becomes apparent that, provided $(1/T_1)_{\mathrm{b}}$ is not a function of b, the relaxation rate $1/T_1$ should be directly proportional to the fraction of the bound water. This dependence has been observed both in simple protein solutions, where it is possible to achieve a detailed understanding of the important relaxation mechanisms (Koenig and Schillinger 1969), and also in some biological tissues. A particularly clear illustration of this is afforded by the work of Lauterbur, Frank, and Jacobson (1976) on oedematous canine lung tissue, undertaken as part of an early feasibility study for n.m.r. imaging. Figure 6.2, taken from their work, shows the striking linearity of the water-content/spin–lattice relaxation rate plot.

Whereas the finding that $1/T_1$ is proportional to b in a particular tissue type over a limited range of water contents is entirely reasonable, the often-made assumption that water content is the sole factor determining T_1 values in different tissues is much more controversial. A typical example of the evidence in favour of this assertion is given in Fig. 6.3, which shows spin–lattice relaxation rates for a number of healthy and tumourous murine tissues determined by the inversion recovery method at 24 MHz using excised tissue samples (Taylor and Bore 1981). Many similar examples are to be found in the literature; for example, Kiricuta, Demco, and Simplaceanu (1973); Kiricuta and Simplaceanu (1975); Inch, McCredie, Knispel, Thompson, and Pintar (1974); Saryan, Hollis, Economou, and Eggleston (1974); and Bovée, Huisman, and Smidt (1974). However, there are reports in which this simple dependence is not observed.

Notice in Fig. 6.3 that, although the trend of decreasing relaxation rate with increasing water content is readily apparent, the experimental scatter is large. This is really what we should expect; an exact proportionality would require the bound relaxation rates $(1/T_1)_{\mathrm{b}}$ to be the same in all

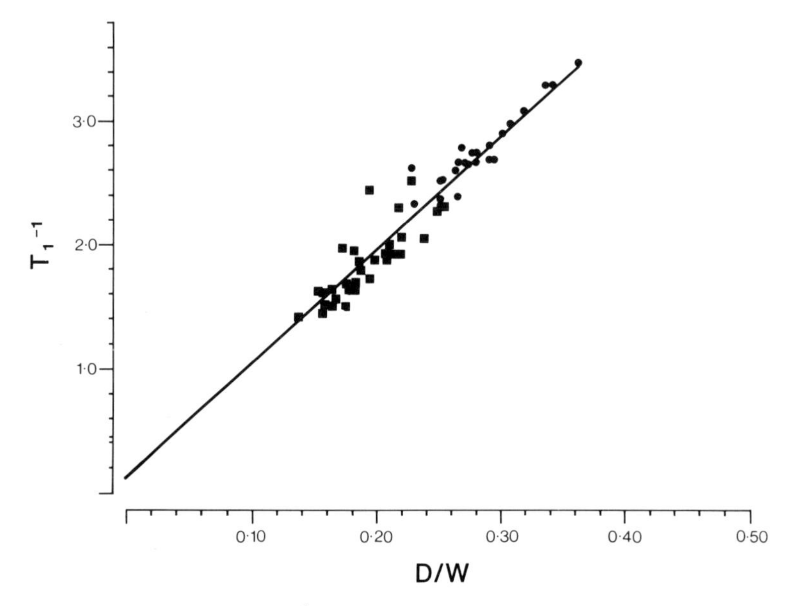

Fig. 6.2. The dependence of ^{1}H spin–lattice relaxation rate, $1/T_1$, on the dry weight to water ratio in normal (●) and oedematous (■) canine lung tissue. Measurements were made at 4 MHz. (From Lauterbur *et al.* 1976.)

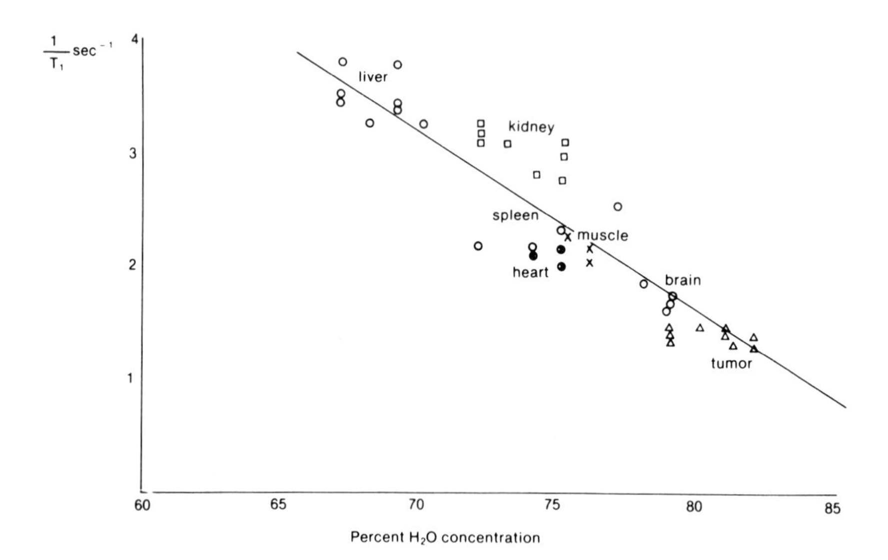

Fig. 6.3. ^{1}H spin–lattice relaxation rate, $1/T_1$, as a function of percentage water content for a number of healthy mouse tissues and tumours at 24 MHz. The solid line is a least-squares best fit to the data. (From Taylor and Bore 1981. Copyright Pergamon Press 1981.)

tissues. Whilst the mechanisms leading to relaxation will be similar in nature, it is highly likely that differences in detail will occur in tissues as diverse as kidney and skeletal muscle, for example.

Some equations for relaxation times were derived in § 2.5. For convenience we repeat the T_1 expression herc. Thus:

$$\frac{1}{T_1}=B\left\{\frac{\tau_c}{1+\omega_0^2\tau_c^2}+\frac{4\tau_c}{1+4\omega_0^2\tau_c^2}\right\}. \tag{6.5}$$

The interaction term B is generally assumed to arise principally from the proton–proton intramolecular dipolar interaction. Motional frequency components at ω_0 and $2\omega_0$ lead to relaxation. These are given, for the case of isotropic rotation, by the first and second terms respectively within the curly brackets of eqn (6.5). This is the appropriate description for bulk water. It is also assumed to be relevant to the free component in the simple two phase model of tissue water. In this case $\tau_c=\tau_f\sim 10^{-12}$ s, $\omega_0\tau_f<1$ and $(1/T_1)_f$ is relatively small. In the bound phase the idea is that the appropriate correlation time τ_b is not that of a freely-rotating water molecule but of the macromolecule to which it is bound. Thus, $\tau_b\gg\tau_f$ and $(1/T_1)_b\gg(1/T_1)_f$. The assumption of isotropic rotation is then questionable, but at least it is possible to get a feel for a possible relaxation mechanism. The fact that the bound relaxation rate is so much greater than the free, perhaps by a factor of ten or so, is the reason for the proportionally larger spread in T_1 relaxation rates. Thus,

$$\Delta\left(\frac{1}{T_1}\right)\sim\left(\frac{1}{T_1}\right)_b\Delta b, \tag{6.6}$$

and a 1 per cent change in water content may give rise to a 10 per cent change in T_1 (see also Figs. 6.2, and 6.3).

T_1 values have been measured for many mammalian tissues over a wide frequency range. Figure 6.4 was compiled from the work of a number of authors and illustrates the frequency dependence of T_1 for various rat tissues. Notice the strong dispersion, a factor of considerable importance given the wide range of frequencies currently used for human n.m.r. imaging.

It was realized at an early stage that there is comparatively little variation in results obtained from the same organs of different mammalian species (Cottam, Vasek, and Lusted 1972), an experience borne out by later investigators (Ling, Foster, and Hutchison 1980). Particularly extensive T_1-determinations have been made on rabbit tissue by Ling, Foster, and Hutchison (1980) and Mallard, Hutchison, Edelstein, Ling, Foster, and Johnson (1980) who made measurements at both 2.5 and 24 MHz. The authors also report the ratio of T_1 values measured at these two frequencies. If water content is the only factor which determines the differences

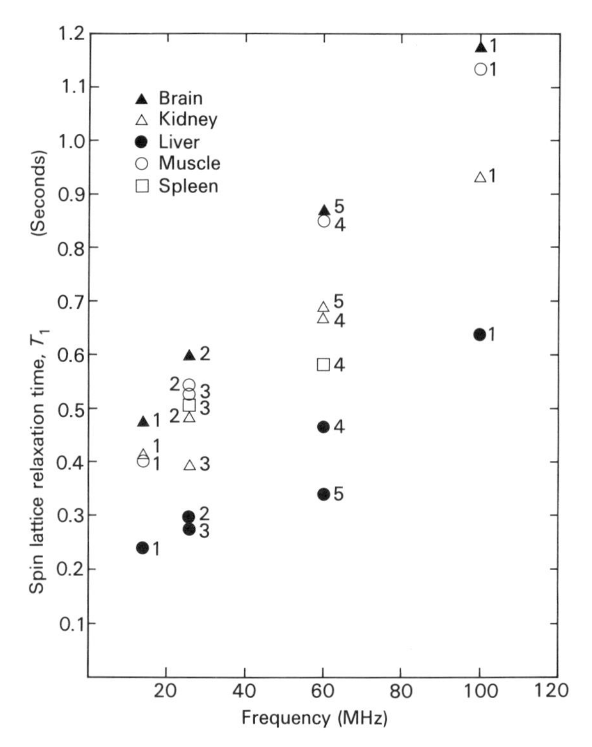

Fig. 6.4. The frequency-dependence of ^{1}H spin–lattice relaxation times for various rat tissues: ▲ brain; △ kidney; ● liver, ○ muscle; □ spleen. Superscripts denote the origin of results: [1] Block and Maxwell (1974); [2] Damadian (1971); [3] Hollis, Saryan, Eggleston, and Morris (1975); [4] Bovée, Huisman, and Smidt (1974); [5] Kiricuta, Jr. and Simplaceanu (1975).

in T_1 relaxation times at a particular frequency, then this ratio should be the same for all tissues. Certainly it is of the same order (T_1(24 MHz)/T_1(2.5 MHz) ~ 2). However, some tissues, notably muscle and nervous tissue, do differ significantly. This is evidence (within the limitations of the two phase model) for differences in $(T_1)_b$. This in turn reflects differences in the nature and extent of local ordering which can become important for T_1 as well as T_2 processes at low frequencies. Very extensive measurements have been made by Damadian and co-workers at 22.5 and 100 MHz on human and other mammalian tissues. Their results fall within the lines, drawn for guidance only, on the rat data plot of Fig. 6.4 and thus lend further support to the assertion that different species do not differ significantly in T_1 properties. Much of Damadian's work has been performed in connection with T_1 measurements in the corresponding tumourous tissues, a subject to which we return in § 6.2.5.

6.2.4. *Spin–spin relaxation*

We turn now to a brief discussion of spin–spin relaxation (see § 2.5.2). The most striking feature of tissue T_2s is their shortness: they are typically 20–100 ms, at least an order of magnitude lower than the corresponding T_1s in the frequency range 4–100 MHz. This suggests that there are additional relaxation mechanisms contributing to the T_2 process.

In order to glean some idea of what is happening we reproduce the expression for $1/T_2$ derived in § 2.5.2. (eqn (2.57)) for dipolar relaxation in the case of isotropic rotation. Thus

$$\frac{1}{T_2}=\frac{\gamma^4\hbar^2 I(I+1)}{5r^6}\left\{3\tau_c+\frac{5\tau_c}{1+\omega_0^2\tau_c^2}+\frac{2\tau_c}{1+4\omega_0^2\tau_c^2}\right\}. \tag{6.7}$$

It must be emphasized that this is an inappropriate model, but at least it gives some basis for qualitative discussion. The new feature of eqn (6.7) compared with the analogous expression for the spin–lattice relaxation rate, eqn (6.5), is the presence of a zero-frequency component proportional to τ_c (see also the discussion of § 2.5.2). In our interpretation of spin–lattice relaxation rates, we were concerned with the correlation time in the bound phase, $\tau_b \sim 10^{-8}$ s, which we suggested might be associated with the tumbling of macromolecules. Motional processes with correlation times $\leqslant 10^{-5}$ s do not contribute significantly to T_1 processes, but in the case of spin–spin relaxation they can be dominant. The relative lack of importance of the ω_0 and $2\omega_0$ components is indicated by the rather weak dependence of T_2 on operating frequency (see, for example, Held, Noack, Pollak, and Melton 1973). This means that images based on T_2 contrast will be qualitatively similar for imaging systems operating at different frequencies; potentially a very attractive feature. T_2-based images can give excellent tissue contrast, and spin echo sequences (usually with long echo delays) are finding favour, particularly for high field systems, where the time-saving relative to inversion recovery sequences is most apparent.

A possible candidate which has been implicated in T_2 processes is τ_e, the time associated with the exchange of a water molecule between bound and free phases. Another possibility is that the motion of the bound water molecules is anisotropic so that some of the dipole–dipole interaction (see § 2.5.1) between the two water protons remains unaveraged and thus provides an efficient spin–spin relaxation mechanism. As was the case with T_1, T_2 increases with water content in an approximately linear fashion (Taylor and Bore 1981) lending further support to the two-phase

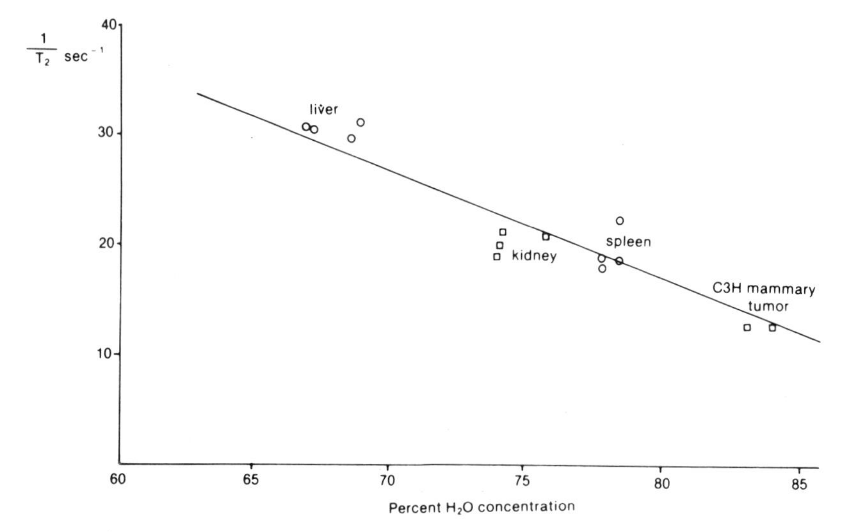

Fig. 6.5. ^1H spin–spin relaxation rate, $1/T_2$, as a function of percentage water content for a number of healthy mouse tissues and tumours at 30 MHz. The solid line is a least-squares best fit to the data. (From Taylor and Bore 1981 Copyright Pergamon Press 1981.)

model. This is illustrated in Fig. 6.5, which shows the 30 MHz ^1H spin–spin relaxation rate as a function of water content for various murine tissues.

A further characteristic of T_2-relaxation is the common occurrence of non-exponentiality. In most instances (Packer 1977), this has been interpreted as a compartmentalization effect, in which case the T_2-relaxation (or free-induction decay, f.i.d.) should be resolvable into two or more exponential components. The two obvious 'compartments' which spring to mind are the intracellular and the extracellular pools. Alternatively, the non-exponentiality could be the consequence of residual dipolar broadening. A detailed understanding of T_2 relaxation has not yet emerged in biological tissues. Many of the attempts to interpret data involve empirical fits which give little feeling for the underlying physical processes. However, what is of importance in the context of this monograph is that the mechanisms of T_1- and T_2-relaxation in tissues are different. Although both would seem to depend largely on the water content, the optimum tissue contrast for a particular clinical condition might be afforded by a pure-water-content image, a T_1-image, a T_2-image, or an image based on some particular combination of these three. Equally, it may require a number of differently-based n.m.r. images to confirm a diagnosis. Some images displaying pure relaxation times (T_1 and T_2) have been published, but this field is only just starting to be explored. Whereas initially there is bound to be some confusion arising from n.m.r. images obtained under different operating conditions (see discussion of § 6.3.1), this great versatility will prove to be one of the great advantages which n.m.r. has over other imaging modalities.

Absolute values of T_2 (rather than T_{2e}) are also of importance in determining the static field homogeneity and the field gradient intensity necessary to achieve the best S/N for a given spatial resolution (see § 2.5.2.1).

Whilst discussing relaxation times, we should mention that the rotating frame relaxation time $T_{1\rho}$ also shows a strong variation between tissue types. Again, Damadian and co-workers have made extensive studies in their efforts to improve methods of tissue discrimination. There has been no detailed discussion of $T_{1\rho}$ in relation to n.m.r. imaging, however.

6.2.5. *Tumour discrimination*

Much of the interest in tissue relaxation times has stemmed from the observation by Damadian (1971) that spin–lattice relaxation times are elevated in tumourous tissues. This was demonstrated using two malignant rat tumours, Novikoff hepatoma and Walker sarcoma, which had T_1 values at 24 MHz of 736 and 826 ms respectively, compared with values in normal tissues extending over the range of 257 to 595 ms. Confirmation that this was a general effect followed rapidly with measurements of T_1 values in animal and human tissues. Table 6.1, taken from the work of Damadian, is typical of the results obtained. Often the human tumour T_1 values turned out to be less markedly elevated than was the case for the rapidly growing laboratory tumours generally used in the animal studies. Nevertheless, with very few exceptions, for example melanoma and possibly cancer of the thyroid (Schara, Šentjurc, Auersperg, and Golouh 1974; de Certaines, Herry, Lancien, Benoist, Bernard, and Le Clech 1982), the tumour T_1 values were longer than those for the corresponding normal tissues. It was appreciated that, since T_1 values for tumourous states of some tissues overlapped with normal states of others, problems of interpretation could arise in the case of metastases.

Soon after Damadian's initial discovery, Hazlewood, Chang, Medina, Cleveland, and Nichols (1972) showed that it was possible to detect cancer at an early stage of development. They were able to distinguish normal, preneoplastic and tumourous murine glands not only on the basis of elevated T_1 values but also by way of their increased T_2 values and diffusion constants. Another important observation by Frey, Knispel, Kruuv, Sharp, Thompson, and Pintar (1972) demonstrated the presence of a tumour systemic effect; they found that non-tumourous tissues in tumourous mice had elevated T_1 values. This indicated that such tissues

Table 6.1

T_1 relaxation times at 100 MHz in normal and malignant human tissues

Tissue	$T_{1\ \text{tumour}}$	$T_{1\ \text{normal}}$	*Probability that difference in means are not significant*
Breast	1.080±0.08 (13)	0.367±0.079 (5)	0.52×10^{-4}
Skin	1.047±0.108 (4)	0.616±0.019 (9)	0.55×10^{-4}
Muscle:			
Malignant	1.413±0.082 (7)	1.023±0.029 (17)	0.50×10^{-5}
Benign	1.307±0.1535(2)		
Esophagus	1.04 (1)	0.804±0.108 (5)	
Stomach	1.238±0.109 (3)	0.765±0.075 (8)	0.40×10^{-2}
Intestinal tract:			
Small bowel	1.122±0.04 (15)	0.641±0.080 (8)[2]	0.27×10^{-5}
		0.641±0.043 (12)[3]	
Liver	0.832±0.012 (2)	0.570±0.029 (14)	
Spleen	1.113±0.006 (2)	0.701±0.045 (17)	
Lung	1.110±0.057 (12)	0.788±0.063 (5)	0.25×10^{-2}
Lymphatic	1.004±0.056 (14)	0.720±0.076 (6)	0.52×10^{-2}
Bone	1.027±0.152 (6)	0.554±0.027 (10)	0.74×10^{-2}
Bladder	1.241±0.165 (3)	0.891±0.061 (4)	0.36×10^{-1}
Thyroid	1.072 (1)	0.882±0.045 (7)	
Nerve	1.204 (1)	0.557±0.158 (2)	
Adipose	2.047 (1)	0.279±0.008 (5)	
Ovary	1.282±0.118 (2)	0.989±0.047 (5)	
Uterus:			
Malignant	1.393±0.176 (2)	0.924±0.038 (4)	
Benign	0.973 (1)		
Cervix	1.101 (1)	0.827±0.026 (4)	
Testes	1.223 (1)	1.200±0.048 (4)	
Prostate	1.110 (1)	0.803±0.014 (2)	
Adrenal	0.683 (1)	0.608±0.020 (5)	
Peritoneum	1.529 (1)	0.476 (1)	
Malignant melanomas	0.724±0.147 (6)		
Tongue	1.288 (1)		
Pericardial layer (mesothelioma)	0.758 (1)		
Kidney		0.862±0.033 (13)	
Brain		0.998±0.016 (8)	
Pancreas		0.605±0.036 (10)	
Heart		0.906±0.046 (9)	

Probability values are reported for series with sample size ⩾3. Errors reported are standard error of the mean (SEM). Number of cases analyzed are indicated in parenthesis. [From R. Damadian *et al.*, *Proc. natn. Acad. Sci. U.S.A.* **71,** 1471 (1974).]

could not be used to establish normal T_1 values as had previously been assumed.

Workers were not slow to appreciate the implications of their results for cancer diagnosis. However, enthusiasm was somewhat dampened when it became clear that elevated T_1 values are not cancer-specific. Increased T_1 values were observed in fetal tissue and regenerating liver, for example (Inch *et al.* 1974; de Certaines, Moulinoux, Benoist, Bernard, and Rivet 1982). Ling, Foster, and Mallard (1979) reported raised T_1 values for tissue adjacent both to malignant tumours and to normal foreign tissue implanted subcutaneously in a rat thigh. In each case, the increase was apparent within 24 hours. However, the T_1 values in the muscle adjacent to the normal foreign tissue subsided over a period of days, eventually returning to lie within the normal range, whereas, in the case of tissue adjacent to the malignant tumour, the T_1 values remained elevated. On the basis of the two-phase model, these results are not unexpected, and can clearly be interpreted in terms of an 'amplification' of differences in water content (see § 6.2.3). Whilst this is undoubtedly the major effect, it would be surprising if changes arising from differences in macromolecular organization were not also of importance. Some evidence for this is provided by the observation of cyclical changes in T_1 for synchronized HELA cell clones (Beall and Hazlewood 1976).

Much effort has been directed towards an improvement in the discrimination between cancerous and normal tissue. Other n.m.r. parameters (T_2, $T_{1\rho}$, and D) have been measured and combined with T_1 values to give a 'malignancy index' which is a more successful predictor than any one parameter taken on its own (Koutcher, Goldsmith, and Damadian 1978).

It is quite clear that T_1 values are increased in tumourous tissues, often to a marked degree with increases of 50–100 per cent being not uncommon. Again, the explanation remains controversial, but the importance, at least initially, lies in the existence of the effect rather than an understanding of its origins. With this very large contrast available, n.m.r. imaging ought to find major application in cancer diagnosis. This is strongly confirmed by the results of the first clinical trials (see below).

Arguments against the use of T_1 measurements for cancer diagnosis, based on the lack of specificity, may be justified in the case of a biopsy specimen but are much less tenable in an imaging context where the tumour is observed *in situ* against a background of normal tissue. Other factors, such as the shape, size and location of the abnormality can then be taken into consideration. In addition, as we have seen, it is possible to obtain a series of images, based on different n.m.r. parameters, which can be displayed separately or else combined to yield an image which offers better discrimination in the manner of Damadian's malignancy index. In fact, far from being a failure of the technique, this lack of specificity means that any condition which affects the state of tissue water, and there are many which fall into this category, is a potential candidate for diagnosis by n.m.r. imaging.

As a final cautionary note, it must be pointed out that the majority of the tissue T_1s have been determined from biopsy samples. Mathur-de Vré, Grimée, and Rosa (1983) have published details of an experimental protocol which ensures repeatability for *in vitro* measurements. However, although the indications are that these *in vitro* values accurately reflect the *in vivo* ones, there is a need to exercise care in this respect.

6.3. N.m.r. imaging: general considerations

6.3.1. *Dependence of image on* ρ, T_1 *and* T_2

The aim of any imaging technique is to measure some physical property, such as ultrasonic absorption, X-ray attenuation or, in the case of n.m.r., mobile proton density, as a function of position and to display this information, suitably encoded, as an image. There is no guarantee that such images will resemble what would be seen by exposure and visual inspection of the object under investigation. Indeed, in the case of medical imaging, it is very much to be hoped that the various imaging modalities will exhibit different forms of tissue contrast. N.m.r. is a somewhat unusual imaging technique in that the physical property employed is, to a certain extent, a matter for choice. Thus, mobile proton density, spin–lattice relaxation time, or spin–spin relaxation time, can be made the basis for an n.m.r. image. Usually, however, all three parameters will be involved to a greater or lesser degree depending on the operating conditions and the particular n.m.r. technique used.

Other parameters may be involved either directly or indirectly. It is clear, for example, that the presence of flow will have an important bearing on the image—this is discussed in § 6.3.6. It is less clear, however, why temperature might be influential. In fact, the relaxation times (particularly T_1) are temperature-dependent, and it has been suggested (Parker, Smith, Sheldon, Crooks, and Fussell 1983) that this dependence may be useful in monitoring the hyperthermic treatment of tumours. Although temperature effects have been observed in phantoms, the resolution obtained (~2 °C for a five-minute scan) was insufficient to justify an immediate application in non-invasive thermometry.

In Chapters 3 and 4 several imaging methods were described which differ in the manner and efficiency with which they achieve spatial encoding of the n.m.r. signal. There are also pulse sequences available which are designed to elicit different types of tissue contrast: for the most part they are the analogues of the high-resolution methods for measuring relaxation times (see § 2.5), namely saturation recovery (SR) and inversion recovery (IR) for T_1, and spin-echo (SE) for T_2. They can be used in combination with most of the proposed imaging methods. The precise manner in which the schemes for spatial and contrast encoding dovetail together depends on the particular method chosen—some examples have been given in Chapter 4 for the cases of projection reconstruction and Fourier imaging.

A method of data collection which was once widely advocated for n.m.r. imaging purposes is steady state free precession (s.f.p.). Its attraction lies in its ability to maintain a sizeable fraction of the total available magnetization in the plane of observation. It found application initially with the sensitive point and sensitive line techniques developed by Hinshaw (see Chapter 3), and more recently with projection-reconstruction methods (see § 4.1). In the limit of 90° pulses and short interpulse intervals the dependence of the n.m.r. signal on the spin density ρ, T_1 and T_2 is given by

$$S \propto \rho(T_1/T_2). \tag{6.8}$$

Latterly the s.f.p. method has fallen out of favour; a principal reason being this dependence on the ratio T_1/T_2. As indicated in § 6.2, water content is the prime factor determining relaxation times in different tissues. Since T_1 and T_2 depend on the water content in similar fashion (compare Figs 6.3 and 6.5), the use of a ratio technique will eliminate much of the available contrast. This argument applies to other similar techniques, for example the driven equilibrium Fourier transform method, or d.e.f.t. A second reason for abandoning such techniques arises from the relatively high duty cycle of the r.f. pulse train. Although this is not a problem at low frequency, significant r.f. power absorption can occur at the high end of the frequency range used for n.m.r. imaging (see § 6.8.3).

Perhaps the simplest pulse sequence to be used is the saturation recovery (SR) or repeated f.i.d. method. Essentially, this consists of 90° pulses, selective or non-selective, which are regularly repeated with an interpulse interval TR which is not sufficient to allow complete recovery of the longitudinal magnetization (Fig. 6.6(a)). After a few cycles a steady state is reached in which the signal (initial amplitude of the f.i.d.) is given by

$$S \propto \rho[1 - \exp(-\mathrm{TR}/T_1(\omega))] \tag{6.9}$$

where the dependence on frequency of T_1 is indicated. Thus the signal is proportional to the spin density but the T_1 contrast can be changed by varying the repetition rate (TR value). This is illustrated in Fig. 6.7(a) which shows the signal as a functon of TR for brain tissue and cerebrospi-

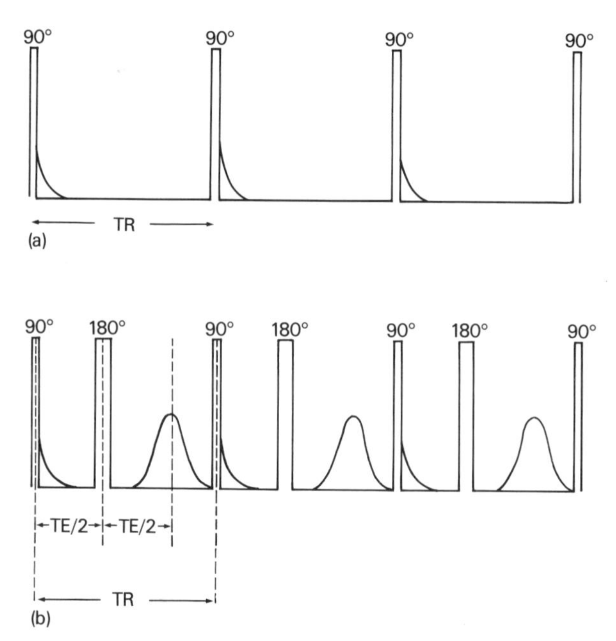

Fig. 6.6. Theoretical (a) and practical (b) saturation recovery (SR) sequences, in which 90° pulses are repeated at intervals TR. In (b) the addition of a 180° pulse causes echo-formation at a time TE after the preceding 90° pulse. N.m.r. signals are shown schematically as solid lines.

nal fluid (c.s.f.). Curves have been calculated using the following parameters: white matter: % $H_2O = 80$, $T_1 = 285$ ms, $T_2 = 90$ ms; grey matter: % $H_2O = 97$, $T_1 = 525$ ms, $T_2 = 100$ ms; c.s.f.; % $H_2O = 100$, $T_1 = 1500$ ms, $T_2 = 600$ ms. These values are intended to simulate the situation at 6.5 MHz. They have also been used in the calculations for Figs. 6.9 and 6.11. If a TR value is chosen which is long compared with the T_1 values of brain tissue and c.s.f., the signal intensity from the two regions will be similar, since they are nearly isodense (same ρ value). A head scan obtained under such conditions would therefore show little tissue contrast (Fig. 6.7(b)). If, however, TR is reduced ($\lesssim T_1$ of brain and c.s.f.) then substantial differences in signal intensity arise (Fig. 6.7(c)) and a high-contrast head image can be generated (Fig. 6.7(d)). Edelstein, Bottomley, Hart, and Smith (1983) have addressed themselves to the correct choice of TR value, taking as their criterion the differential signal-(i.e. contrast)-to-noise ratio obtainable per unit time. They conclude that TR should be chosen to be equal to the shortest T_1 value which is of interest; at 5.1 MHz they suggest that a value in the range 0.1 to 0.15 s would be appropriate for most regions of the body. Since these short intervals are

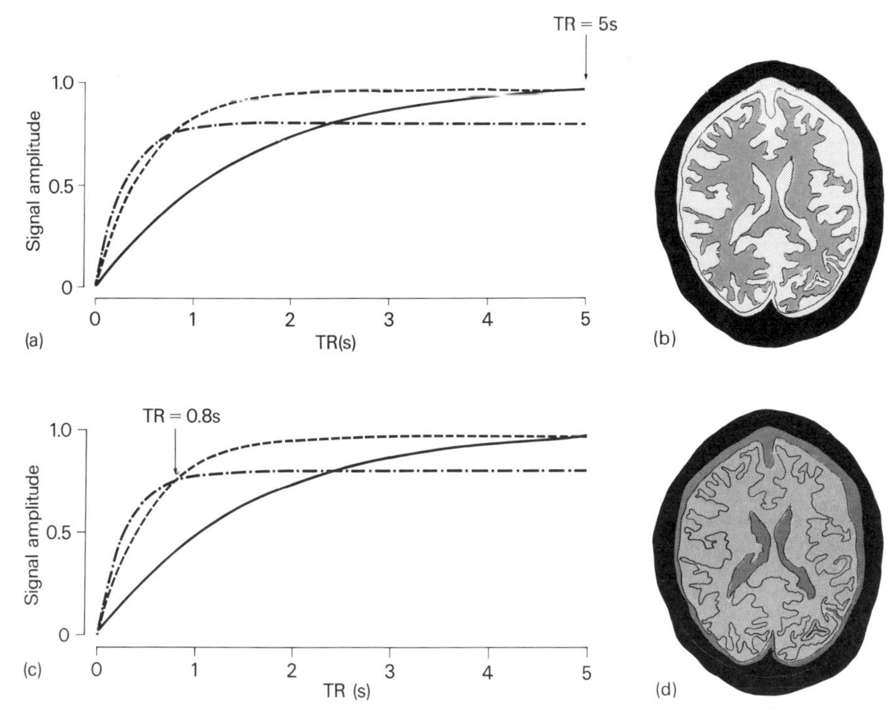

Fig. 6.7. (a), (c) SR signal amplitudes for brain tissue (white matter —·—; grey matter - - -) and c.s.f. (——) as a function of TR. Signals corresponding to TR > T_1 (brain, c.s.f.), (a) and TR ~ T_1 (brain, c.s.f.) (c) are indicated. (b), (d) Schematic SR n.m.r. head scans corresponding to TR values indicated by arrows in (a) and (c) respectively.

comparable with the spin–spin relaxation time it is necessary to prevent coherence build-up by, for example, introducing gradient 'spoiler' pulses. Excellent contrast can be achieved by this method. As well as producing high-quality visually attractive images much interest has been expressed in the possibility of using absolute T_1 images to make the procedure quantitative, allowing, for example, intercomparison between scans obtained on different machines. Laudable though this objective is, it is fraught with difficulty. There is the problem of what is actually meant by T_1—it is certainly not an absolute parameter since, as we have seen, it has a strong frequency-dependence. Although the T_1-relaxation process in tissues is usually exponential and can therefore be described by a single relaxation time, if there is heterogeneity to the degree that one pixel spans more than one tissue type, multicomponent relaxation will be observed. T_1-determinations based on insufficient measurements (often only two are used) will not detect this and will give values which are dependent on the interpulse interval chosen. Even in the case of a single T_1 the technical problems associated with measuring it still have to be faced. First, it is necessary to use a model which is appropriate to the sequence actually in use. The expression of eqn (6.9) will generally not be relevant for practical sequences, which almost universally employ a spin-echo to separate acquisition from gradient and r.f. pulses (this also has the effect of reducing the effects of static field inhomogeneity). A typical SR sequence is illustrated in Fig. 6.6(b). The corresponding expression for the signal is:

$$S \propto \rho \exp(-\mathrm{TE}/T_2)[1-\exp(-\mathrm{TR}/T_1)(2\exp(\mathrm{TE}/2T_1)-1)], \quad (6.10)$$

where TE is the (total) time between 90° pulse and echo. Note that for $\mathrm{TE} \ll T_1$, T_2 this reduces to eqn (6.9). (In practice, TE is normally made fairly short, typically ~10 ms, so this condition is reasonably fulfilled). The extra terms arise from a T_2 decay $(\exp(-\mathrm{TE}/T_2))$ and a shift in the origin of the T_1 recovery $(2\exp(\mathrm{TE}/2T_1)-1)$. The recovery process remains exponential with a time constant T_1 so that the addition of a spin-echo does not significantly change things in this case. However, like its high-resolution counterpart, the SR method relies on the precise setting of the 90° pulse condition. Unfortunately, it is not possible to achieve this over the whole field of view, due to the r.f. inhomogeneity of the transmitter coil (see § 5.3.4). Additionally, if the sequence is used with a planar imaging method (i.e. selective pulses are employed) then the use of any slice profile other than square, either by design or as a result of imperfect selection (see § 3.3.3), will lead to errors on this count. As TR is shortened, contributions from the wings of the slice profile become progressively more important, leading to an effective increase in slice thickness—clearly a highly undesirable effect.

Another pulse sequence widely used for n.m.r. imaging studies is the inversion recovery or IR sequence, the basic version of which is illustrated in Fig. 6.8(a). The signal intensity is given in this case by:

$$S \propto \rho[1-2\exp(-\mathrm{TI}/T_1(\omega))]. \quad (6.11)$$

Figure 6.9 analagous to Fig. 6.7 for the SR method, illustrates the effect of different choices of TI on tissue contrast. There is, however, a new feature arising from the initial sign change of the magnetization, namely that for very short TI values one can get an apparent reversal of tissue contrast (Fig. 6.9(b)). This arises because, particularly in the case of Fourier methods, one often works with the power spectrum, so that the sign of the signal is lost. However, if steps are taken to preserve it, then no confusion can arise. The region around the zero crossing given by $\mathrm{TI} = T_1 \log_e 2$ is, of course, to be avoided. Normally, optimal contrast is achieved for somewhat longer TI values, 0.4 s being a

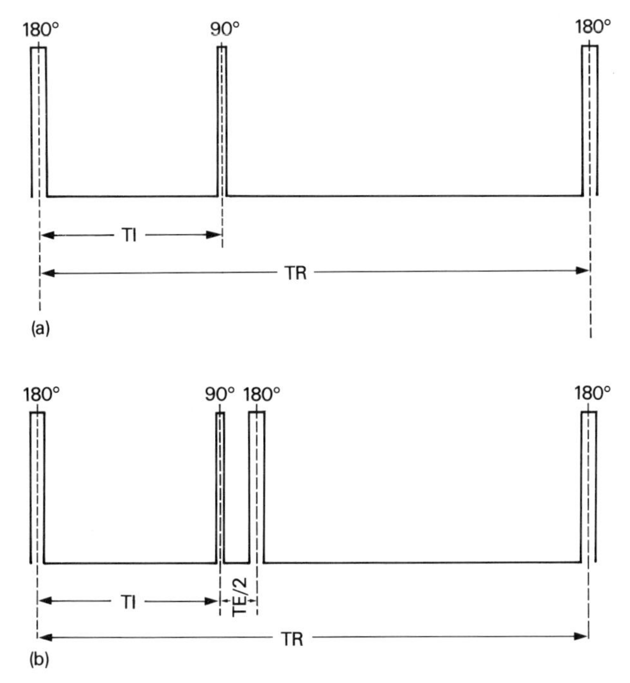

Fig. 6.8. Theoretical (a) and practical (b) inversion recovery (IR) sequences in which a 180° pulse inverts the z-magnetization, which is then allowed to recover towards its equilibrium value. The recovery process is inspected with a 90° pulse applied at a time TI following the inversion. In (b) an additional 180° pulse is used to create a spin-echo as in Fig. 6.6(b).

typical choice at 6 MHz, rather longer at higher frequencies. The IR sequence is capable of yielding excellent contrast, clearly delineating white from grey matter in brain, for example. However, the cycle time TR is necessarily long compared to the SR method. It is possible to use IR to determine quantitative T_1 values. In high-resolution n.m.r. it is the method of choice for this purpose, so the sequence clearly merits careful consideration. Again, we need to take account of the actual sequence used; Figure 6.8(b) is a typical example. Ideally, TR would be chosen to be about five times the longest T_1 present, to allow complete recovery between repetitions. However, this leads to unacceptable prolongation of the imaging time, particularly at high frequency, and it is necessary to come to some compromise. In this case the expression for the signal is:

$$S \propto \rho \exp(-\mathrm{TE}/T_2)[1-(1+\exp(-\mathrm{TE}/T_2))[2\exp(\mathrm{TE}/2T_1)-1]\exp(-\mathrm{TI}/T_1) + \exp(-\mathrm{TE}/T_2)[2\exp(\mathrm{TE}/2T_1)-1]\exp(-\mathrm{TR}/T_1)]. \quad (6.12)$$

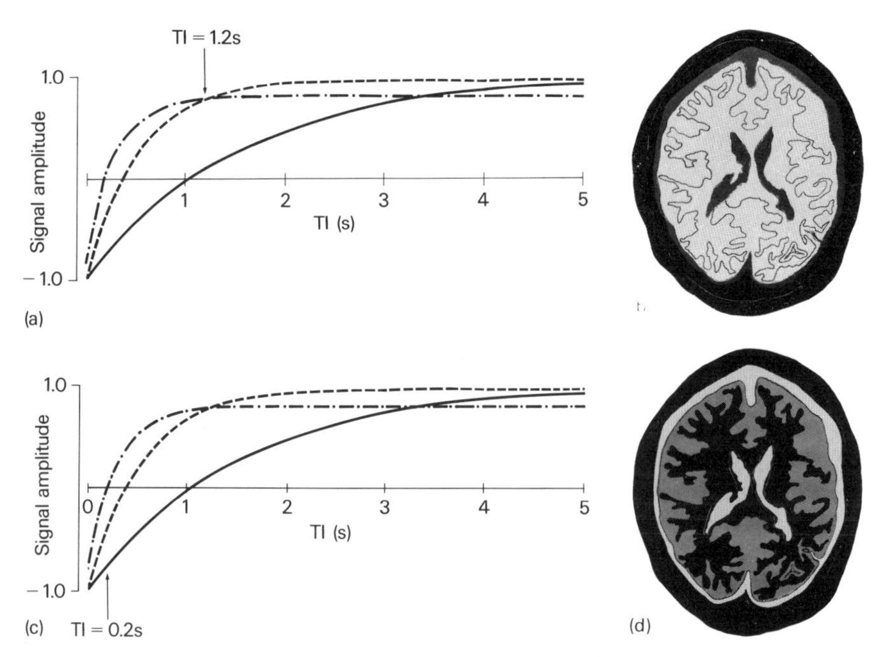

Fig. 6.9. (a), (c) IR signal amplitudes for brain tissue (white matter —·—; grey matter - - -) and c.s.f. (——) as a function of TI in the limit TR → ∞, TE → 0. Signals corresponding to TI ~ T_1 (brain, c.s.f.) (a) and TI < T_1 (brain, c.s.f.) (c) are indicated. (b), (d) Schematic IR n.m.r. head scans corresponding to TI values indicated by arrows in (a) and (c) respectively. Note reversal of contrast.

As TE → 0 this reduces to:

$$S \propto \rho[1 - \exp(-\mathrm{TI}/T_1) + \exp(-\mathrm{TR}/T_1)], \qquad (6.12a)$$

and if in addition TR → ∞ it further simplifies to give Equation (6.11). If TE is not short, an unwanted element of T_2-contrast can enter which in some circumstances can cancel the T_1-contrast that the sequence is designed to elicit. In spite of the complexity, the process is still single exponential, with a time constant T_1. However, if an attempt is made to determine T_1 from a single IR measurement in combination with a pure spin density image, using eqn (6.11), gross inaccuracies can result. It is possible to use a T_1 look-up table calculated from the full expression, and this method has been applied with some success (Pykett, Rosen, Buonanno, and Brady 1983). However, a multipoint determination is much to be preferred, if only because it can confirm the single exponentiality of the relaxation. Comments regarding pulse imperfections are also applicable to the IR method.

The spin-echo (SE) pulse sequence (Fig. 6.10(a)) generates contrast on

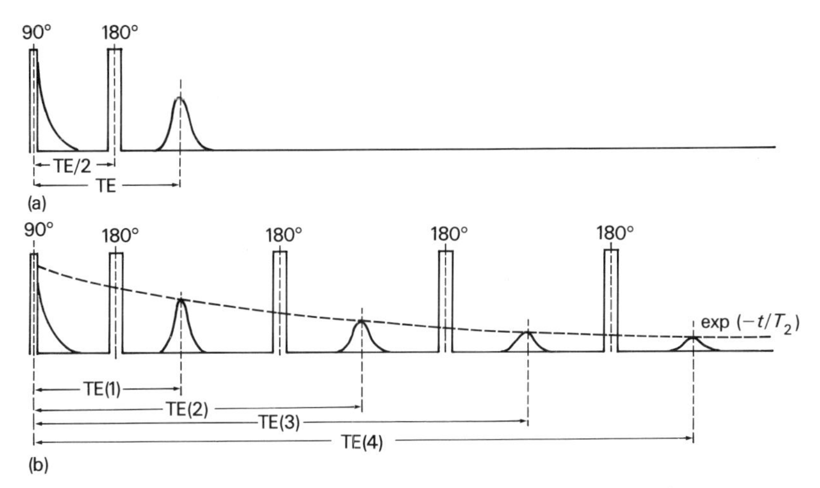

Fig. 6.10. (a) Spin-echo (SE) sequence in which a 180° pulse following a 90° pulse causes a spin-echo at time TE. (b) Multiple spin-echo (MSE) sequence in which repeated 180° pulses successively recall signal to give echoes at times TE(1) = TE; TE(2) = 2TE, TE(3) = 3TE and TE(4) = 4TE. The n.m.r. signal is shown as a full line and the T_2 decay envelope $\exp(-t/T_2)$ in (b) as a broken line.

the basis of the spin–spin relaxation time T_2. To a first approximation the signal is given by

$$S \propto \rho \exp(-\mathrm{TE}/T_2). \tag{6.13}$$

Figure 6.11 shows the T_2-decay of brain tissue and c.s.f., and illustrates the contrast available for different choices of echo-delay TE. The SE sequence has been particularly successful in the delineation of multiple sclerosis lesions (see § 6.5.2). It can also be used to determine quantitative T_2 values, either from scans with different TE values or, more conveniently, by using a multiple spin-echo (MSE) sequence in which the signal is refocused by 180° pulses to give a train of echoes in the manner of Carr and Purcell (Fig. 6.10(b)) (see also § 2.5.2.2). A separate image can then be derived from each echo. This method was first introduced by the team at the University of California, San Francisco, who generated two echoes. However, the signal can be recalled many more times within a total time about three times the natural T_2. Four echoes are commonly employed (see Fig. 6.29), but up to thirty-two separate MSE images have been reported (see front cover of Bruker Medical Report 83/1). Of course, successive images become noisier as the signal progressively decays: T_2 calculations should therefore weight the data accordingly. Typically, TE might have a value of 20 ms, giving times for four echoes (see Fig. 6.10(b)) of TE(1) = 20 ms, TE(2) = 40 ms, TE(3) = 60 ms, and

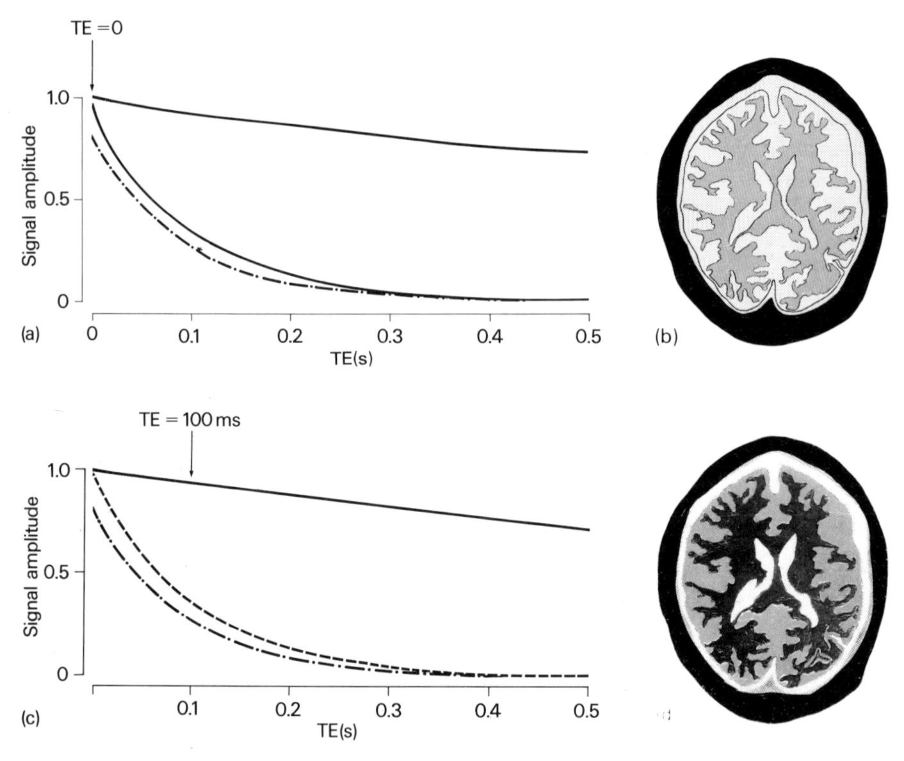

Fig. 6.11. SE signal amplitudes for brain tissue (white matter —·—; grey matter ---) and c.s.f. (——) as a function of TE. Signals corresponding to TE $< T_2$ (brain, c.s.f.) (a), and TE $\sim T_2$ (brain, c.s.f.) (c), are indicated. (b), (d) Schematic head-scans illustrating contrast arising from TE values indicated by arrows in (a) and (c) respectively.

TE(4) = 80 ms. Again, the sequence is susceptible to pulse errors and careful consideration needs to be given to possible cumulative effects of a train of imperfect 180° pulses. Note, too, that the T_2 process is generally not exponential in biological tissues (see § 6.2.4), particularly in the short time-limit. Although the concept of a single T_2 is valuable, it should be treated with caution. The MSE method is nevertheless growing in popularity, since it generates high-contrast images in an efficient manner.

The basic sequences for generating tissue contrast have been discussed above. However, it is quite possible to use them in combination or derive more sophisticated variants. The appearance of the image, as well as depending on the pulse sequence, will also depend on the manner in which the grey scale is selected. Often this is done in rather arbitrary fashion by altering the window controls (see § 5.6) to product a visually-attractive result. However, this does not necessarily guarantee optimum contrast for the particular clinical condition for which one is screening. A great deal of work is required in this area before routine operating procedures are established (see, for example, Droege, Wiener, Rzeszotarski, Holland, and Young 1983). Until n.m.r. screening becomes routine it will demand considerably more expertise from operator and interpreting radiologist than has its X-ray CT counterpart.

6.3.2. *Choice of imaging plane and resolution*

In general, whether by choice or physical constraint, an image of the entire object is not normally required. It is therefore necessary to come to a decision over which particular region to select and how best to divide it into a necessarily finite number of image points. The usual choice is a planar slice, which can be selected with the aid of a simple magnetic field gradient oriented normal to the slice. Figure 6.12(a) illustrates this for an axial geometry. Definition within the slice is achieved through the use of two further mutually-orthogonal field gradients. Herein lies one of the great advantages of n.m.r. imaging: it is possible to interchange the roles of these field gradients in order to define sagittal (Fig. 6.12(b)) and coronal (Fig. 6.12(c)) slices. By using the gradients in combination, intermediate slice orientations can be achieved. Note that, provided the desired plane lies within the region of magnet uniformity, no patient movement is required, and switching between slice orientations is entirely electronic. This versatility is extremely useful when attempting to establish the full extent of any pathological changes which may be present.

An important consideration is the thickness of the imaging plane. Whereas there is no theoretical obstacle to making the plane as thin as we like through the simple expedient of increasing the field gradient amplitude, the amount of material contributing to the image decreases in proportion, and the S/N per unit time suffers (see § 5.7). On the other hand, there is no point in increasing the plane thickness too much beyond the dimensions of the smallest structure one hopes to see, since partial volume effects then become dominant (see Fig. 6.13.). Typically, slices are chosen to have a thickness of 5–15 mm, although slices as thin as 3 mm have been used for n.m.r. head scans.

This leads on to a discussion of the manner in which the image slice is divided into individual image points. Normally, this is done in simple fashion by imposing a regular matrix on the slice, thereby dividing it into a set of equal volume elements or 'voxels'. We measure the physical parameter, spin density, T_1, etc., corresponding to each volume element, convert it to a grey level or colour and display it on a two-dimensional matrix in which each two-dimensional element or 'pixel' has uniform intensity. Of course, in the final image presentation, the data is often

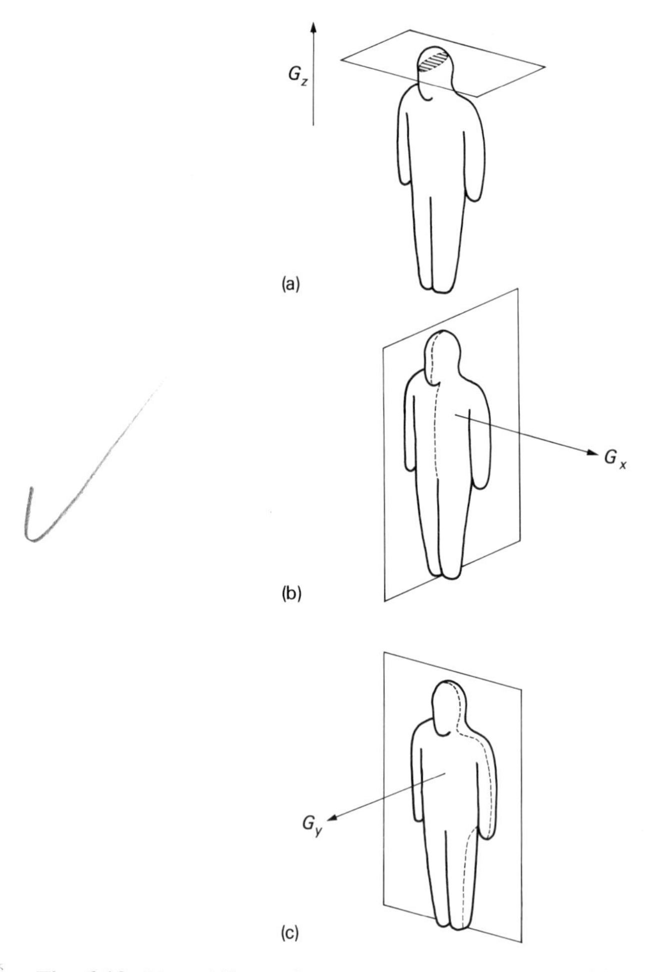

Fig. 6.12. Use of linear field gradient to select axial (a), sagittal (b), and coronal (c) slices.

interpolated and smoothed to the point at which the discreteness of the individual pixels is no longer visible. The pixels are normally chosen to be square, i.e. the in-plane spatial resolution is the same in both dimensions. However, this is certainly not always the case; in a line-scanning method, for example, the width of the image lines is likely to exceed the separation of image points along the line. At first sight it might seem appropriate to make the in-plane spatial resolution equal to the slice thickness. However, the majority of n.m.r. studies, especially whole-body scans, are axial and there is often much greater spatial correlation between adjacent

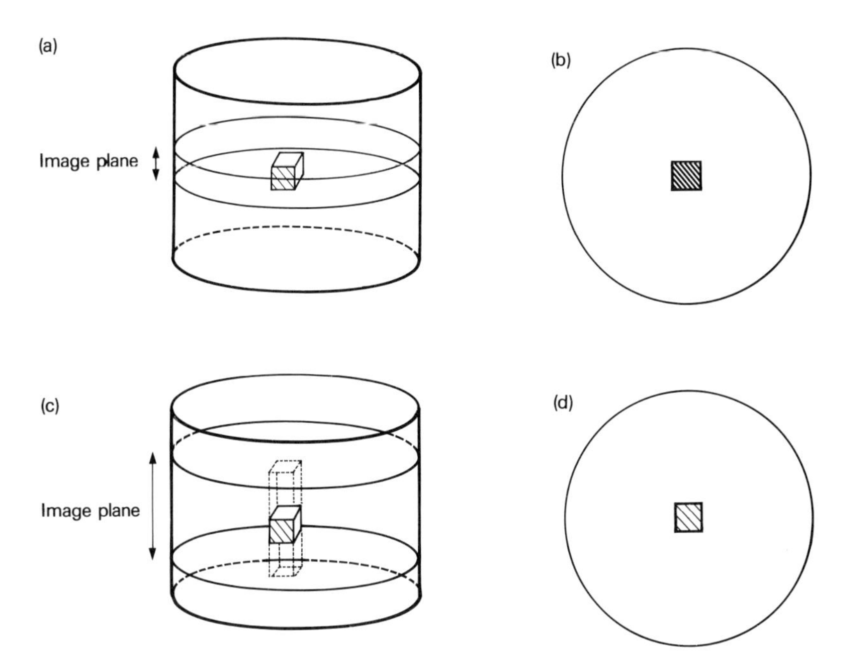

Fig. 6.13. Demonstration of partial volume effects. (a) Object consisting of a single water-filled cube whose side is equal to the image plane thickness; (b) image of (a) showing 100 per cent contrast; (c) same object as in (a) but image plane much thicker; (d) the spin density is assumed to be uniformly distributed over the plane thickness and image contrast is thus greatly reduced.

slices than there is within a slice. Many of the important vessels—for example, the aorta and vena cava in the trunk, carotid arteries in the head—run, for most of their course, along the body axis, and many of the organs and their internal structures are similarly disposed. This is, of course, true *a fortiori* for studies confined to limbs. Thus, for head scans, in-plane resolution is often chosen to be ~1–4 mm (c.f. typical slice thickness of 5–10 mm). In most cases, the total number of pixels across the image field is between 32 and 512, a value of 128 or 256 being typical. (It is generally a power of two, as required by the fast Fourier transforms used in data processing.) In the case of whole-body sections this number is not usually increased in proportion to the size of the imaging field, and there is consequently a degradation of spatial resolution relative to head scanning. This increase in pixel size (slice thickness is also often increased) helps to compensate for the loss in sensitivity incurred by the increased receiver coil dimensions necessary to accommodate the body cross-section. A large number of pixels would, in any event, impose more severe demands on data processing and collection; an increased number of projections in the case of the projection reconstruction method, for example. However, this in itself should not be seen as too much of a problem since, for present generation X-ray CT scanners, 512- or even 1024-point displays are the norm. The important consideration is the relationship between voxel size and image S/N per unit time. This was discussed in § 5.7, where it was shown that imaging time depends on the inverse sixth power of the linear dimension. Thus, each twofold improvement in spatial resolution costs a factor of 64 in imaging time if the S/N is to be maintained.

6.3.3. Imaging times

Although there is no theoretical bound to the resolution achievable by n.m.r. imaging, the inherently low sensitivity of the method (see §§ 2.3.2 and 5.7) imposes a time limitation.

It is possible to trade off S/N against imaging time to some extent. However, all of the current techniques, with the exception of echo-planar imaging and the related recalled-echo scheme (see § 4.2.4), have associated with them minimum performance times which result from the need to record a number of f.i.d.s in, for example, different gradient orientations, before image reconstruction can take place. This number increases in direct proportion to the total number of pixels required. With a typical cycle time TR of 1 s a 128×128 image will take about two minutes to produce by the projection reconstruction or Fourier imaging method operating in the 'minimum performance time' limit (i.e. with no signal averaging). This normally leads to acceptable S/N for head slices of typical thickness (5–10 mm). If an inversion recovery sequence is used to accentuate spin–lattice relaxation contrast, then longer delays are necessary and imaging times are somewhat increased. It should be noted at this point that the images used to demonstrate potential clinical applications in the subsequent sections, derive from a number of different sources and have been obtained by a variety of different techniques. Imaging times have been quoted where they are available and it will be seen that for head sections, they span a fairly limited range of from 30 s to 10 min, a figure of two minutes being 'typical'.

The minimum performance time problem is most apparent in the case of three-dimensional projection reconstruction or Fourier imaging (see below) where the above figures must be increased by the number of pixels in the third dimension, leading to imaging times which are several tens of minutes.

This brings us to another point, namely what is the true resolution of an imaging system? In the case of a static object, assuming the field gradients satisfy the criteria laid down in § 2.5.2.1, one could reasonably claim that it was equal to the true pixel dimension (i.e. before any interpolation is carried out). In a living subject, however, there are motional processes which are rapid in comparison with the imaging times and a given point in a particular organ may well undergo a complex motion taking it through several pixels on the (fixed) imaging matrix. The problem is especially acute in the case of whole-body sections where respiratory, cardiac, and peristaltic motions are all present. Even the brain undergoes regular pulsatile motion, although this is more restricted and is consequently rather less of a problem. A number of techniques, including e.c.g. and respiratory gating are available to alleviate this problem. Nevertheless, for body scans, some degradation of resolution will normally be apparent. It is, however, possibly to effectively 'freeze out' the motion by imaging on a much faster time scale (typically 30 ms for a complete image) using the echo-planar technique. See § 4.3 for a description of this method.

6.3.4. Three-dimensional n.m.r. imaging

Although, in the majority of n.m.r. studies, slice section has been employed, the full three-dimensional experiment has been developed and offers a number of advantages, notably, that one has a complete data set for a given volume, the human head, for example. Three-dimensional studies have been carried out using both projection-reconstruction and Fourier-based methods (see Chapter 4). The resolution may be isotropic, or it can be reduced in one dimension to give the effect of a series of contiguous slices. (The imaging time is thereby correspondingly reduced.) This is more efficient than a series of single-slice studies since all spins are simultaneously excited. It is also preferable to the use of multislice studies (see § 4.2.4) in which the slice profile is generally not sufficiently 'clean' to allow excitation of contiguous slices.

Of course, if time permits, the retention of isotropic resolution has many attractions. Careful positioning of the subject is then unnecessary and problems of registration are avoided, thus facilitating comparison with images obtained by other techniques. Image planes at any level or orientation can be selected for display at leisure without the need for patient recall and with obvious benefit to the diagnostician. With so much information available it is clear that there is considerable potential in the development of three-dimensional display systems, and it is to be anticipated that considerable effort will be expended in this area.

Against these undoubted advantages must be weighed the increased time necessary for acquisition of image data, typically forty minutes for a human head at 3 mm resolution (in all three dimensions). Image processing time is also significantly lengthened although this of less importance.

6.3.5. *Use of contrast agents*

Contrast-enhancing agents have been of great benefit to diagnostic radiology, and there is little doubt that they will ultimately play an important role in n.m.r. imaging, too. Applications include differentiation of 'isomagnetic' tissues and the checking of organ function: for example, a contrast agent which is excreted in urine can be used to demonstrate urological anatomy and renal function. Similarly, an intravenous (i.v.) agent may be useful in showing vascular anatomy, tissue perfusion, and brain pathology at the site of leakage across the blood brain barrier. Non-functioning foci can be distinguished through their failure to show contrast enhancement. It should also be possible to demonstrate biochemical function by imaging natural metabolites or suitably labelled metabolic analogues—see the discussion of chemical-shift imaging in § 6.7.

It is possible to envisage a number of different types of n.m.r. contrast agent. The simplest would be the local increase of mobile spin density through the administration of relatively large quantities of fluid. An example of this type of experiment is the use of mineral oil to highlight the stomach and intestines in a series of studies on the guinea pig abdomen (Newhouse, Pykett, Brady, Burt, Goldman, Buonanno, Kistler, and Pohost 1982). This is really the n.m.r. analogue of radio-opaque dye methods, such as the barium meal, designed to raise the local electron density and consequently the X-ray absorption. It is also possible to envisage the reverse experiment in which existing signal-producing spins are replaced with those of another nuclear species, for example by using heavy water. (Deuterons have a resonance frequency which is only about 15 per cent that of protons and are therefore not excited in a proton imaging instrument.)

Methods which change the water content or balance in tissues will also be of interest. Hricak, Crooks, Sheldon, and Kaufman (1983*a*) have demonstrated pronounced effects on the images of human kidneys following a 24 h period of dehydration. Beall (1982) has used the diuretic furosemide (5 mg/animal) to remove water from normal tissue in murine mammary glands, highlighting tumours which would otherwise not be clearly delineated, due to the signal from the 'halo' of oedematous tissue which often surrounds diseased areas. The oestrogen analogue clomiphene (20 μg/animal) was also effective in a similar context, though in a reverse role: normal cells with oestrogen receptors were stimulated to take up water in this case whereas tumour cells again failed to respond. Tissue contrast was thereby increased more than tenfold.

Changes in proton density, as well as directly affecting the signal intensity, will also give rise to relaxation time changes (see § 6.2.3). These are potentially an even greater source of tissue contrast. This leads us to consider another category of n.m.r. contrast agents—those which act through their direct effect on the relaxation times. Paramagnetic centres with one or more unpaired electrons are particularly efficient in this regard, due to the strength of the electronic dipole moment (see § 2.1). Typical paramagnetic substances including simple molecular gases, ions, and stable free radicals are detailed in Fig. 6.14. The theory of solvent relaxation by paramagnetic centres has been given by Solomon (1955), and Bloembergen and Morgan (1961), who showed that this contribution to the relaxation rate depends on the concentration of the paramagnetic

(a)

N̈ :Ö:	:Ö: N: Ö:	:Ȯ::Ȯ:
Nitric oxide	Nitrogen dioxide	Molecular oxygen

(b)

Transition Series cations		Electron distribution in 3d orbitals	Spin Quantum number
V^{+2}	$3d^1$	↑	1/2
Ti^{+2}	$3d^2$	↑ ↑	2/2
Cr^{+3}	$3d^3$	↑ ↑ ↑	3/2
Cr^{+2},Mn^{+3}	$3d^4$	↑ ↑ ↑ ↑	4/2
Mn^{+2},Fe^{+3}	$3d^5$	↑ ↑ ↑ ↑ ↑	5/2
Fe^{+2},Co^{+3}	$3d^6$	↑↓ ↑ ↑ ↑ ↑	4/2
Co^{+2},Ni^{+3}	$3d^7$	↑↓ ↑↓ ↑ ↑ ↑	3/2
Ni^{+2}, Cu^{+3}	$3d^8$	↑↓ ↑↓ ↑↓ ↑ ↑	2/2
Cu^{+2}	$3d^9$	↑↓ ↑↓ ↑↓ ↑↓ ↑	1/2
Cu^{+1}	$3d^{10}$	↑↓ ↑↓ ↑↓ ↑↓ ↑↓	0

(c)

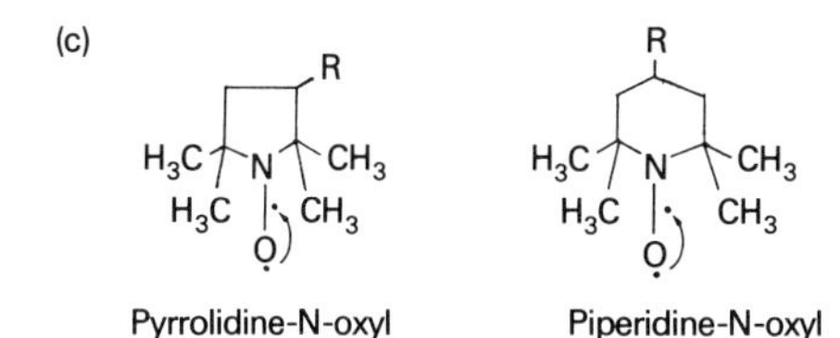

Fig. 6.14. (a) Electron dot diagram showing unpaired electrons for nitric oxide, nitrogen dioxide, and molecular oxygen, all of which exhibit strong paramagnetic behaviour. (b) Electron subshell diagrams of ions in the first transition series of the periodic table. The larger the number of unpaired electrons the higher the spin quantum number and the stronger the paramagnetic behaviour. (c) Pyrrolidine-*N*-oxyl (i) and piperidine-*N*-oxyl (ii) nitroxide stable free radicals. Stability is attributed to the steric hindrance of the methyl groups and delocalization (indicated by the arrow) of the unpaired electron (dot) between oxygen and nitrogen atoms. (After Brasch 1983.)

centre, the strength of its magnetic moment, the proximity of the solvent to the centre, as well as on a number of correlation times: one describing the electron–spin relaxation, one describing the tumbling rate of the centre, and one describing the contact time of the centre with the solvent. The details need not concern us here, though the dependence on magnetic moment should be noted. It indicates that the more unpaired spins in the centre the better. Hence, cations near the middle of the transition series, such as Fe^{3+} and Mn^{2+} which both usually (though not always, e.g. Fe(I) in $Fe(CN_6^{3-})$) have five unpaired electrons, are likely to be especially effective. Potentially, an even greater enhancement is available if lanthanide or actinide ions are used. Gadolinium III, for example, generally has seven unpaired electrons.

Trial studies using both Mn^{2+} and Fe^{3+} have been reported. In particular, orally-administered ferric chloride (0.06 per cent; Gore 1982; Gore, Doyle, and Pennock 1981) and ferric ammonium citrate (Wesbey, Brasch, Engelstad, Moss, Crooks, and Brito 1983) were found to give good enhancement of the gastrointestinal tract. Manganese is selectively retained by the liver; an intravenous injection of 20 μmol/kg Mn^{2+} into a rabbit leads to a halving of the liver T_1 (Gore *et al.* 1982). At the level of 50 μmol/kg, Mn^{2+} has also allowed good visualization of ischaemic regions induced in the canine heart by ligation of a coronary artery (Goldman, Brady, Pykett, Burt, Buonanno, Kistler, Newhouse, Hinshaw, and Pohost 1982*a*; Goldman, Brady, Pykett, Burt, Newhouse, Buonanno, Kistler, Hinshaw. Pohost 1982*b*; see also Lauterbur, Dias, and Rudin 1978; Hollis, Bulkey, Nunnally, Jacobus, and Weisfeldt 1978). Contrast arose from the reduction of the relaxation times in the normal, adequately perfused, myocardium. Excellent correlation was obtained between the ischaemic regions delineated by this method and by thallium-201 scintigraphy and triphenyl tetrazolium chloride-stained sections.

A major problem with the direct use of paramagnetic ions as contrast agents, arises from their comparative toxicity. One measure of the efficiency of a contrast agent is the ratio of its LD50 to a diagnostically useful dose; for simple salts values ~10:1 are typical (see Brasch 1983, Table III, for LD50 values and details of metabolism). One way of overcoming this obstacle is to use insoluble contrast agents: barium sulphate, for example, is a safe and widely-used radiological contrast agent, whereas Ba^{2+} itself is highly toxic. A suitable insoluble n.m.r. contrast might be gadolinium oxalate which has been shown to be useful in opacifying the stomach and gut (Runge, Stewart, Clanton, Jones, Lukehart, Partain, and James 1983*b*).

Rather than administer the ion in the form of a simple salt, it is also possible to 'package' it using a chelator such as ethylenediaminetetraacetic acid (EDTA) or diethylenetriaminepentaacetic acid (DTPA), giving greatly reduced toxicity. For example, Cr-EDTA, a practical i.v. agent which is renally excreted, does not undergo metal exchange *in vivo*, and doses of up to 3 mmol/kg/day can be tolerated in dogs and rats (Runge *et al.* 1983*b*). In fact, chelation therapy is an established means of treating metal poisoning, and n.m.r. imaging could provide a means of assessing its efficacy if the ion overload is sufficiently acute. In iron-storage disease of the liver, for example, iron-tissue levels are typically 1.8 mg/g and in Wilson disease liver copper levels as high as 700 mg/g have been described. Runge, Clanton, Smith, Hutchison, Mallard, Partain, and James (1983*a*) have reported an *in vitro* study in which T_1 determinations on mouse plasma and homogenized tissue were used to assess the efficacy of the iron chelator desferoxamine (1.2 mg/g) following i.v. injection of iron dextran (0.5 mg/g).

The use of ion chelators can confer other advantages to the contrast agent. For example, the rotational correlation time is increased, rendering, the relaxation process more efficient and enabling lower doses to be used. Huberty, Engelstad, Wesbey, Moseley, Young, Tuck, Brito, Hattner, and Brasch (1983) have been able to demonstrate renal parenchyma and collecting system enhancement for Gd^{3+}, Mn^{2+} and Fe^{3+} doses as low as 0.001 mmol/kg with a variety of chelating agents. Once the ion has been chelated it is possible to take the process a stage further by adding species such as monoclonal antibodies which give the relaxation agent tissue specificity. Brady, Rosen, Gold, Khaw, Fallon, Goldman, Ter Penning, Yasuda, and Haber (1983) have used Mn^{2+} bound to the monoclodal antibody antimyosin via the bifunctional chelator DTPA to determine the extent of infarcted myocardium in canine hearts. Doses as low as 2 mg of antibody (50 μg Mn^{2+}) were sufficient to cause substantial reduction of T_1 (from 1068 to 245 ms at 61.4 MHz) in the infarcted region, allowing it to be distinguished readily from the normal myocardium, whose T_1 was reduced much less dramatically (from 818 to 690 ms). The method, though costly, is of general applicability—the use of tumour specific antibodies, can be envisaged, for example. Another possibility for the targeting of contrast agents is through the use of liposomes.

Stable free radicals (s.f.r.s) are another group of compounds which can be used as relaxation-enhancing agents. The nitroxide s.f.r.s have attracted particular attention in this regard (Brasch 1983; Brasch, London, Wesbey, Tozer, Nitecki, Williams, Doemeny, Tuck, and Lallemand 1983*a*; Brasch, Nitecki, Brant-Zawadzki, Enzmann, Wesbey, Tozer, Tuck, Cann, Fike, and Sheldon 1983*b*), pyrrolidine and piperidine nitroxide s.f.r. derivatives being typical (see Fig. 6.14). Both series of radicals derive their stability from a combination of the steric hindrance afforded by the four methyl groups on the α-carbons and by electron delocalization to the centre of the molecule. Brasch *et al.* (1983*a*) have described the use of the water-soluble piperidinyl nitroxide s.f.r. derivative *N*-succinyl-4-amino-2,2,6,6-tetramethylpiperidine-1-oxyl (t.e.s.) in a study of experimentally-induced cerebritis in dogs. Using unenhanced n.m.r., a high-intensity ring was seen surrounding a low-intensity necrotic core, whereas with t.e.s.-(0.9 g/kg i.v.)-enhanced n.m.r. the necrotic centre was enhanced and the overall size of the anomalous region was much greater. Signal intensity increased by 45 per cent, twenty minutes after t.e.s. injection, subsiding after a period of 1 h, indicating that, in contrast to simple paramagnetic ions, t.e.s. has a biological retention-time ideally suited to imaging applications. No enhancement was observed contralaterally, showing that t.e.s. crosses the blood brain barrier only at the site of disease, and suggesting that it has a pathophysiology similar to common radiographic contrast agents. The same group (Brasch *et al.* 1983*b*) have also described the use of t.e.s. for n.m.r. urography, finding that in rats the threshold dose for observable contrast enhancement was ~0.04 g/kg. One problem with s.f.r. contrast agents is their susceptibility to reduction, which leads to a decreased efficiency. Griffiths, Rosen, Rauckman, and Drayer (1983) have shown that a single passage through the lungs is sufficient to reduce up to 97 per cent of a nitroxide s.f.r. when given at low dose, though the process is saturable and the distribution and elimination follow expected kinetics at higher dose-levels.

Interestingly, the paramagnetism of dissolved molecular oxygen can be used to selectively enhance cardiac tissues. For example, when a subject was given pure oxygen to inhale, increased signals were observed from the left ventricle (Young, Bailes, Collins, and Gilderdale 1982*a*). In measurements on human volunteers and dogs with indwelling catheters, Bydder, Goatcher, Hughes, Orr, Pennock, Steiner, and Tripathi (1982*b*) have determined the increase in left ventricular T_1 rate as the proportion of inspired oxygen was increased from 21 to 100 per cent. They found it to be 11.6 per cent in humans and 9.6 per cent in greyhound dogs.

These few examples serve to illustrate the potential of n.m.r. contrast agents. However, the field remains very much in its infancy and much more work needs to be done on the development of more specific and efficient agents and on the pharmaco-kinetics of their distribution and elimination. As we have seen, many relaxation enhancing agents rely on the use of paramagnetic centres. Conveniently, these can be measured directly, even at the low concentrations in which they are used, by n.m.r.'s more sensitive sister technique of electron–spin resonance. Alternatively, conventional isotope tracer techniques are available. Given that (at least) three correlation times enter the expression for the paramagnetic enhanced relaxation rate, the effect is likely to be strongly dispersive. It will thus be important to conduct trials at or near the frequency at which the agents are to be used clinically (rather than in a conventional high-field analytical instrument). Additionally, the influence of molecular environment on correlation time means that there is no simple substitute for *in vivo* trials. See § 6.5.2 for an example of the use of Gd-DTPA as an enhancing agent for brain tumours.

6.3.6. *Flow imaging*

There is great interest in the possibility of n.m.r. flow imaging and its use for studies of the vascular system. Much research effort has been expended in the search for a general method. However, the imaging of a vector rather than a scalar quantity poses considerable difficulties, and the problem is greatly simplified if one has *a priori* knowledge of the flow direction, as would be the case for a principal artery or vein, for example.

In fact, many standard n.m.r. images show the evidence of vascular flow. This is particularly apparent in the case of imaging schemes based on steady state free precession, which requires a certain time for the equilibrium (steady state) signal to become established. If during this period, blood flows out of the imaging plane, its signal is lost. Arteries therefore often appear as circles or ellipses (depending on whether or not they run perpendicular to the imaging plane) of reduced signal intensity. Measurement of the lumen is a relatively simple matter. In favourable cases it is also possible to visualize the arterial wall as a surrounding ring of intermediate signal strength (Newhouse *et al.* 1982).

In general, the effect of rapid flow will result in loss of signal intensity due to the disruptive effect on the imaging sequence. This causes the vascular system to be highlighted (negatively) and allows problem areas to be easily identified without recourse to intravenous contrast agents. At lower rates of flow, signal is not completely lost; indeed, for low velocities, it can be increased. In the case of a SR experiment, for example, this arises because fresh, fully polarized, blood entering the imaging plane gives rise to a stronger signal than that from the partially saturated blood which it replaces. Similar effects can be seen in IR scans where flow alters the effective relaxation rate. Flow effects have also been observed in multislice studies. They are most pronounced in the outermost slices. Thus Herfkens, Higgins, Hricak, Lipton, Crooks, Lanzer, Botvinick, Brundage, Sheldon, and Kaufman (1983*b*) have reported 'paradoxical' flow enhancement for multislice studies of the abdomen, in which the aorta shows the effect in the most cranial slice and the vena cava in the most caudal. Presumably this arises from the entry of unpolarized blood into the imaging volume from top (aorta) and bottom (vena cava) respectively.

The magnitude of flow effects depends on the flow velocity and the nature of the imaging sequence and its timing parameters. For images showing some saturation, there is generally enhancement at low flow rates. This rises to a maximum and thereafter falls in an approximately linear fashion with increasing flow rate until a cut-off is reached, beyond which no signal from the region of flow is visible (Crooks and Kaufman 1984). As indicated, flow effects are often observed with standard imaging sequences and, with little modification, these same sequences can be used to obtain quantitative estimates of flow. In order to circumvent any difficulties which might arise from the pulsatile nature of the blood flow, sequences are commonly gated from the electrocardiogram (e.c.g.) (see Singer 1982).

A simple method based on an extension of the ideas outlined above has been applied by Singer and Crooks (1983) to determine flow velocities in principal vessels. It involves saturation of the spins in the cross-section of interest, followed by observation of the same region after a short variable delay. Signal arises principally from fully polarized spins entering the selected slice. There is, however, some contamination from the signals of fast-relaxing static components. A method developed by Feinberg (1983) is somewhat more effective in rejecting signal from stationary spins. It uses a selective 90° pulse followed by a selective 180° refocusing pulse in a non-overlapping downstream slice.

If phase sensitive detection (see § 2.4.2) is employed then flow velocity can be determined by virtue of the additional phase encoding it generates. Thus, in a properly designed echo experiment, signal from the static spins is refocused by a 180° pulse and all components are in phase at the echo maximum. If, however, spins undergo a displacement between evolution in the gradients on opposite sides of the 180° pulse, this will manifest itself as a phase shift at the echo maximum which is proportional to the flow velocity. Measuring the phase shifts therefore gives a direct measure of flow. The method works for flow in or through the selected slice, though an extended sequence is required to separate the two effects if present simultaneously. Van Dijk (1984*a*) has used such methods to measure both blood flow and the motion of the heart itself.

These few examples illustrate the potential for flow determinations inherent in n.m.r. imaging. Major developments are to be anticipated in this area.

6.4. Small-scale imaging

6.4.1. Historical perspective

The most widespread use of conventional n.m.r. instruments has been for chemical analysis, structural elucidation, and the study of motional processes at the molecular level. The high-resolution spectrometers needed for these experiments require extremely good magnetic field homogeneity (see § 2.6) which can only be achieved over very limited regions, and this in turn has limited sample volumes to about 0.1–1 ml for proton-only work and 1–10 ml for the less sensitive nuclei with larger chemical-shift ranges.

From the first report by Lauterbur in 1973, the early n.m.r. imaging experiments were therefore limited by the pole gaps of existing magnets to sample sizes of a few cubic centimetres. Since that time, the trend has been to move up in scale through intermediate-sized systems capable of small animal and human limb studies. (Andrew, Bottomley, Hinshaw, Holland, Moore, and Simaroj 1977) towards the ultimate goal of whole-body imaging first achieved by the f.o.n.a.r. method of Damadian, Goldsmith, and Minkoff in 1977, and followed shortly by the line-scanning technique of Mansfield, Pykett, Morris, and Coupland (1978).

If it is assumed that the same number of pixels is retained in each case, then the increased pixel volume leads to higher picture quality for large-scale systems, although this is partially offset by the need to operate at lower frequencies (see § 5.7). When surveying the literature, however, one should beware of drawing comparisons between recent head and whole-body sections, and small-scale images which may well be representative of n.m.r. imaging at a much earlier stage of its development. Also, since the relaxation times are frequency dependent, the spin–lattice relaxation time, in particular, showing a strong dispersion, it is not possible to make direct comparisons between values obtained on instruments operating at widely different frequencies.

Although in one sense they can be seen as steps along the path to whole-body studies, small- and intermediate-scale imaging have an important role to play in the animal studies which are an essential part of the clinical development of the technique. In addition, n.m.r. imaging promises to make important contributions in other, non-medical areas such as food technology, for example. It is worth bearing in mind that it is equally possible by going to higher frequencies (e.g. 600 MHz) to extend n.m.r. imaging to the study of much smaller systems, perhaps down to a spatial resolution of 5–10 μ. The potential of such an n.m.r. microscope remained entirely unexplored until Lauterbur conducted trials with an experimental 90 MHz system in 1983. A 1.7-mm-diameter 15-turn solenoidal coil was used to observe objects of roughly 1 mm diameter; the spatial resolution achieved was 20 μ in a scan time of 10 min.

6.4.2. *Human studies*

The first images to demonstrate *in vivo* human anatomy were cross-sections through Maudsley's finger, obtained in 1976 by Mansfield and

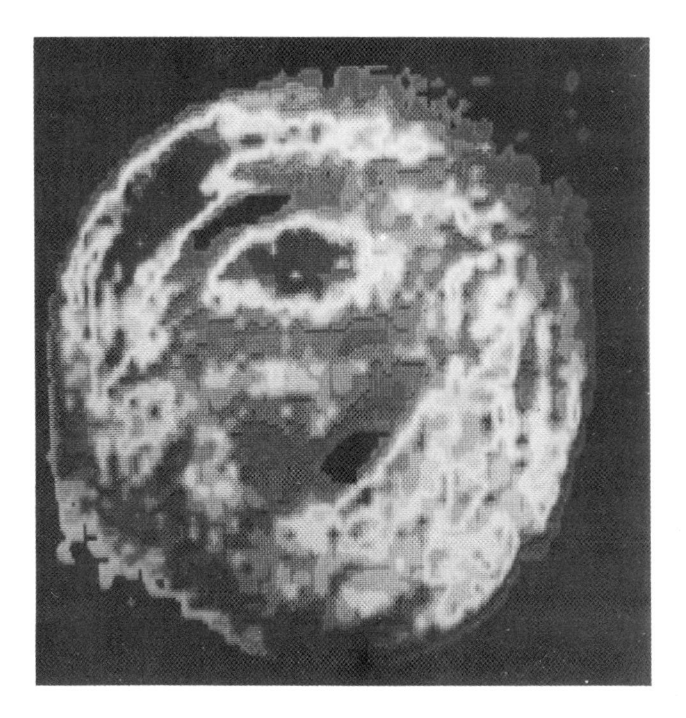

Fig. 6.15. Cross-sectional image through a human finger obtained by the line-scanning method of Mansfield and Maudsley (1977) using a repetition interval of 0.5 s.

Maudsley with a line-scanning technique using selective irradiation (Mansfield and Maudsley 1976, 1977). An example is shown in Fig. 6.15. Each line of the image was averaged 48 times with an interpulse delay of 0.5 s giving a total imaging time of 23 min. Many anatomical features can be recognized including bone (the mid phalanx), bone marrow, flexor and extensor tendon sheaths, digital arteries, and nerves. Even at this early stage, the importance of T_1 effects was appreciated, and interpulse delays were varied to elicit maximum tissue contrast. The reduction in the apparent T_1 of blood by virtue of its motion through the image plane was also well understood. No attempt was made to select the image plane, which was consequently defined by the receiver coil geometry and corresponded to a slice approximately 1 cm thick. The long imaging times required, even in a relatively thick slice ($\sim$10 mm), reflect both the comparative inefficiency of line-scan imaging and the low operating frequency (15 MHz). For a more up-to-date assessment of resolution and imaging times achievable with optimally-designed small-scale systems; see Mansfield and Morris (1982).

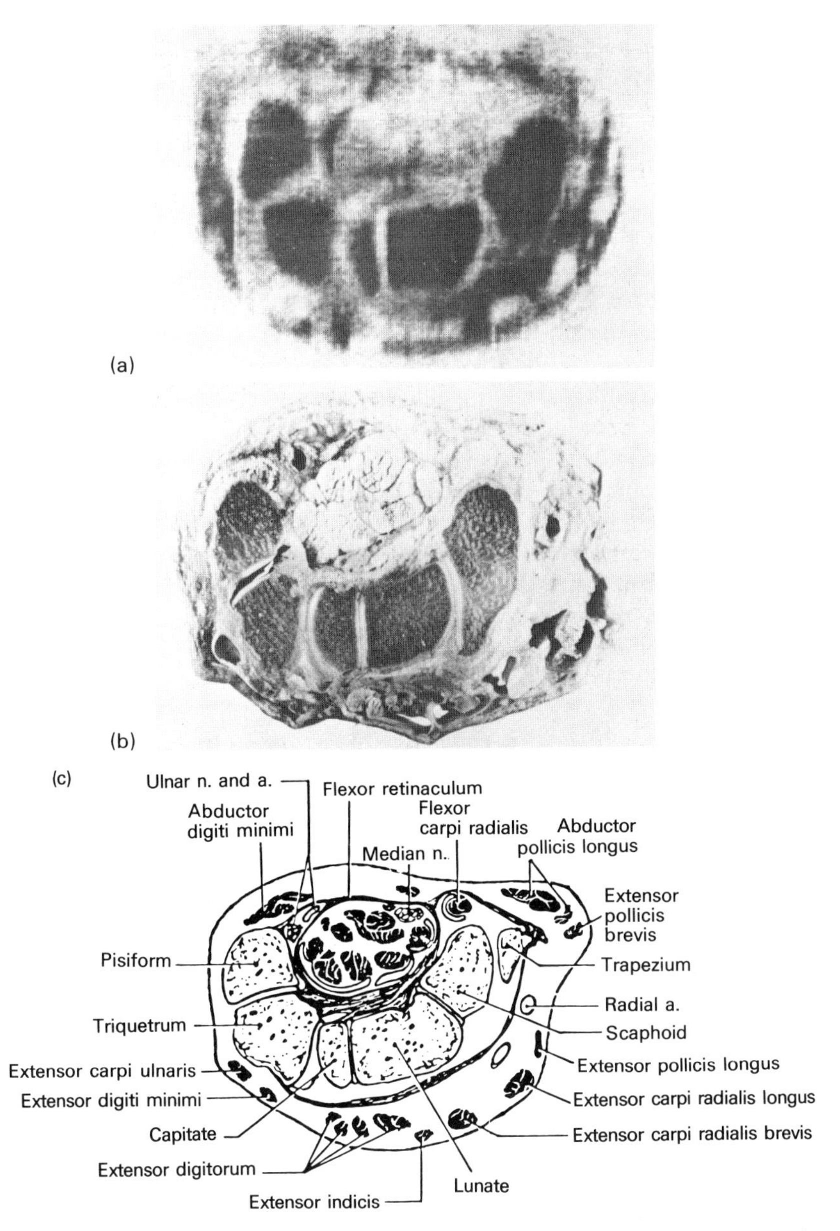

Fig. 6.16. (a) *In vivo* wrist cross-section obtained by the multiple sensitive point method at 30 MHz. Dark image areas correspond to regions of high mobile proton density. The slice thickness was 3 mm, in-plane resolution 0.4 mm, and scan time 9 min. (b) Corresponding cadaver section. (c) Annotated anatomical sketch for comparison. (From Hinshaw *et al.* 1977. Reprinted by permission from *Nature* **270,** 722. Copyright © 1977 Macmillan Journals Limited.)

Figure 6.16 shows an *in vivo* cross-section through a human wrist obtained at 30 MHz with a purpose-built instrument designed for intermediate-scale imaging (up to 8 cm) using the multiple sensitive point method (Andrew *et al.* 1977; Hinshaw, Bottomley, and Holland 1977). (Note that the grey scale has been reversed relative to Fig. 6.15; black corresponds to high signal intensity.) The in-plane resolution of 0.4 mm and slice thickness of 3 mm achieved through the use of oscillating field gradients (see § 3.2.3), give a beautiful representation of human wrist anatomy: compare the n.m.r. image (a) with the photograph of a similar cadaver section (b) and the labelled diagram (c). In many people's eyes this was the first clear indication that n.m.r. imaging could achieve the necessary tissue contrast and resolution to become a competitive imaging modality. An interesting feature of the image is that, with the exception of the one from subcutaneous fat, the highest signal intensity comes from the cancellous interior of the carpal bones. In contrast, other n.m.r. sections through the forearm (Hinshaw, Andrew, Bottomley, Holland, Moore, and Worthington 1979) show very little signal from the compact bone of the radius and ulna, but a signal of intermediate strength from their central marrow-filled cavities. The use of an operating frequency appropriate to the scale of the imaging study, reduced the scan times to some 10–12 min. However, further improvement can be made by changing from a line to a planar technique. As discussed in § 6.3.6, steady-state free precession which is part and parcel of the multiple sensitive point technique used, is particularly sensitive to motion. Although human limbs can be clamped fairly efficiently, some movement artefacts are nevertheless apparent in Fig. 6.16(a). These take the form of low-intensity vertical stripes, one pixel wide, corresponding to missing imaging lines. Blood-vessels show as regions of low signal intensity, for example, the radial artery visible to the right of the scaphoid bone.

Other small scale 'human images' have been obtained from specimens removed at surgery or biopsy allowing direct comparison both with pre-operative large-scale imaging results and also with subsequent histopathological and biochemical assays. Bovée *et al.* (1980) have investigated thin slices of mammary tissue obtained from mastectomy specimens. They were able to locate with accuracy breast tumours using an imaging scheme based on Hinshaw's sensitive point method (see § 3.2.3).

6.4.3. *Animal studies*

A number of cross-sectional images of experimental animals were reported during the early development of n.m.r. imaging. They were chosen to demonstrate morphological detail and gave an idea of the degree of tissue contrast to be expected from later human studies. As discussed in § 6.2.3, there is little variation in n.m.r. relaxation properties between the different mammalian species, so one might expect a rather close correlation between animal and human images. The only drawback is that the strong frequency-dependence of T_1 often requires the use of different imaging parameters and precludes direct comparison of relaxation-time measurements in the two regimes.

Sometimes these early studies were performed *in vivo*; often, in view of the comparatively long imaging times then required, the animals were sacrified prior to imaging. See, for example, the cross-sectional study of rabbit anatomy by Hinshaw and co-workers (Hinshaw, Andrew, Bottomley, Holland, Moore, and Worthington 1978). With the use of higher operating frequencies and planar techniques, it is now possible to obtain images of small experimental animals with the same relative resolution (i.e. the same number of pixels spanning the object) and having the same quality as their human counterparts, in times which are broadly comparable. The natural rhythyms—respiratory, cardiac, and peristaltic—are nevertheless considerably faster than is the case with humans, and methods which are especially susceptible to motional artefacts will have difficulty in coping. In such cases, to achieve the highest image-quality (in terms of genuine resolution), it is generally still necessary to sacrifice the animal prior to imaging (Newhouse *et al.* 1982). This is true of Fig. 6.17 which shows a cross-sectional image through the mid abdomen of a cat. It was obtained at a frequency of 61.4 MHz, corresponding to a (superconducting) magnetic field of 1.44 T, with a prototype system built by the Technicare Corporation. A steady-state free precession projection-reconstruction method was used, and plane definition was provided by oscillating field gradients at a frequency ~5 Hz. Imaging time for a 128×128 matrix was 2.3 min. The spinal column, surrounded by the paraspinal muscles, can be seen towards the bottom centre of the image. The kidneys are also clearly seen as regions of intermediate signal strength. Their internal structure is apparent in the form of discrete bands corresponding to the outer cortex, inner cortex, and medulla. The intrarenal collecting system was not visualized, but the ureters can be seen running out of the medial apices of the kidneys.

The image of Fig. 6.18 shows a cross-section through the mid abdomen of a guinea pig obtained under similar conditions to Fig. 6.17. The kidneys, including the renal pelvis, and spinal system are again seen. In addition, the abdominal aorta and vena cava are well resolved as discs of low signal intensity lying immediately anterior to the spinal column. The loops of the small intestine can be seen as circles of low intensity, highlighted by the bright regions of surrounding peritoneal fat. In some

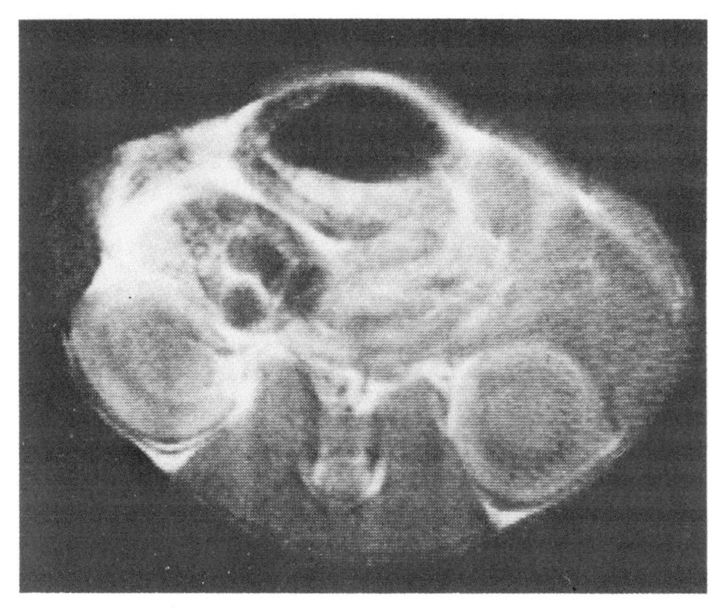

Fig. 6.17. 61.4 MHz ^{1}H n.m.r. cross-section through the mid abdomen of a cat. The kidneys display discrete bands corresponding to outer cortex, inner cortex, and medulla. (From Newhouse *et al.* 1982.)

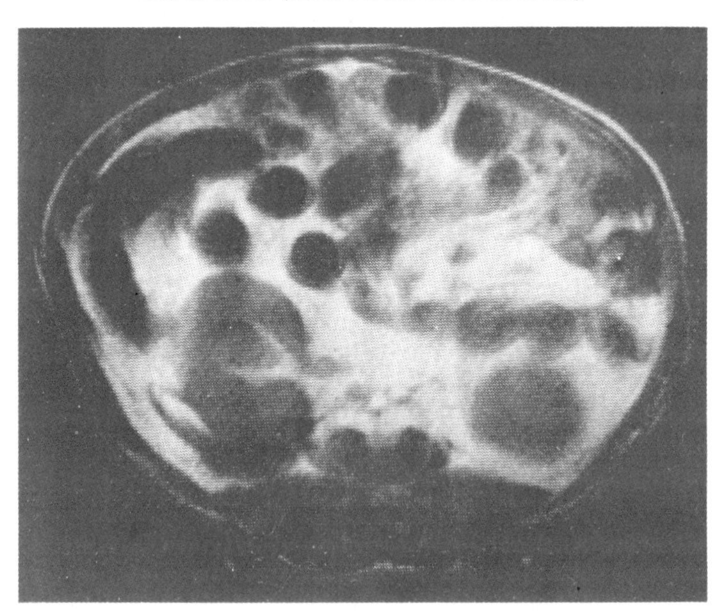

Fig. 6.18. 61.4 MHz ^{1}H n.m.r. cross-section through the mid abdomen of a guinea pig. A crescent-shaped retroperitoneal haematoma is seen posterior to the right kidney. (From Newhouse *et al.* 1982.)

cases the lumen of those intestines running normal to the image plane can be seen as rings of intermediate signal strength. Intestinal sections running in other directions are also apparent but are less clearly visualized.

The usefulness of animal models for the study of human disorders is well established; controls are readily available in the form of identical litter mates, the inducement of the clinical condition can be standardized, and samples can be taken as the disease progresses. In the human situation, diagnoses are often made at a late stage when the patient presents with severe symptoms, and tissue samples are only available at the time of surgery or autopsy. A simple example of an animal model is shown in Fig. 6.18: a retroperitoneal haematoma, visible in the image as a crescent-shaped region of low signal strength, was created near the right kidney by injection into the peritoneal space of 3 cc of blood aspirated from the heart (Newhouse *et al.* 1982).

Figure 6.19 shows cross-sections, obtained under similar conditions to Figs 6.17 and 6.18, at two levels through the head of a gerbil 1 h and 24 h following ligation of the left carotid artery—the well-known gerbil stroke

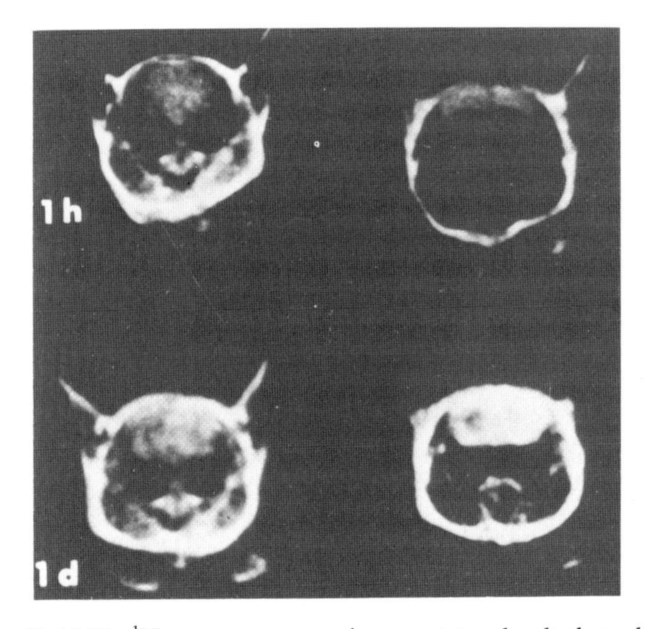

Fig. 6.19. 61.4 MHz ^{1}H n.m.r. transverse images at two levels through the head of a gerbil with left common carotid artery ligation. No image abnormalities are noted one hour post-operatively (top), but by 24 h (bottom) left hemisphere swelling with mass effect and right downward displacement of the midline is noted. The left cortical mantle is thickened and of markedly increased intensity. (From Buonanno *et al.* 1982.)

model (Levine and Payan 1966; Buonanno *et al.* 1982; Buonanno, Pykett, Brady, Vielma, Burt, Goldman, Hinshaw, Pohost, and Kistler 1983.). No asymmetries in signals from the brain were observed in the 1 h scans but after 24 h the left cortical mantle (right side of image) thickens and increases considerably in intensity. This can be reconciled with the T_1 and T_2 values, which were shown in separate experiments at 150 MHz to have increased by 15 and 30 per cent respectively. No changes were observed in the hemisphere contra-lateral to the ligature. Similar changes were recorded in rats, normally after about 12 h, but sometimes as early as 2 h following ligation. Note that an *increase* in signal intensity depends on the use of a steady-state free precession method; an inversion recovery sequence, for example, would yield a corresponding decrease. Studies of experimentally-induced infarction have also been reported in primates (Spetzler, Zabramski, Kaufman, and Yeung 1983). See also Asato, Handa, Hashi, Hatta, Komocke, and Yazaki (1983) for details of a study involving cryogenically-induced brain oedema, and Herfkens, Sievers, Kaufman, Sheldon, Ortendahl, Lipton, Crooks, and Higgins (1983*a*) for ^{1}H n.m.r. studies of infarcted muscle.

Crooks and co-workers at UCSF (Crooks, Hoenninger, Arakawa, Kaufman, McRee, Watts, and Singer 1980; Hansen, Crooks, Davis, De Groot, Herfkens, Margulis, Gooding, Kaufman, Hoenninger, Arakawa, McRee, and Watts 1980; Herfkens, Davies, Crooks, Kaufman, Price, Miller, Margulis, Watts, Hoenninger, Arakawa, and McRee 1981; Davis, Kaufman, Crooks, and Miller 1981; Davis, Kaufman, and Crooks 1982; Stark, Bass, Moss, Bacon, McKerrow, Cann, Brito, and Goldberg 1983) have undertaken extensive cross-sectional studies of both normal rats and rats with various types of lesion; for example, pyogenic and sterile abcesses produced respectively by injection of fecal matter and turpentine, haematoma produced by injection of uncoagulated whole blood (Herfkens *et al.* 1981; see Fig. 6.20), and various types of benign and malignant tumours both implanted and spontaneous (Davis *et al.* 1981, 1982). A number of images were obtained under different operating conditions allowing pure T_1 and T_2 images to be calculated. All the different lesions could be discriminated from the background tissue in one or other of the relaxation-time images; many could be distinguished in both and some in the standard image as well. Originally, the team hoped to demonstrate that measurements of T_1, T_2 and possibly also spin density, taken together, would provide sufficient information to characterize uniquely a particular lesion. Unfortunately, their results did not support their hopes, and some overlap between abscesses, haematoma, and tumours occurred. An interesting aspect of this work was that a good correlation was found between the relaxation rates $1/T_1$, $1/T_2$ and the water content, both for normal tissues and the different types of lesion, providing *in vivo* support for the *in vitro* results discussed in § 6.2.3.

A number of tumour models were studied by this same group. In one series of experiments, #3924A hepatoma was injected into the liver of ACI/N rats and any palpable extra hepatic masses which developed were imaged over a period of up to six weeks. Relaxation times, at the imaging frequency of 15 MHz, were found to be 0.91 ± 0.32 s (T_1) and 59 ± 12 ms (T_2) for the tumours, compared with values for normal liver of 0.43 ± 0.4 s and 41 ± 7 ms respectively. In spite of the 30 per cent variation, present even for this uniform tumour model, all tumours with an area greater than 8 mm^2 were seen. Again, there was an overlap in relaxation-time values with those measured in other lesions and some normal tissues, notably brain. The development of hepatoma in Wistar rats, monitored by n.m.r. imaging, has also been reported by Bottomley (1979, 1981). The increased T_1 of the growing tumour enabled it to be seen as a region of high signal intensity with the multiple sensitive point method used.

Figure 6.21 is taken from a second tumour model study by the UCSF team, and shows the development, over a six-week period, of an adeno-carcinoma implanted in the lower right thigh of a rat (Davis *et al.* 1982). All tumours were discriminated from background tissue without difficulty:

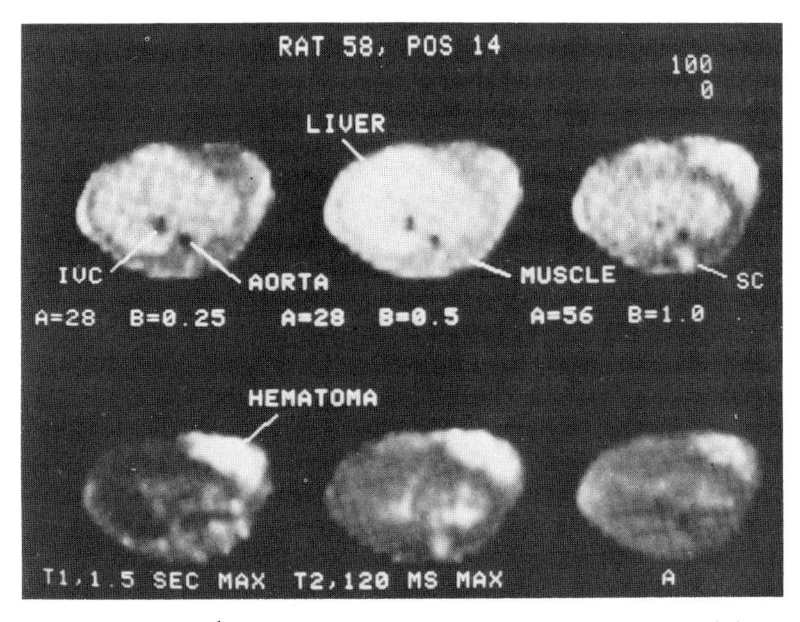

Fig. 6.20. Transverse ^{1}H n.m.r. images at 15 MHz through the upper abdomen of a live supine rat with a subcutaneous haematoma in its left flank. In the top row of images A and B, parameters correspond to TE and TR respectively (see § 6.3.1). The lower row of images show calculated T_1, T_2 and pure spin density sections. Although the intensity of the haematoma is not unique in the intensity images it is the brightest region in both T_1 and T_2 images. (From Herfkens, *et al.* 1981.)

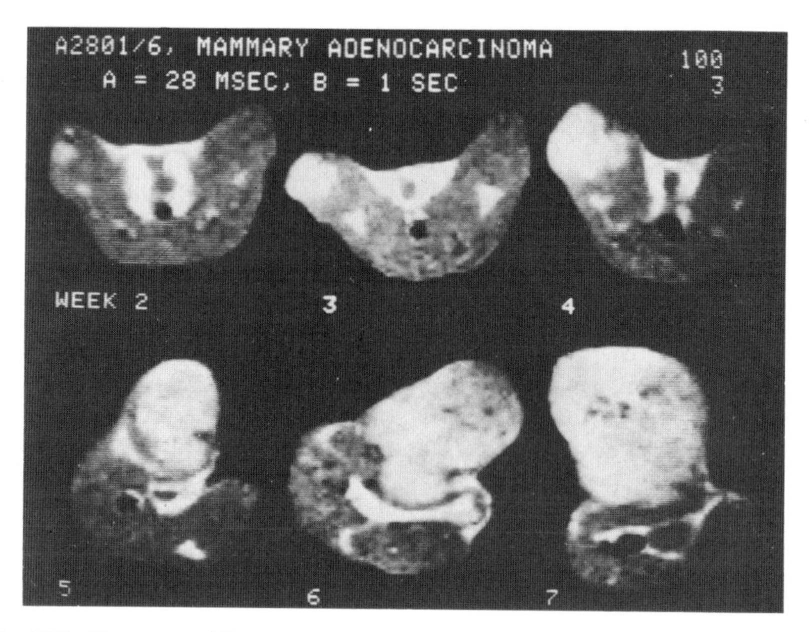

Fig. 6.21. Transverse [1]H n.m.r. images at 15 MHz through a mammary adenocarcinoma implanted in the lower right leg of a rat. The images show the growth of the tumour over a six-week period. (From Davis *et al.* 1982.)

although tumour T_1 values overlapped those for the normal muscle and T_2 values those for subcutaneous fat, the combination of T_1 and T_2 was tumour-specific. Necrotic regions which developed in some of the tumours displayed especially-elevated T_1 and T_2 values.

The problems caused by the short time-scale of motional processes in small experimental animals were mentioned above; for example, the basal rate for a rabbit heart is approximately 250 beats per minute. In spite of this, Mansfield and co-workers (Ordidge, Mansfield, and Coupland, 1981; Ordidge, Mansfield, Doyle, and Coupland 1982*a,b*) have been able effectively to 'freeze' these cardiac motions using the echo-planar sequence (§ 4.3; see also discussion of § 6.3.3). Full justice can only be done to this work by viewing the images in rapid succession, as a film loop for example, when the eye signal averages the stationary regions and the cardiac motion becomes graphically apparent. Figure 6.22 shows six high-speed images (frames) of a rabbit thorax taken at different stages throughout the cardiac cycle. The image matrix consists of 32×32 real points interpolated to 64×64 for display. The slice thickness was 8.6 mm, in-plane resolution about 0.75 mm and data collection time 32 ms per image. More recently, higher quality dynamic images of a similar nature

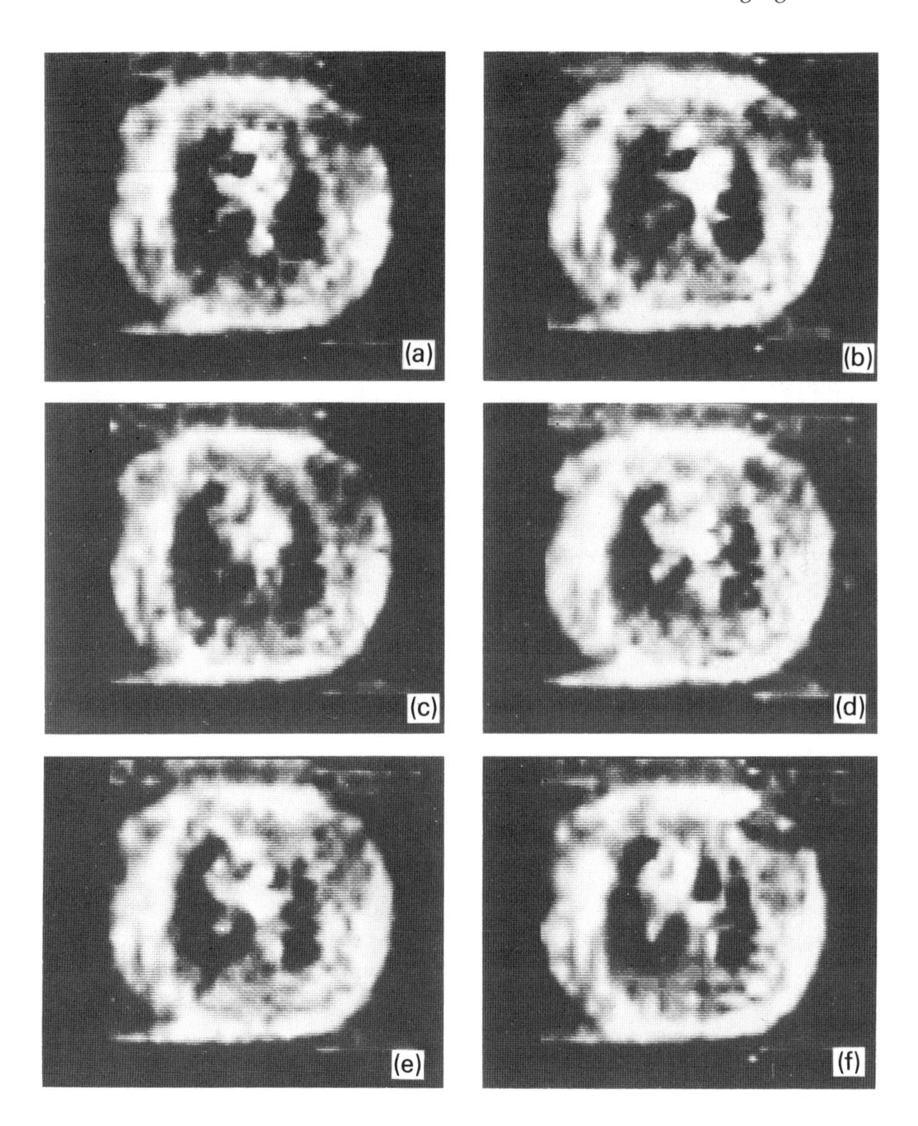

Fig. 6.22. Six single shot frames spanning a complete cardiac cycle from an n.m.r. movie loop showing the cross-section view through the thorax of a live rabit. Lung fields and heart, including chambers, are seen in all images: (a) right chamber open (dark zone); (b) right chamber and overall size of heart collapsing; (c) close to systole; (d) systole; (e) left ventricle beginning to open; (f) left ventricle in full diastole. Data collection time for each frame was 32 ms; each point representing a $0.75 \times 0.75\ \mathrm{mm}^2$ area in a slice of thickness 8.6 mm. (From Ordidge, Mansfield, Doyle, and Coupland 1982*b*).

have been obtained from a piglet (Doyle, Rzedzian, Mansfield, and Coupland 1983). Thoracic studies of three children with high respiratory and heart rates have also been carried out using this method (Rzedzian, Mansfield, Doyle, Guilfoyle, Chapman, Coupland, Chrispin, and Small 1983). By collecting data from a number of contiguous axial slices, dynamic images could also be constructed in sagittal, coronal, and intermediate plane orientations, giving a beautifully complete picture of cardiac motion.

Although the present instrument can only accommodate small-sized animals (1/4 human scale), work nears completion on the whole-body system and the implications for adult cardiac studies are clear: as well as major anatomical defects, irregularities in the heart motion, characteristic of myocardial infarction, will be apparent, and frame-by-frame analysis will allow the measurement of the ejection fraction, a valuable indicator of cardiac performance.

Finally, it should be noted that animal studies are expecially useful in the evaluation of n.m.r. contrast agents (see § 6.3.5, and, for examples, Brasch *et al.* 1983*a*).

6.4.4. *Other uses of n.m.r. imaging*

It is undoubtedly true that the principle application of n.m.r. imaging will be in diagnostic medicine, and it is to this area that by far the greatest research effort has been directed. Nevertheless, there are other disciplines in which n.m.r. in general, and n.m.r. imaging in particular, can make a substantial impact. Food science and technology is one such. Many early images were taken of fruits and vegetables, often for no good reason other than convenience of size, ready availability, or their undeniable aesthetic appeal, as exemplified by the lemon cross-section which adorned the front cover of *Nature* in 1977 (Hinshaw *et al.*) and which is reproduced in Fig. 6.23. However, such studies may well be of more practical value in assessing the correct conditions and timing for picking, storing, and marketing—areas in which even small improvements can reap large financial returns.

Seeds are also of great interest, from fundamental studies of germination, through problems of storage, where moisture content is of crucial importance, to the selection of the appropriate times for processing. (The milling process is, for example, critically dependent on the state of the grain and hence on its water content.) Proton n.m.r. has already been of value in improving the oil-bearing yield of certain seed crops by selecting those seeds with the highest oil contents for re-planting, and ^{13}C n.m.r. is starting to achieve similar results for protein content (Rutar 1982): n.m.r. imaging would add a new dimension to such studies.

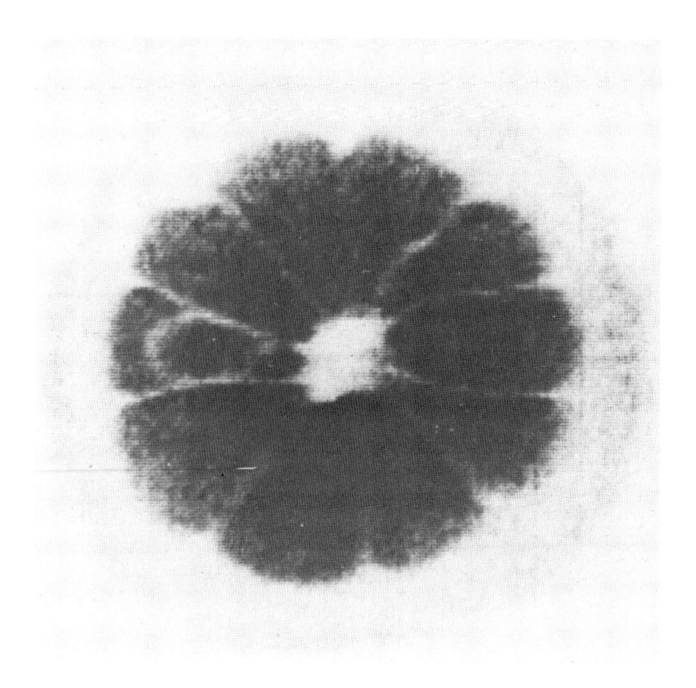

Fig. 6.23. N.m.r. cross-section through an intact lemon. Conditions as for Fig. 6.16. (From Hinshaw *et al.* 1977. Reprinted by permission from *Nature* **270** (5639) front cover illustration. Copyright © 1977 Macmillan Journals Limited.

There are possible applications, too, in the meat processing industry, particularly if quality can be correlated with n.m.r. parameters. In general, water content is of vital importance throughout the processed food industry, from the condition of individual processed vegetables to the seepage of moisture through pie cases!

Measurements of water content and diffusion may also have application in the building industry, both in the design of new materials and in the preservation of old ones. In this latter respect, n.m.r. imaging has recently been used to study the replacement of sea-water by polyethylene glycol, one stage in the preservation of timbers raised from Henry VIII's flagship, the *Mary Rose.* Studies of the materials themselves, rather than the water they contain, present difficulties since the n.m.r. resonances of solids are characteristically broad (see § 2.5.2). Some rubbers do give liquid-like spectra and are therefore amenable, and it is possible to combine the line-narrowing techniques of solid state n.m.r. to extend n.m.r. imaging in this direction (Wind and Yannoni, 1979). The detection of flaws in structural components would be an important area of application for such a system. Unfortunately, since r.f. fields do not penetrate metals to any great degree, their study is precluded. Studies of the new plastic and epoxy composite materials, many of whose failure mechanisms are not properly understood, would be a possibility, however.

It is also possible to use n.m.r. to detect plastic explosives which have a tell-tale T_1/T_2 characteristic!

This list is by no means exhaustive: only time will tell what areas will draw the most benefit from n.m.r. imaging.

6.5. Head scans

6.5.1. Normal studies

The improvement in quality of n.m.r. head scans from the first such image published in 1978 (Clow and Young) by Young's imaging team (then part of EMI), through the early multiple sensitive point and projection reconstruction results of Moore and colleagues at Nottingham (Holland, Moore, and Hawkes 1980; Moore, Holland, and Kreel 1980; Moore and Holland 1980; Hawkes, Holland, Moore, and Worthington 1980) to present-day images produced on commercial instruments, has been truly remarkable. Much, though by no means all, of the preliminary clinical evaluation of n.m.r. imaging has been concerned with head rather than whole-body studies, largely because they make less stringent demands on the imaging system: there are no major motional problems, the region over which the magnetic field homogeneity needs to be maintained is less, and the smaller receiver coils are more easily tuned. Another reason for the interest in head-scanning is the experience already derived from the use of X-ray CT systems which have revolutionized neuroradiology since their introduction by EMI in 1971 (Hounsfield 1973). Whilst this existing experience undoubtedly facilitated the development of n.m.r. techniques, it also invited comparison with the X-ray studies on precisely those grounds considered to be the latter's forte. Encouragingly, and perhaps somewhat surprisingly, in view of the earlier emphasis on the complimentary rather than the competitive nature of n.m.r. imaging, it is well capable of living up to these high standards.

In the images presented below lengthy anatomical descriptions have not been entered into, but a few illustrative features have been selected for discussion. The faithfulness with which the structure of the brain is represented by n.m.r. is nevertheless quite remarkable and, in order that this can be appreciated by those unfamiliar with brain anatomy, Fig. 6.24

Fig. 6.24. (b)–(f) Labelled post-mortem axial sections through the human head at the levels indicated in (a). (h)–(j) Labelled post-mortem axial sections through the human head at the levels indicated in (g). (From Bo, Meschan, and Krueger 1980.)

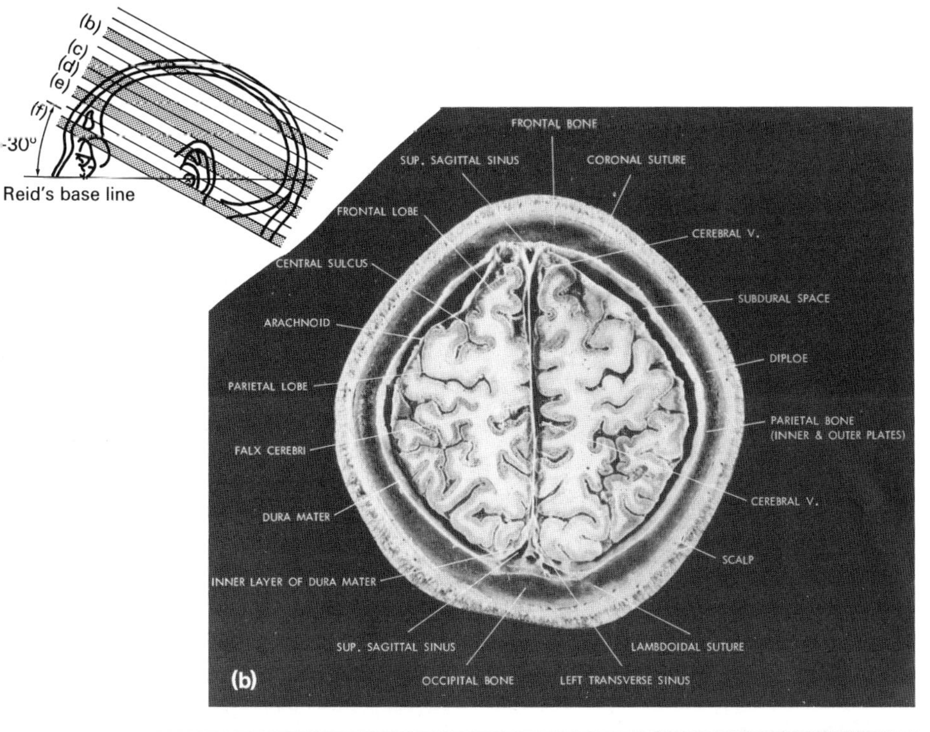
(b)
(c)
(d)
(e)
(f)
30°
Reid's base line
FRONTAL BONE
SUP. SAGITTAL SINUS
CORONAL SUTURE
FRONTAL LOBE
CEREBRAL V.
CENTRAL SULCUS
SUBDURAL SPACE
ARACHNOID
DIPLOE
PARIETAL LOBE
PARIETAL BONE (INNER & OUTER PLATES)
FALX CEREBRI
CEREBRAL V.
DURA MATER
SCALP
INNER LAYER OF DURA MATER
SUP. SAGITTAL SINUS
LAMBDOIDAL SUTURE
OCCIPITAL BONE
LEFT TRANSVERSE SINUS
(b)

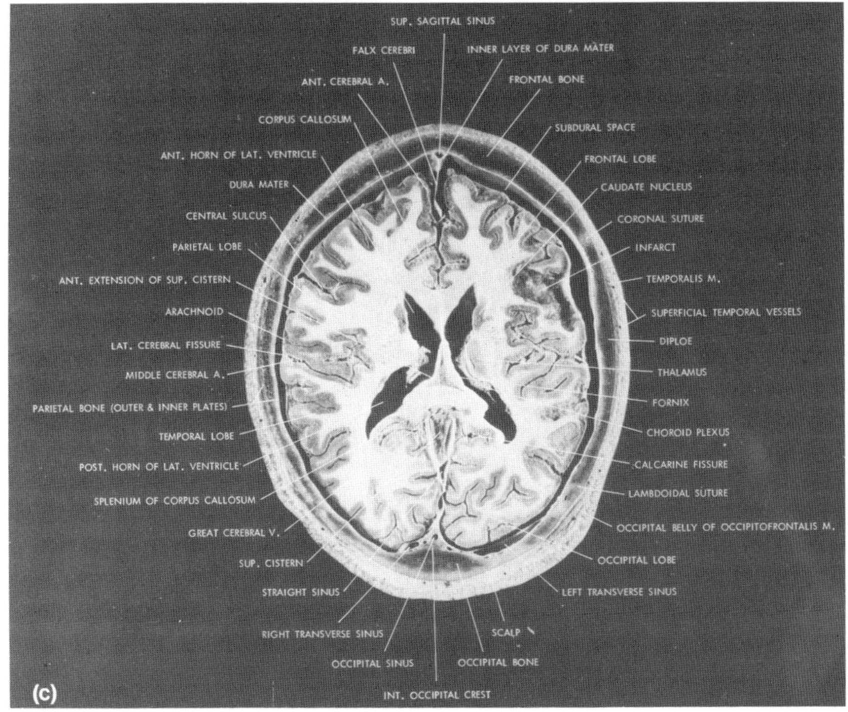
SUP. SAGITTAL SINUS
FALX CEREBRI
INNER LAYER OF DURA MATER
ANT. CEREBRAL A.
FRONTAL BONE
CORPUS CALLOSUM
SUBDURAL SPACE
ANT. HORN OF LAT. VENTRICLE
FRONTAL LOBE
DURA MATER
CAUDATE NUCLEUS
CENTRAL SULCUS
CORONAL SUTURE
PARIETAL LOBE
INFARCT
ANT. EXTENSION OF SUP. CISTERN
TEMPORALIS M.
ARACHNOID
SUPERFICIAL TEMPORAL VESSELS
LAT. CEREBRAL FISSURE
DIPLOE
MIDDLE CEREBRAL A.
THALAMUS
PARIETAL BONE (OUTER & INNER PLATES)
FORNIX
TEMPORAL LOBE
CHOROID PLEXUS
POST. HORN OF LAT. VENTRICLE
CALCARINE FISSURE
SPLENIUM OF CORPUS CALLOSUM
LAMBDOIDAL SUTURE
GREAT CEREBRAL V.
OCCIPITAL BELLY OF OCCIPITOFRONTALIS M.
SUP. CISTERN
OCCIPITAL LOBE
STRAIGHT SINUS
LEFT TRANSVERSE SINUS
RIGHT TRANSVERSE SINUS
SCALP
OCCIPITAL SINUS
OCCIPITAL BONE
INT. OCCIPITAL CREST
(c)

Fig. 6.24(a)–(c)

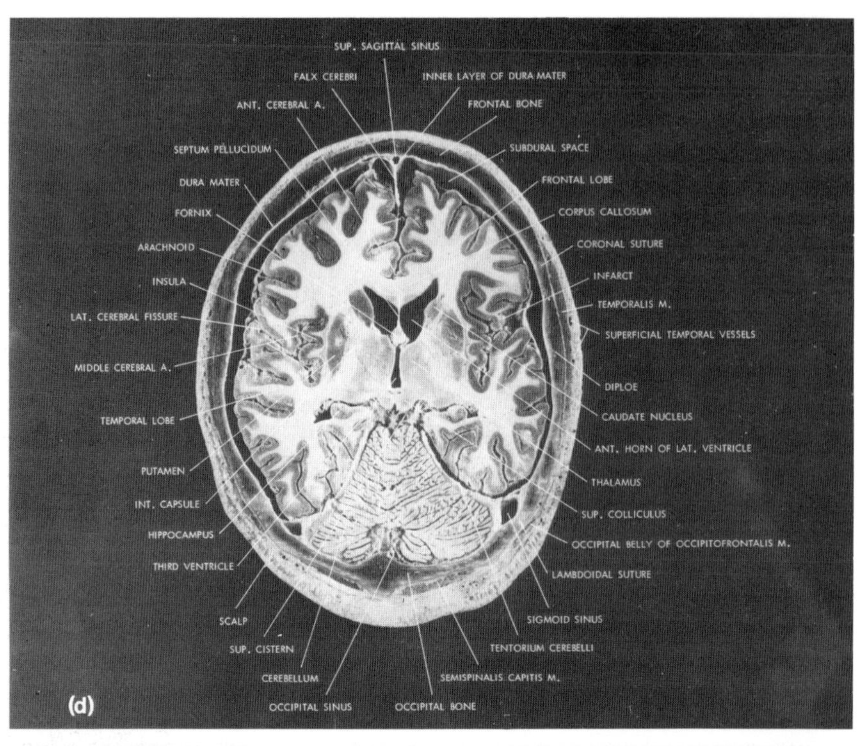
SUP. SAGITTAL SINUS
FALX CEREBRI
INNER LAYER OF DURA MATER
ANT. CEREBRAL A.
FRONTAL BONE
SEPTUM PELLUCIDUM
SUBDURAL SPACE
DURA MATER
FRONTAL LOBE
FORNIX
CORPUS CALLOSUM
ARACHNOID
CORONAL SUTURE
INSULA
INFARCT
TEMPORALIS M.
LAT. CEREBRAL FISSURE
SUPERFICIAL TEMPORAL VESSELS
MIDDLE CEREBRAL A.
DIPLOE
TEMPORAL LOBE
CAUDATE NUCLEUS
ANT. HORN OF LAT. VENTRICLE
PUTAMEN
THALAMUS
INT. CAPSULE
SUP. COLLICULUS
HIPPOCAMPUS
OCCIPITAL BELLY OF OCCIPITOFRONTALIS M.
THIRD VENTRICLE
LAMBDOIDAL SUTURE
SCALP
SIGMOID SINUS
SUP. CISTERN
TENTORIUM CEREBELLI
CEREBELLUM
SEMISPINALIS CAPITIS M.
OCCIPITAL SINUS
OCCIPITAL BONE
(d)

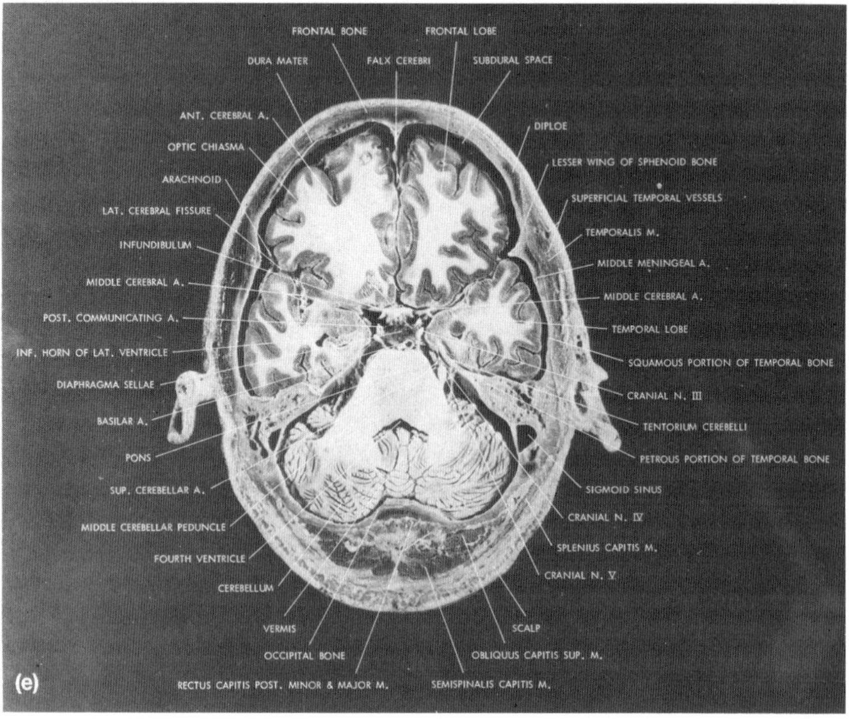
FRONTAL BONE
FRONTAL LOBE
DURA MATER
FALX CEREBRI
SUBDURAL SPACE
ANT. CEREBRAL A.
DIPLOE
OPTIC CHIASMA
LESSER WING OF SPHENOID BONE
ARACHNOID
SUPERFICIAL TEMPORAL VESSELS
LAT. CEREBRAL FISSURE
TEMPORALIS M.
INFUNDIBULUM
MIDDLE MENINGEAL A.
MIDDLE CEREBRAL A.
MIDDLE CEREBRAL A.
POST. COMMUNICATING A.
TEMPORAL LOBE
INF. HORN OF LAT. VENTRICLE
SQUAMOUS PORTION OF TEMPORAL BONE
DIAPHRAGMA SELLAE
CRANIAL N. III
BASILAR A.
TENTORIUM CEREBELLI
PONS
PETROUS PORTION OF TEMPORAL BONE
SUP. CEREBELLAR A.
SIGMOID SINUS
MIDDLE CEREBELLAR PEDUNCLE
CRANIAL N. IV
SPLENIUS CAPITIS M.
FOURTH VENTRICLE
CRANIAL N. V
CEREBELLUM
VERMIS
SCALP
OCCIPITAL BONE
OBLIQUUS CAPITIS SUP. M.
RECTUS CAPITIS POST. MINOR & MAJOR M.
SEMISPINALIS CAPITIS M.
(e)

Fig. 6.24(d)(e)

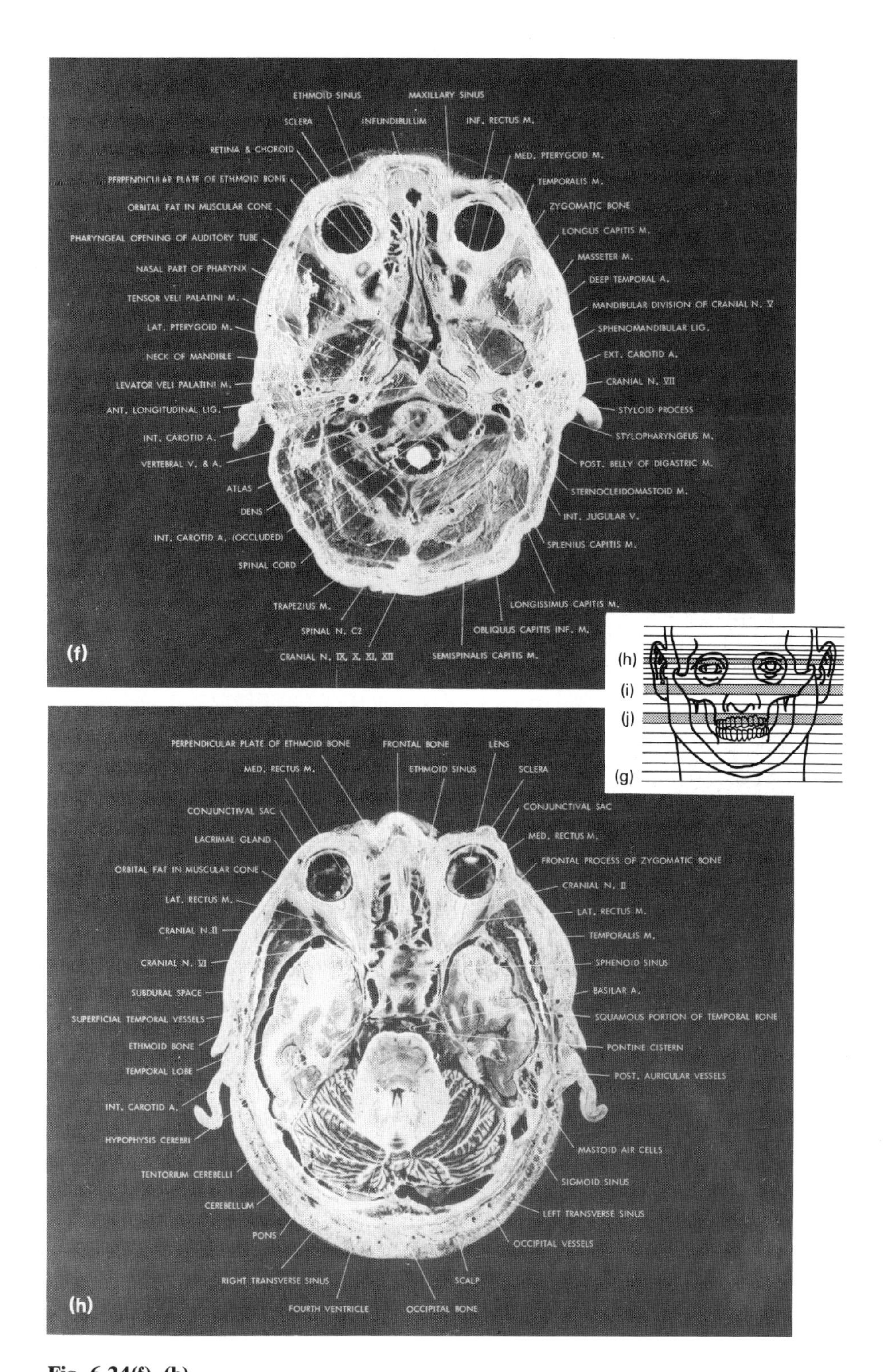
ETHMOID SINUS
MAXILLARY SINUS
SCLERA
INFUNDIBULUM
INF. RECTUS M.
RETINA & CHOROID
MED. PTERYGOID M.
PERPENDICULAR PLATE OF ETHMOID BONE
TEMPORALIS M.
ORBITAL FAT IN MUSCULAR CONE
ZYGOMATIC BONE
PHARYNGEAL OPENING OF AUDITORY TUBE
LONGUS CAPITIS M.
NASAL PART OF PHARYNX
MASSETER M.
DEEP TEMPORAL A.
TENSOR VELI PALATINI M.
MANDIBULAR DIVISION OF CRANIAL N. V
LAT. PTERYGOID M.
SPHENOMANDIBULAR LIG.
NECK OF MANDIBLE
EXT. CAROTID A.
LEVATOR VELI PALATINI M.
CRANIAL N. VII
ANT. LONGITUDINAL LIG.
STYLOID PROCESS
INT. CAROTID A.
STYLOPHARYNGEUS M.
VERTEBRAL V. & A.
POST. BELLY OF DIGASTRIC M.
ATLAS
STERNOCLEIDOMASTOID M.
DENS
INT. JUGULAR V.
INT. CAROTID A. (OCCLUDED)
SPLENIUS CAPITIS M.
SPINAL CORD
LONGISSIMUS CAPITIS M.
TRAPEZIUS M.
SPINAL N. C2
OBLIQUUS CAPITIS INF. M.
(f)
CRANIAL N. IX, X, XI, XII
SEMISPINALIS CAPITIS M.
(h)
(i)
(j)
(g)
PERPENDICULAR PLATE OF ETHMOID BONE
FRONTAL BONE
LENS
MED. RECTUS M.
ETHMOID SINUS
SCLERA
CONJUNCTIVAL SAC
CONJUNCTIVAL SAC
LACRIMAL GLAND
MED. RECTUS M.
ORBITAL FAT IN MUSCULAR CONE
FRONTAL PROCESS OF ZYGOMATIC BONE
CRANIAL N. II
LAT. RECTUS M.
LAT. RECTUS M.
CRANIAL N. II
TEMPORALIS M.
CRANIAL N. VI
SPHENOID SINUS
SUBDURAL SPACE
BASILAR A.
SUPERFICIAL TEMPORAL VESSELS
SQUAMOUS PORTION OF TEMPORAL BONE
ETHMOID BONE
PONTINE CISTERN
TEMPORAL LOBE
POST. AURICULAR VESSELS
INT. CAROTID A.
HYPOPHYSIS CEREBRI
MASTOID AIR CELLS
TENTORIUM CEREBELLI
SIGMOID SINUS
CEREBELLUM
LEFT TRANSVERSE SINUS
PONS
OCCIPITAL VESSELS
RIGHT TRANSVERSE SINUS
SCALP
(h)
FOURTH VENTRICLE
OCCIPITAL BONE

Fig. 6.24(f)–(h)

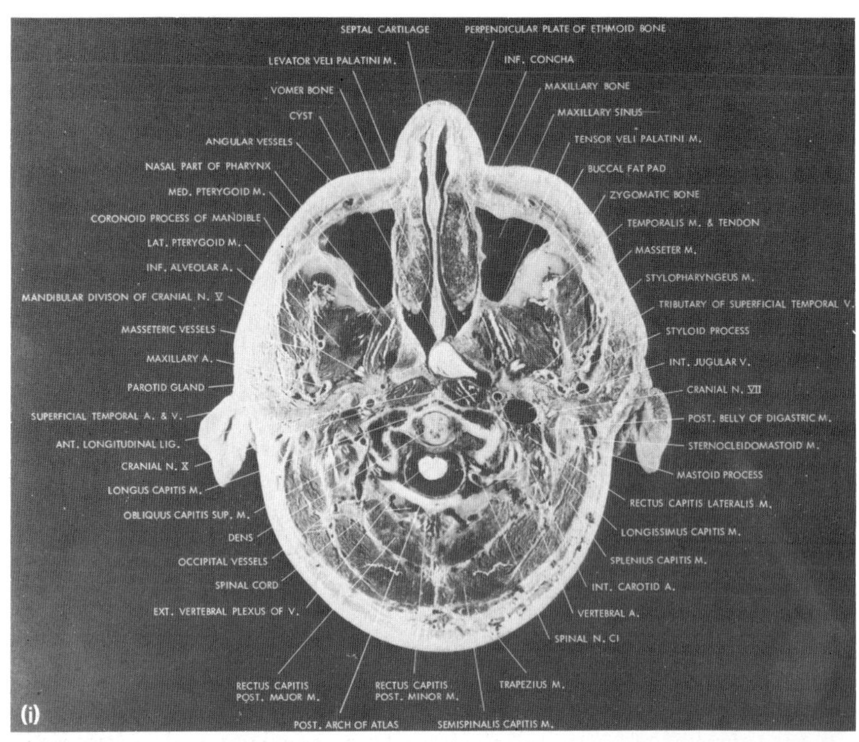
SEPTAL CARTILAGE
PERPENDICULAR PLATE OF ETHMOID BONE
LEVATOR VELI PALATINI M.
INF. CONCHA
VOMER BONE
MAXILLARY BONE
CYST
MAXILLARY SINUS
ANGULAR VESSELS
TENSOR VELI PALATINI M.
NASAL PART OF PHARYNX
BUCCAL FAT PAD
MED. PTERYGOID M.
ZYGOMATIC BONE
CORONOID PROCESS OF MANDIBLE
TEMPORALIS M. & TENDON
LAT. PTERYGOID M.
MASSETER M.
INF. ALVEOLAR A.
STYLOPHARYNGEUS M.
MANDIBULAR DIVISON OF CRANIAL N. V
TRIBUTARY OF SUPERFICIAL TEMPORAL V.
MASSETERIC VESSELS
STYLOID PROCESS
MAXILLARY A.
INT. JUGULAR V.
PAROTID GLAND
CRANIAL N. VII
SUPERFICIAL TEMPORAL A. & V.
POST. BELLY OF DIGASTRIC M.
ANT. LONGITUDINAL LIG.
STERNOCLEIDOMASTOID M.
CRANIAL N. X
MASTOID PROCESS
LONGUS CAPITIS M.
RECTUS CAPITIS LATERALIS M.
OBLIQUUS CAPITIS SUP. M.
LONGISSIMUS CAPITIS M.
DENS
SPLENIUS CAPITIS M.
OCCIPITAL VESSELS
INT. CAROTID A.
SPINAL CORD
VERTEBRAL A.
EXT. VERTEBRAL PLEXUS OF V.
SPINAL N. C1
RECTUS CAPITIS POST. MAJOR M.
RECTUS CAPITIS POST. MINOR M.
TRAPEZIUS M.
POST. ARCH OF ATLAS
SEMISPINALIS CAPITIS M.
(i)

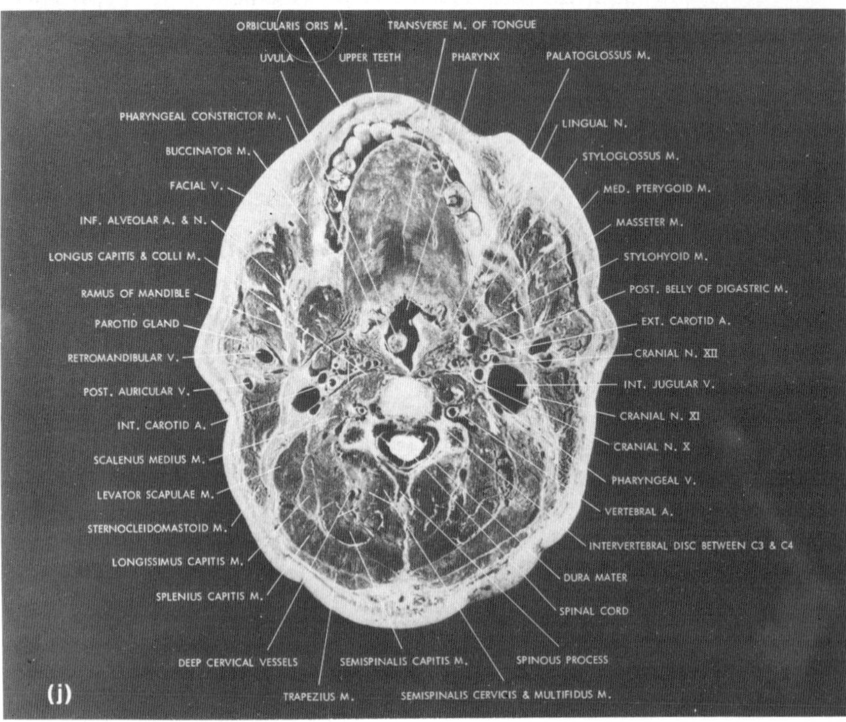
ORBICULARIS ORIS M.
TRANSVERSE M. OF TONGUE
UVULA
UPPER TEETH
PHARYNX
PALATOGLOSSUS M.
PHARYNGEAL CONSTRICTOR M.
LINGUAL N.
BUCCINATOR M.
STYLOGLOSSUS M.
FACIAL V.
MED. PTERYGOID M.
INF. ALVEOLAR A. & N.
MASSETER M.
LONGUS CAPITIS & COLLI M.
STYLOHYOID M.
RAMUS OF MANDIBLE
POST. BELLY OF DIGASTRIC M.
PAROTID GLAND
EXT. CAROTID A.
RETROMANDIBULAR V.
CRANIAL N. XII
POST. AURICULAR V.
INT. JUGULAR V.
INT. CAROTID A.
CRANIAL N. XI
SCALENUS MEDIUS M.
CRANIAL N. X
LEVATOR SCAPULAE M.
PHARYNGEAL V.
STERNOCLEIDOMASTOID M.
VERTEBRAL A.
INTERVERTEBRAL DISC BETWEEN C3 & C4
LONGISSIMUS CAPITIS M.
DURA MATER
SPLENIUS CAPITIS M.
SPINAL CORD
DEEP CERVICAL VESSELS
SEMISPINALIS CAPITIS M.
SPINOUS PROCESS
TRAPEZIUS M.
SEMISPINALIS CERVICIS & MULTIFIDUS M.
(j)

Fig. 6.24(i)(j)

has been included. This shows a series of annotated photographs of post-mortem sections through the head at the levels indicated in Figs. 6.24(a)(g).

The ability of n.m.r. imaging to determine the level and orientation of an image slice electronically has already been discussed in § 6.3.2. The prototype Fourier imaging system developed by Siemens (Zeitler and Schittenhelm 1981), for example, could select any image plane lying within a 30-cm-diameter sphere. This ability is illustrated in Fig. 6.25,

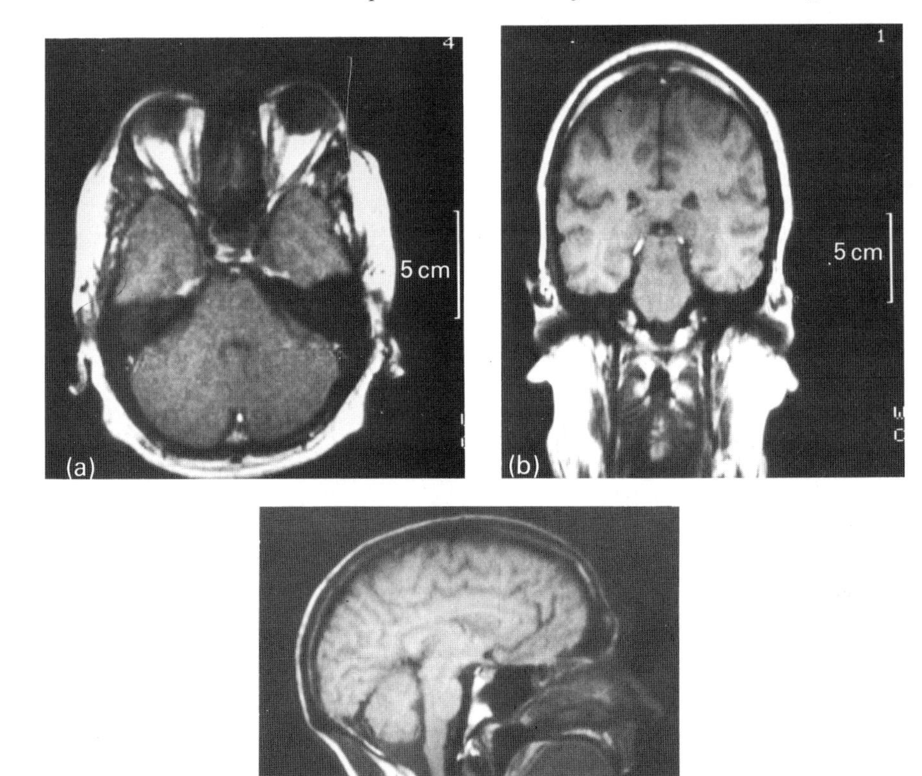

Fig. 6.25. Axial (a), coronal (b), and sagittal (c) proton n.m.r. images of the human head obtained at 64 MHz (1.5 T) using a rapidly-repeated SE method. The slice thickness was 6 mm. For (a) and (b) TE was 40 ms and TR 0.2 s; for (c) TE was 30 ms and TR 0.3 s. With such short values of TR, T_1 is also influential in determining image contrast. (Supplied courtesy of Siemens Ltd. 1984.)

which shows 6 mm axial (a), coronal (b), and sagittal (c) head scans obtained, from a healthy volunteer using a rapidly-repeated SE sequence at 64 MHz. The repetition period (TR) was either 0.2 or 0.3 s and the echo delay (TE) was either 30 or 40 ms (see caption for details). Similarly, the spatial resolution is also under electronic control and can be improved through the simple expedient of increasing the field gradient strengths. Usually, the field of view is reduced correspondingly to keep the number of image points constant. Figure 6.26(b) shows a 1-mm-resolution image, corresponding to the central region of a mid sagittal slice, an example of which is shown in Fig. 6.26(a).

For sequences with long TR values, such as used in the production of Fig. 6.26, it is possible to take advantage of the 'dead' time following signal acquisition to excite slices at other levels. In this manner a number of images (typically up to fifteen) can be simultaneously recorded. (Crooks, Arakawa, Hoenninger, Watts, McRee, Kaufman, Davis, Margulis, and De Groot 1982). Multi-slice head studies are illustrated in Figs. 6.27 and 6.28 for inversion recovery (TI = 400 ms) and spin-echo (TE = 45 ms) sequences respectively. This multislice capability should not be

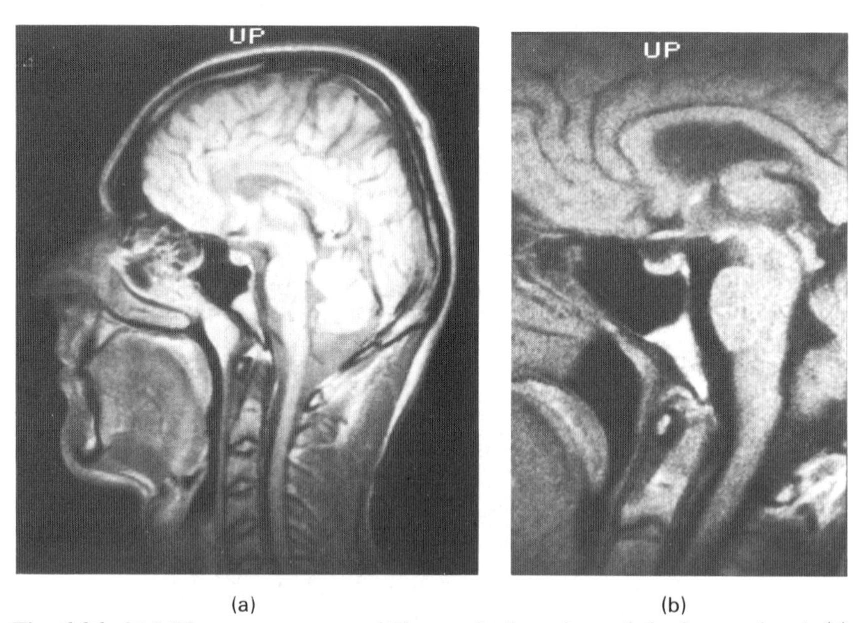

Fig. 6.26. 21 MHz proton n.m.r. midline sagittal sections of the human head. (a) At normal resolution, with TE = 30 ms and TR = 1.6 s; (b) high-resolution image of brainstem achieved by increasing field gradient amplitudes; TE = 37 ms, TR = 1.8 s. With the long TR values used here, T_2 is primarily responsible for image contrast. (Supplied courtesy of Siemens Ltd. 1984.)

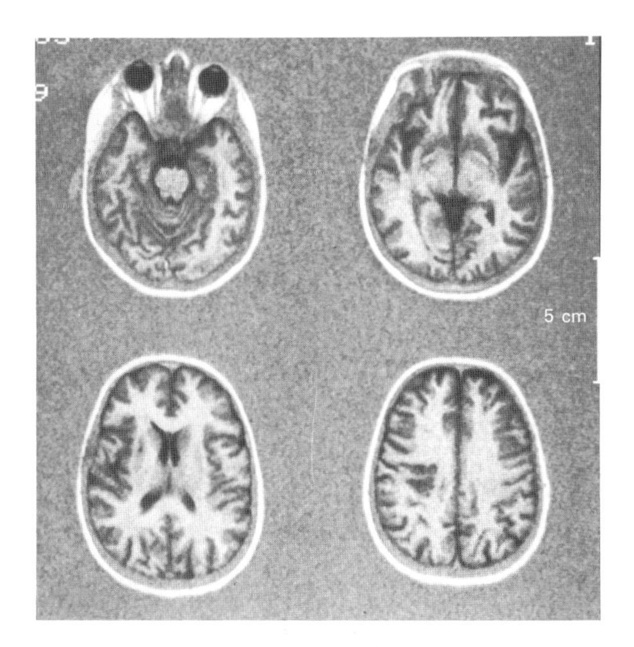

Fig. 6.27. Multislice n.m.r. axial head scans (slice thickness 6 mm) obtained at 21 MHz using an IR method (TI = 400 ms; TR = 2.2 s). Note exquisite white/grey contrast afforded by this technique. (Supplied courtesy of Siemens Ltd. 1984.)

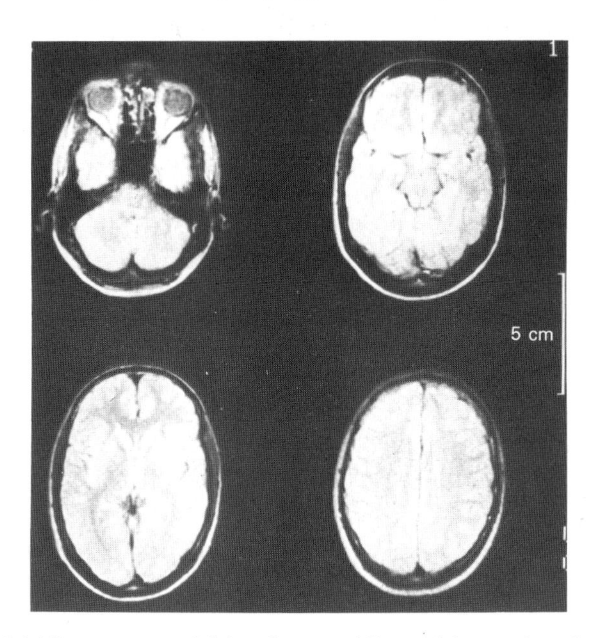

Fig. 6.28. Multislice n.m.r. axial head scans (slice thickness 6 mm) obtained at 21 MHz using a SE sequence (TE = 45 ms; TR = 1.6 s). Note reversal of white/grey contrast relative to Figure 6.27. (Supplied courtesy of Siemens Ltd. 1984.)

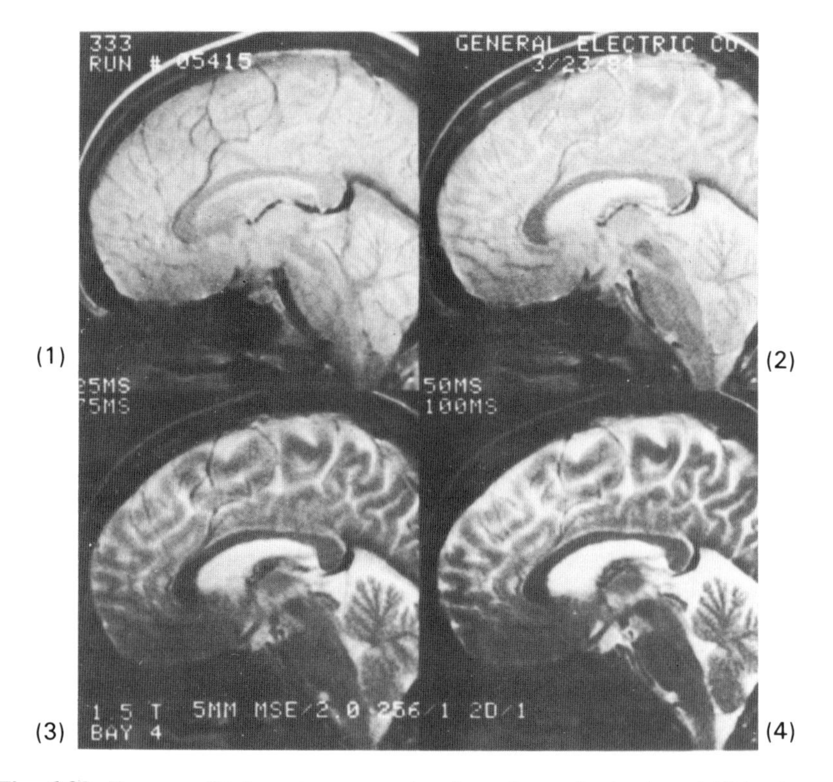

Fig. 6.29. 5 mm sagittal proton n.m.r. head sections obtained at 1.5 T using a multiple spin echo (MSE) Fourier method with TR = 2 s. Four echoes were recorded with echo delays of 25 ms TE (1), 50 ms TE (2), 75 ms TE (3) and 100 ms TE (4). (Reproduced with permission from General Electric Co. 1984.)

confused with the multiple spin-echo (MSE) sequence discussed in § 6.3.1. The latter generates a number of images at the *same* level but with different spin-echo (TE) delays. An example is shown in Fig. 6.29. These 5 mm sagittal MSE images show the cerebral vasculature especially well for short echo delays whereas the grey/white matter contrast is better appreciated at longer delays. If required, it is possible to combine multislicing and multiple-echo methods—for example, the 0.35 T prototype system described by Crooks *et al.* (1982) was used to generate five slices each with two echo-delays (at 28 and 56 ms) in the same time normally required for a single slice. As noted in § 6.3.4, full three-dimensional studies of the human head are also a feasible and attractive proposition.

We turn now to a consideration of tissue discrimination. In many cases the most striking feature of n.m.r. head scans, when compared with their

X-ray CT counterparts, is the excellent contrast between white and grey matter. This was clearly demonstrated for the first time by the Hammersmith team, using a projection reconstruction method with an IR sequence (Doyle *et al.* 1981). It is also apparent in the IR multislice study of Fig. 6.27 where the white matter is seen both centrally and peripherally to subcortical level. Note that the grey matter gives a signal of intermediate strength and is therefore grey in the image, whereas the white matter produces a more intense signal and thus appears white. The surprise comes when we look at Figs. 6.28 and 6.29, where it will be seen that this contrast is reversed. This is not merely an inversion of the grey scale, which itself can be a source of confusion (see, for example, Figs 6.16 and 6.23), but corresponds to a genuine change in the relative signal intensities. It is a prime example of the problems discussed in § 6.3.1, and underlines the point that it is essential to provide detailed information concerning the pulse sequence and acquisition parameters if images are to be compared.

In this instance it will be shown at some length how such differences can be rationalized. First consider the detailed compositions of white and grey matter which are shown in Table 6.2. It will be noted that, whereas the total proton contents are virtually identical (10.56 and 10.58 per cent respectively), they originate from differing proportions of the main proton-containing constituents. Leaving aside the proteins which are likely to have extremely short T_2 values (~10 μs; see § 2.5.2) and will therefore be lost in the time between excitation and observation, we see that grey matter has a water content which is some 10 per cent greater than that of white matter. As far as the proton concentration is concerned, the balance is almost entirely redressed by the 10 per cent greater

Table 6.2
Composition of cerebral white and grey matter.

	Total %	% H	% C	% O
WHITE MATTER				
Protein	11.29	0.82	6.24	2.58
Lipid	16.06	1.78	11.56	2.20
Water	71.60	7.96		63.64
		10.56	17.80	68.42
GREY MATTER				
Protein	10.55	0.76	5.75	2.35
Lipid	6.30	0.72	4.34	0.96
Water	81.90	9.10		72.80
		10.58	10.09	76.11

From Gore (1982).

lipid content of white matter. If we were to obtain a pure proton spin density image, we should therefore expect to see virtually no white/grey contrast. This is indeed the case: repeated f.i.d. images with an interpulse interval of 4 s, allowing plenty of time for soft tissue T_1-relaxation to occur, are largely featureless (Young *et al.* 1982*a*). It is not clear, however, to what extent the myelin contributes to the image. Membrane lipids, in contrast to depot lipids (triglycerides) tend to have short T_2 values, in the range 0.1–5 ms. Since virtually all imaging methods now include a refocusing period, any membrane lipid signal will be selectively suppressed. Interestingly, proton chemical-shift imaging studies of anaesthetized cats (see § 6.7) show no evidence of myelin, though lipids are seen in other regions, for example in retrobulbar fat.

When T_1-relaxation discrimination is introduced, either by reducing the interpulse delay (SR sequence) or by using an inversion recovery (IR) sequence, one begins to see a contrast between white and grey matter: the shorter 'effective' T_1 of white matter causing it to recover more rapidly and give rise to a greater signal amplitude as demonstrated in Fig. 6.27. What then is the explanation for the reversed contrast of Fig. 6.28? Since the interpulse delay (TR) employed was 1.6 s, T_1-relaxation effects are unlikely to be of major importance. The answer lies in the relatively-long refocusing period (TE) of 45 ms during which spin–spin relaxation occurs with a time constant T_2. Grey matter, with its higher water content, has a rather longer T_2 than white matter. Its signal is therefore less attenuated by the echo-delay (see eqn (6.13) and Fig. 6.11) and it appears correspondingly brighter in the image. A similar argument also applies for Fig. 6.29.

In an inversion recovery experiment the choice of the interval TI between 180° and 90° pulses profoundly influences the appearance of the image. Figure 6.30 which shows IR images at 6.5 MHz for TI = 500 ms (a), the usually preferred value of 400 ms (b), 300 ms (c) and 200 ms (d), illustrates the point (Young *et al.* 1982*a*). As can be seen, there is a contrast inversion which occurs between (c) and (d). The explanation for this has been given in § 6.3.1 where it was shown that the magnetization passes through zero when $\mathrm{TI} = T_1 \log_e 2 (\simeq 0.69 T_1)$. Clearly, if $\mathrm{TI} \leqslant 0.69 T_1$(white), then the signal from the grey matter has a greater amplitude (though negative sign) than that for the white matter, which approaches the null point more rapidly. This is the case in Fig. 6.30(d). Under normal conditions, one arranges that $T \geqslant 0.69 T_1$(grey) so that both signals are positive as in Fig. 6.30 (a), (b), and (c). Note that a 6.5 MHz head image is likely to be 'unstable' for TI values lying between 200 and 300 ms.

Before progressing to a discussion of brain pathology, attention is directed to some further important differences between n.m.r. and X-ray

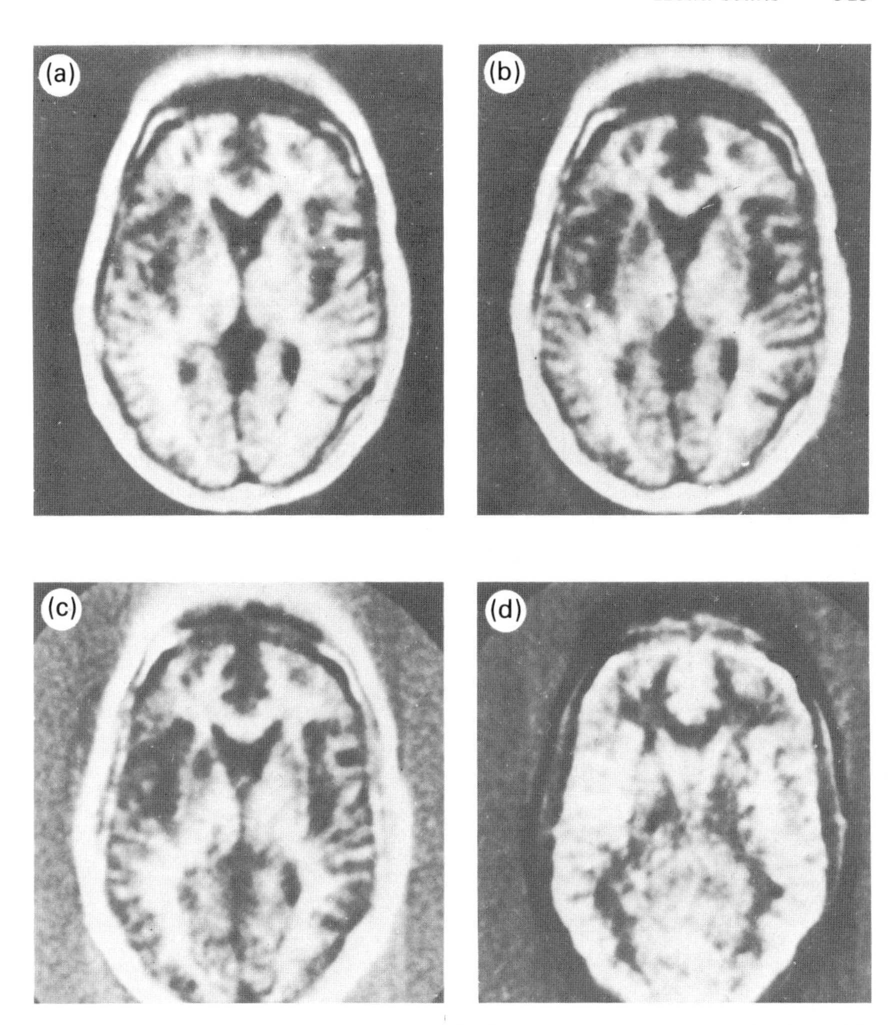

Fig. 6.30. 6.5 MHz inversion-recovery (IR) n.m.r. images of the brain, with TI = 500 ms (a), 400 ms (b), 300 ms (c), and 200 ms (d). As TI is decreased, changes in contrast are seen; note particularly the reversal of contrast in (d). (From Young *et al.* 1982*a*.)

CT scans of the normal brain. For example, the ring of high intensity seen round the periphery of the head corresponds not to the skull, as it does in the case of X-ray imaging, but rather to the subcutaneous fat of the facial tissues. The bone itself has a very weak n.m.r. signal and in consequence appears as an underlying dark ring. Occasionally, bone marrow is visible as a high-intensity region within. This is clearly seen in Figs 6.25; 6.26(a), for example. Cerebrospinal fluid (c.s.f.) has a particularly long T_1 value of

1.5 s at 6.5 MHz (Young *et al.* 1982) compared with a typical range for soft tissues of 0.2–0.5 s. Its T_2 is also long. Under standard imaging conditions, designed to provide good soft tissue contrast, there is usually insufficient time for relaxation recovery, and the ventricular system is well visualized as a low-intensity structure (Figs 6.25, 6.26, and 6.27). If the interpulse interval is increased in a SR experiment, the c.s.f. starts to appear bright, and, in view of its slightly greater proton content, should

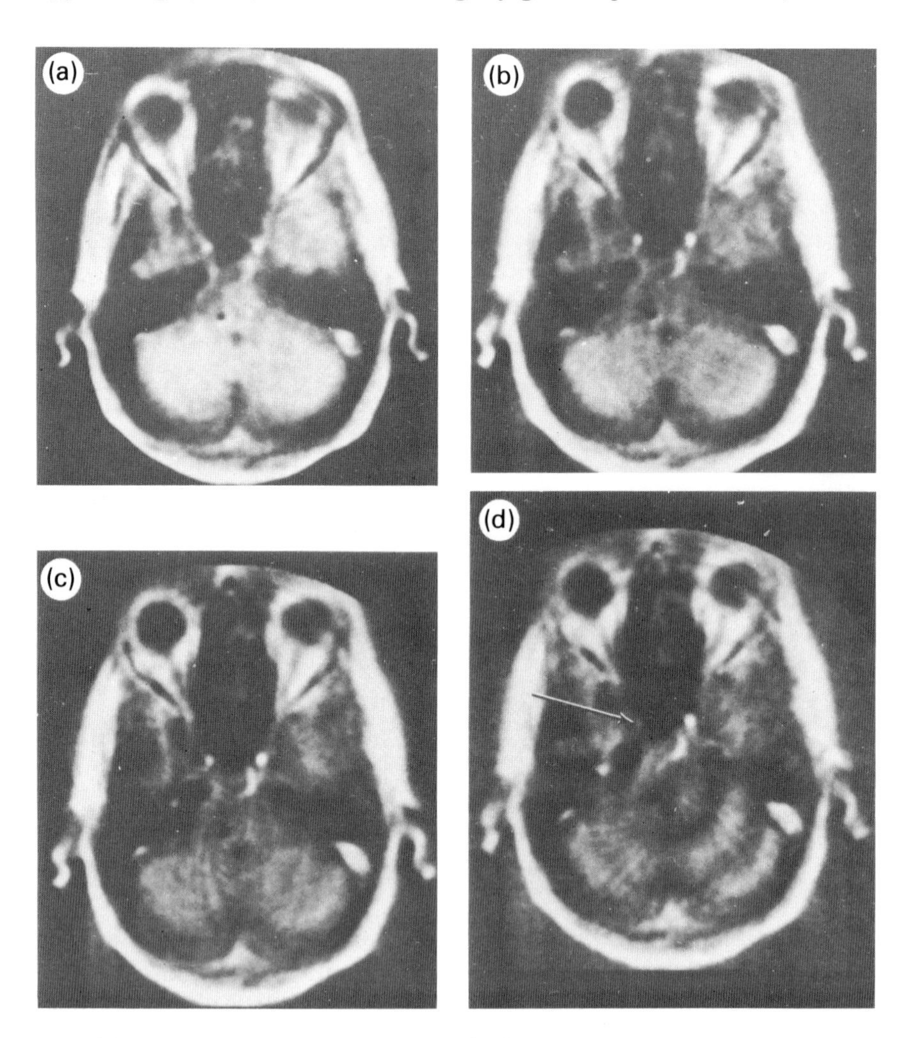

Fig. 6.31. 6.5 MHz saturation recovery (SR) images of the brain, with TR = 1000 ms (a), 300 ms (b), 150 ms (c) and 75 ms (d). As the repetition rate is increased, blood vessels are highlighted. Note, however, that in (d) the left internal carotid artery (arrow) is less clearly seen. (From Young *et al.* 1982*a*.)

ultimately become predominant. However, even with a delay of 4 s it is barely distinguished by this method. SE images with long TE intervals do, however, show the ventricular system as a bright region due to the long T_2 of c.s.f.. See, for example, the image recorded from the last spin echo (TE = 100 ms) in Fig. 6.29. Pure T_1 and T_2 images also give good visualization of the ventricular system (Zeitler and Schitenhelm 1981).

It is interesting to compare n.m.r. orbital sections (see, for example, Fig. 6.25(a) and Fig. 6.27) with their X-ray CT counterparts. In the n.m.r. images the orbits are clearly seen as low-intensity circles highlighted to the rear by the very intense signal from the retrobulbar fat in which the optic nerve shows as a dark line. With X-ray CT the fat has a low attenuation and appears as a dark region crossed by the high-intensity line of the optic nerve. X-ray images at this level are notorious for artefacts arising from the posterior fossa. These are not present in the n.m.r. sections, which show the cerebellum and brainstem clearly outlined by the c.s.f.-filled basal cisterns. The carotid arteries are well seen at this level and can be highlighted by decreasing the interpulse interval in repeated f.i.d. images, as illustrated in Fig. 6.31. The signal from the blood within the arteries remains strong, as a result of the short effective relaxation time caused by the motion. Asymmetries in flow can also be observed and will clearly be of major importance in assessing the danger of brain damage due to loss of blood supply (stroke). It is also possible to make quantitative blood-flow measurements in the head. For example, Singer and Crooks (1983) have used a presaturation method to highlight flow in several of the principal vessels and to determine flow rate in the left and right external jugular veins (see § 6.3.6 for a discussion of this and other n.m.r. flow measurements).

6.5.2. Pathology

Since the first report in 1980 of intracranial pathology demonstrated by n.m.r. (Hawkes *et al.*), many imaging teams, both academic and industrial, have published their own preliminary clinical assessments of the method. Whilst it is still too soon to make definitive comparisons between n.m.r. and other head-scanning techniques, the prognosis, at least for the common pathologies so far investigated, seems excellent. In this section we shall be concerned almost exclusively with brain lesions, though it is important to note that n.m.r. may also have a role to play in the detection of organic brain diseases. For example, Smith and co-workers at Aberdeen (Smith 1982) have been able to follow the treatment of alcoholic brain disease through the observed changes in spin–lattice relaxation time. They speculate that other organic diseases such as presenile dementia and schizophrenia, may also be susceptible to n.m.r. study.

Space-occupying lesions may be seen directly, if they have proton densities or relaxation times which differ significantly from those of the surrounding material. If this contrast is insufficient, or they are located in an adjacent slice, their presence can often be inferred through the mass effect which leads to displacement of surrounding structures. Additionally, the white/grey contrast afforded by n.m.r. gives local boundaries for assessing such effects. In many cases of uncertainty the symmetry of left and right hemispheres provides the all important clue. The characterization of lesions in terms of their n.m.r. properties, in much the same way that EMI or Hounsfield units are assigned for X-ray studies, is of vital importance. However, as we have seen, standardization is made rather more difficult both because of the diversity of imaging methods with their different dependencies on n.m.r. parameters, and also through the use of different operating frequencies, on which the spin–lattice relaxation time, in particular, is strongly dependent (see § 6.2.3). If images showing pure relaxation times or spin densities can be extracted this problem is alleviated, at least on the first of these counts. In their preliminary clinical assessment Hawkes *et al.* (1980) were able to demonstrate the presence of a variety of tumours, including an internal bone-tumour of the vault, a glioma, a chromophobe adenoma, and a craniopharyngioma. All of these were visualized as regions of abnormally-high signal with the steady-state free precession based projection-reconstruction method used (see eqn (6.8)). Coronal and sagittal sections were used in addition to the axial ones to help in determining the extent of the tumour. Tumours invading the sinuses were clearly seen, because of the dramatic increase in signal intensity that they cause in this region and obstructions caused by mucosa or pus were similarly apparent.

The Hammersmith team have studied a large number of brain tumours using SR, IR, and SE sequences (Doyle *et al.* 1981; Young *et al.* 1982*b*; Bydder, Steiner, Young, Hall, Thomas, Marshall, Pallis, and Legg 1982*c*; Bailes, Young, Thomas, Straughan, Bydder, and Steiner 1982; Bydder, Steiner, Thomas, Marshall, Gilderdale, and Young 1983). In agreement with the early *in vitro* observations of Damadian (1971), the T_1 values are found to be substantially elevated, both for intrinsic tumours and for metastases. This renders the lesions visible as regions of low intensity on standard IR scans (TI = 400 ms; 6.5 MHz) and as high-intensity regions on short IR (TI = 200 ms; 6.5 MHz) scans. The tumour is sometimes surrounded by an oedematous ring, particularly if it is malignant. Although this is something of a nuisance from the point of view of accurate size-assessment, it is nevertheless usually possible to choose scan conditions which allow some separation of the two regions. In any event, it has the advantage of presenting a larger target for detection, which may be valuable if the T_1 elevation of the tumour is less pronounced. This is true,

Table 6.3
Proton T_1-relaxation times at 6.5 MHz for brain tumours.

Tumour		*Number of cases*	T_1 *range* (ms)
Benign	Meningioma	3	520–600
	Acoustic neuroma:		
	Noncystic	2	570–700
	Cystic	1	1070
	Prolactinoma	1	580
Malignant	Glioma	3	750–1520
	Metastasis	7	510–1370

From Bydder, Steiner, Young, Hall, Thomas, Marshall, Pallis, and Legg 1982*c*.

for example, in cases of adenocarcinoma of the nasopharynx (Bydder *et al.* 1983). T_1 values obtained *in vivo* at 6.5 MHz are shown in Table 6.3 for a number of tumours. These may be compared with normal values for white and grey matter of 290 ms and 525 ms respectively, whence the prolongation is evident.

Spin-echo sequences also provide excellent tumour discrimination via the associated increase in T_2 value. An example is given in Fig. 6.32 which shows a large glioblastoma seen as a region of high signal intensity in the front left hemisphere. It was obtained with a prototype imaging system built by Siemens (Loeffler and Oppelt 1981) using a SE sequence.

N.m.r. is particularly well-suited to imaging of the posterior fossa, since it is untroubled by bone artefacts which plague X-ray CT scans at this level. A number of reports (see, for example, Bydder *et al.* 1983; McGinnis, Brady, New, Buonanno, Pykett, De La Paz, Kistler, and Taveras 1983) have demonstrated the ability of n.m.r. to detect a wide range of tumours in this region and it may well become the method of choice for such applications. An example is shown in Fig. 6.33.

It is possible that the use of n.m.r. contrast agents (see § 6.3.5) will further enhance the ability of the method to delineate tumours. Carr, Brown, Bydder, Weinmann, Speck, Thomas, and Young (1984) have used gadolinium diethylenetriaminepentaacetic acid given intravenously at a dose of 0.1 mmol kg^{-1} for this purpose. Figure 6.34 shows an example from this study. After contrast administration, the astrocytoma grade IV displays ring enhancement marking the boundary between tumour and oedema.

Vascular lesions are also of major importance. Cerebral infarction is generally better visualized by n.m.r. than by X-ray CT (Doyle *et al.* 1981; Young *et al.* 1982*b*; Bydder *et al.* 1982*a*; Buonanno *et al.* 1982, 1983;

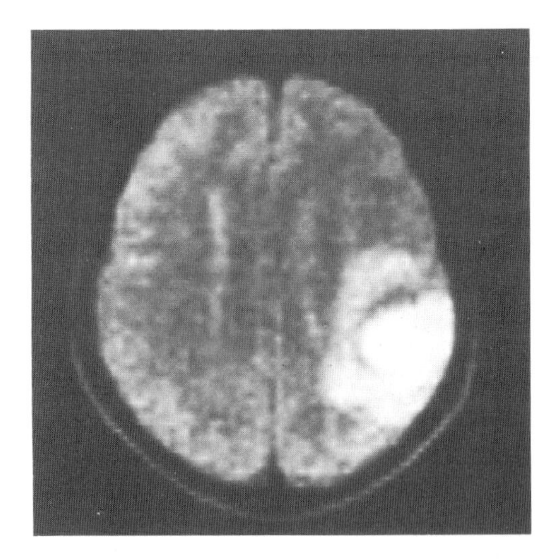

Fig. 6.32. Axial SE head scan at 5 MHz showing a large glioblastoma (white mass in left hemisphere of the brain). (Courtesy of Siemens Ltd. 1981.)

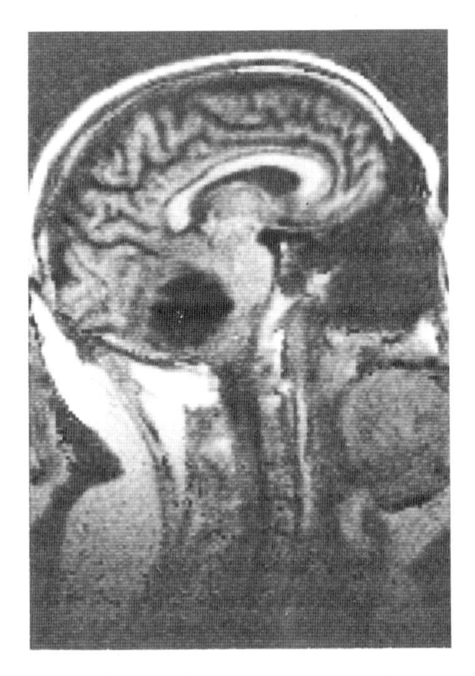

Fig. 6.33. Sagittal IR (TI = 400 ms) head scan at 6.5 MHz showing large posterior fossa tumour (dark mass). (Courtesy of Picker International Ltd. 1983.)

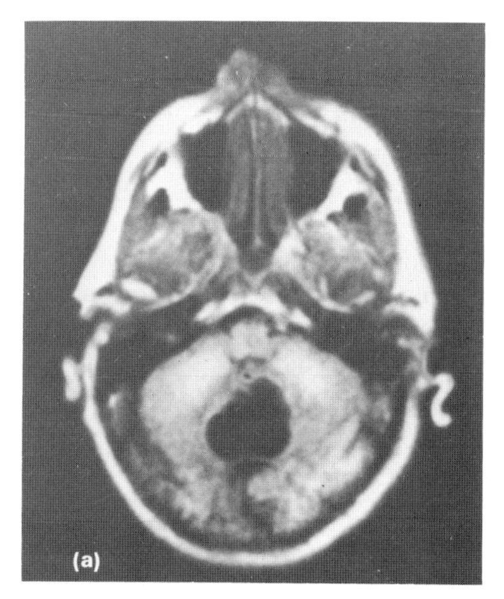

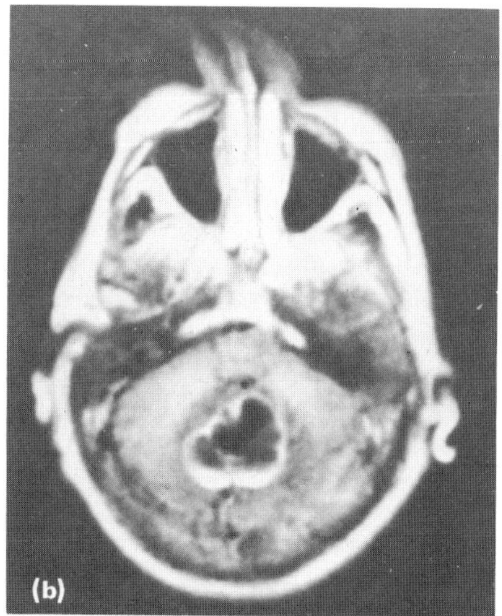

Fig. 6.34. Axial transverse inversion recovery (TI = 500 ms; TE = 40 ms; TR = 1.5 s) head scan of a patient with a grade IV astrocytoma before (a) and after (b) i.v. Gd-DTPA (0.1 mmol/kg) injection. Ring enhancement of the tumour is clearly evident. (From Carr *et al.* 1984.)

see also the animal stroke models discussed in § 6.4.3). It is characterized by a loss in white/grey contrast and an increase in spin–lattice relaxation rate. Mass effects are also evident. Hemispheric infarcts are usually peripheral and wedge-shaped (Bydder *et al.* 1982*c*). Their full extent can be determined using three dimensional techniques (see, for example, Pykett, Buonanno, Brady, and Kistler 1983). They have been seen as early as six hours following the onset of stroke and have been serially monitored for periods of up to two months (Sipponen, Kaste, Ketonen, Sepponen, Katevuo, and Sivula 1983). In the early stages white/grey contrast is preserved, but this is soon lost as the progressive oedema destroys the tissue. In lacunar infarcts (Bydder *et al.* 1983) it is possible for the infarct to be confined to either grey or white matter and thus preserve structural contrast. Although many of the early studies were carried out using IR sequences, spin-echo sequences are also effective in demonstrating infarctions via the prolongation of spin–spin relaxation time which is also present (Bailes *et al.* 1982; Bryan, Willcott, Schneiders, Ford, and Derman 1983). An exception to the rule of T_1-prolongation occurs in the case of haemorrhagic infarcts which at certain stages in their evolution have long T_1s on the exterior but short T_1s in the blood-filled interior.

In cases of intracerebellar haemorrhage (Doyle *et al.* 1981; Young *et al.* 1981, 1982; Bydder *et al.* 1983) it is common to find a peripheral short T_1 region, corresponding to clotted blood surrounding a central region with longer T_1. Spin-echo images typically show increased intensity due to the long T_2 of blood (Bailes 1982; Bydder 1983; Crooks, Ortendahl, Kaufman, Hoenninger, Arakawa, Watts, Cannon, Brant-Zawadzki, Davis, and Margulis 1983).

Aneurysms can be distinguished from brain (Hawkes *et al.* 1980; Doyle *et al.* 1981) through the shortening of the spin–lattice relaxation times associated with the clotting of blood or by flow effects; an example is shown in Fig. 6.35, where the aneurysm appears as a bright spot with the inversion recovery method employed. An example of arteriovenus malformation is shown using SR (a) and SE (b) methods in Fig. 6.36.

The ability of n.m.r. imaging to generate coronal, and sagittal, as well as axial sections without recourse to secondary reconstruction is of immense value in spinal cord studies, especially in the region of the cranio-vertebral junction. The point of entry of the cord into the skull, the foramen magnum, is well demarcated by the high-intensity signal, seen under most scanning conditions, arising from the fat in the marrow of the petrous and occipital bone (see, for example, Fig. 6.26(b)). The cord itself is clearly seen at all levels without need of intrathecal contrast agent. The extent of any abnormalities present can be easily mapped and

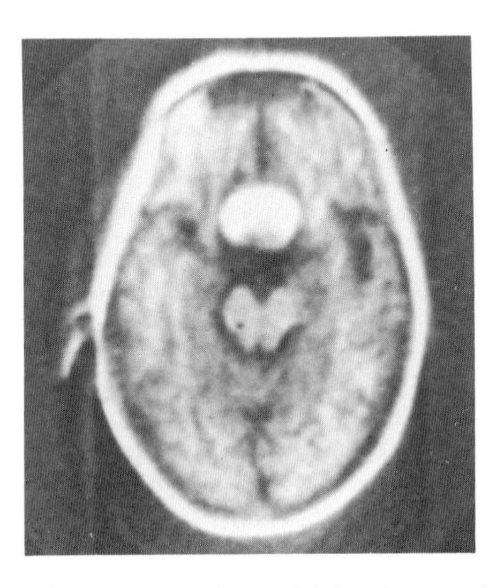

Fig. 6.35. Axial IR (TI = 400 ms, TR = 1.4 s) head scan of a patient with an aneurysm of the anterior communicating artery. The white region within the lesion corresponds to clotted blood. (From Doyle *et al.* 1981*b*.)

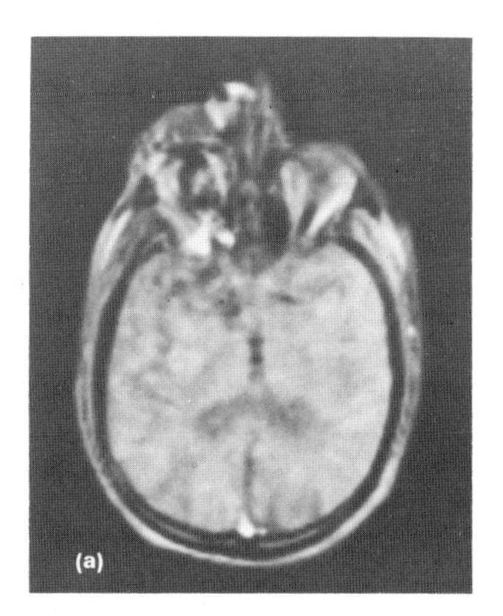

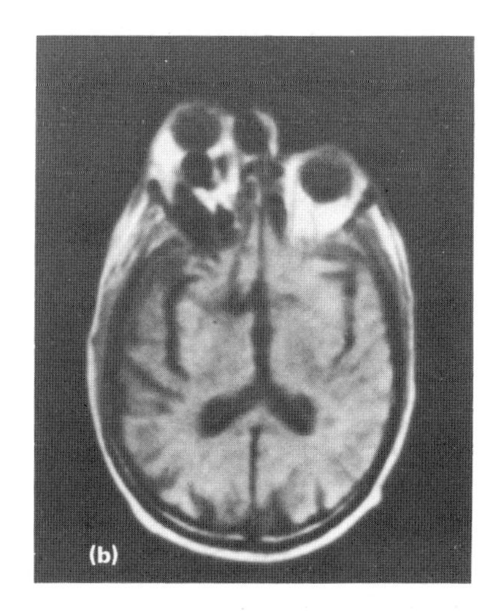

Fig. 6.36. Axial SR (a) and SE (b) head scans at 6.5 MHz of a patient with arteriovenous malformation. (Courtesy of G. M. Bydder, Hammersmith Hospital, London 1984.)

tumours are generally distinguished from c.s.f.-filled lesions on the basis of relaxation time differences. N.m.r. is likely to become the modality of choice for the diagnosis of Arnold Chiari malformation, which may well be possible on an outpatient basis. In such cases the herniation of the cerebellar tonsils is readily apparent (see, for examples, De La Paz, Brady, Buonanno, New, Kistler, McGinnis, Pykett, and Taveras 1983; Bydder *et al.* 1983; Modic, Weinstein, Pavlicek, Starnes, Duchesneau, Boumphrey, and Hardy 1983). The extent of the central cystic cavity in cases of syringomyelia can be delineated, and hydromyelia is similarly well diagnosed (De La Paz *et al.* 1983).

The ability to distinguish white and grey matter so effectively led Doyle *et al.* (1981) to suggest that n.m.r. imaging may be particularly useful in the demyelinating diseases, of which multiple sclerosis is an important example. This was amply demonstrated in 1981 (Young *et al.*) by the publication of a series of n.m.r. head scans showing multiple sclerosis lesions. Figure 6.37 is taken from this work and shows a comparison of X-ray CT and n.m.r. results. In addition to the two posterior periventricular lesions seen in the X-ray scans (large arrows), the n.m.r. image also shows six smaller lesions on the lateral margin of the lateral ventricles (small arrows). The lesions are seen as well-defined regions of low signal intensity in this imaging mode, corresponding to the increased spin–lattice relaxation times associated with the loss of white matter. Young *et al.* (1981) developed a set of diagnostic criteria, including a minimum size

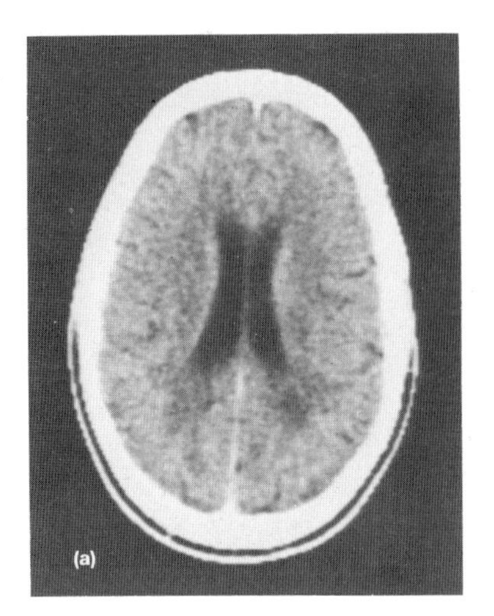

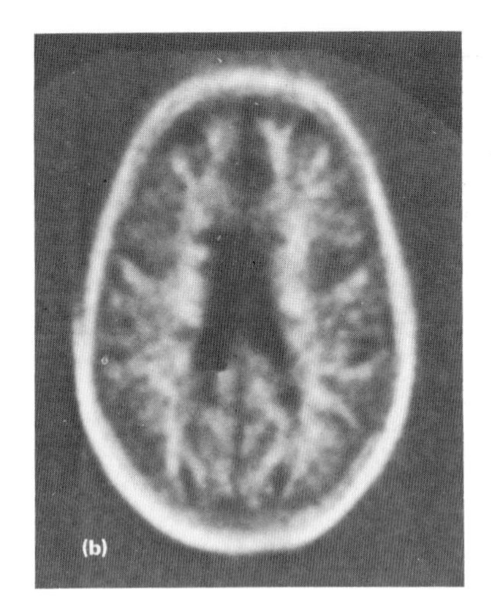

Fig. 6.37. Comparison between X-ray CT (a) and n.m.r. (b)—axial head scans at the midventricular level in a patient with multiple sclerosis. The n.m.r. scan was obtained using an IR sequence (TI = 400 ms, TR = 1.4 s) at 6.5 MHz. (From Young *et al.* 1981.)

limit of 4 mm × 3 mm, which effectively excluded possible suspect areas in normal volunteers. Applying these criteria, Young *et al.* were able to detect a total of 131 multiple sclerosis lesions in ten patients using IR n.m.r. compared with only 19 using contrast-enhanced X-ray CT.

SE sequences have proved to be of particular value in the delineation of multiple sclerosis (m.s.) lesions. Using a TE value of 80 ms at 6.5 MHz, Bailes *et al.* (1982) found that the lesions appeared brighter than both brain and c.s.f. This greatly reduced the danger of false positives from partial volume artefacts and allowed peripheral as well as periventricular lesions to be confidently assigned. Using this approach, Young, Randell, Kaplan, James, Bydder, and Steiner (1983) were able to reduce their minimum size criterion to 3 × 3 mm, finding 47 lesions in five patients, compared with 33 by IR n.m.r., and 8 using post-contrast X-ray CT. An example of the use of SE methods for detecting m.s. lesions is shown in Fig. 6.38, where the facility for sagittal as well as axial sections is also used to good advantage. Apart from the pioneering work at the Hammersmith, m.s. has also been studied by a number of other teams; for example, at Massachusetts General Hospital (Buonanno *et al.* 1982) and the University of California at San Francisco (Crooks, Mills, Davis, Brant-Zawadzki, Hoenninger, Arakawa, Watts, and Kaufman 1982). Multiple sclerosis was the first proven clinical application for n.m.r. imaging and an n.m.r. system has now been supplied solely for this purpose.

N.m.r. imaging of the paediatric brain has also proved to be a fruitful

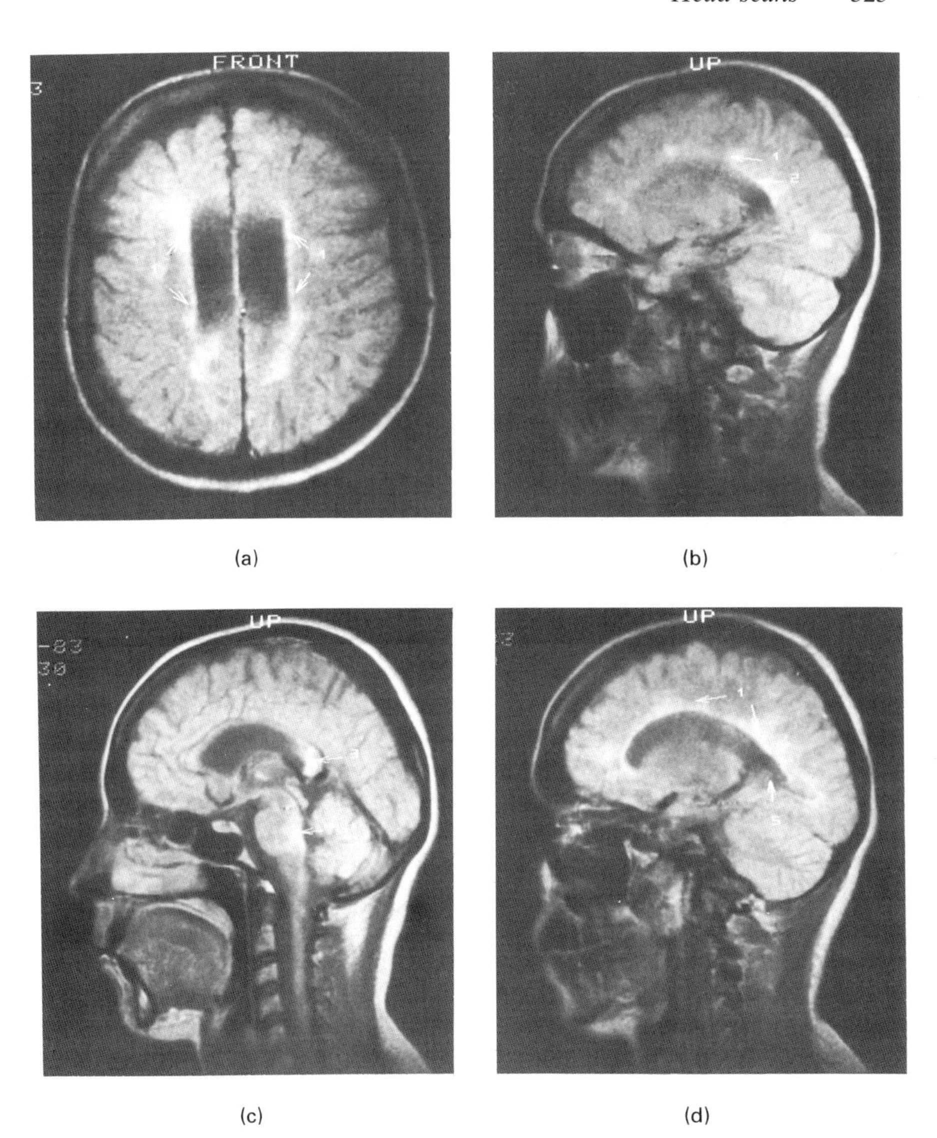

(a) (b)

(c) (d)

Fig. 6.38. Multiple sclerosis lesions (arrowed) highlighted using a SE sequence (SE = 30 ms, TR = 1.6 s) at 21 MHz. Note use of both axial and sagittal sections to identify extent of lesions, and contrast reversal relative to Fig. 6.37. (Courtesy of Siemens Ltd. 1983.)

area of research, again pioneered by the Hammersmith team (Levene, Whitelaw, Bubowitz, Bydder, Steiner, Randell, and Young 1982; Johnson, Pennock, Bydder, Steiner, Thomas, Hayward, Bryant, Payne, Levene, Whitelaw, Dubowitz, and Dubowitz 1983; see also the spectroscopic studies of paediatric brain discussed in § 2.7.1). The process of

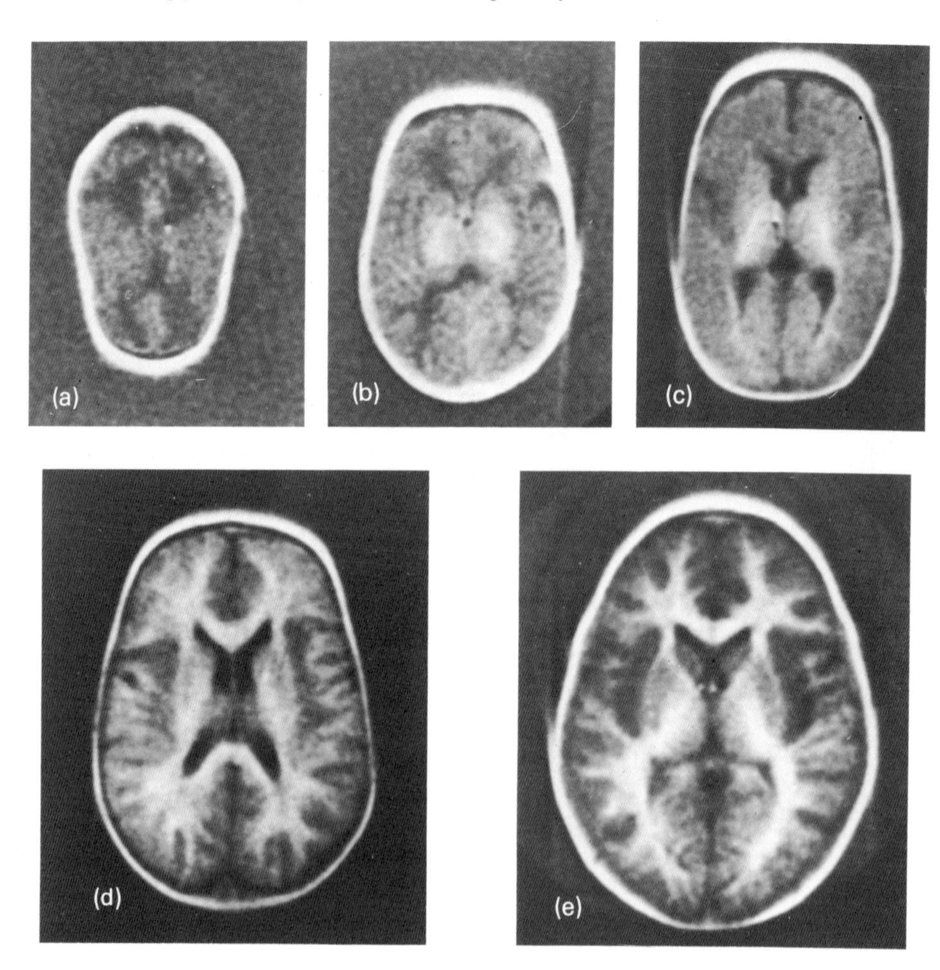

Fig. 6.39. Normal IR (TI = 600 ms, TR = 1.8 s) scans at 6.5 MHz of paediatric brain demonstrating the process of myelination (a) 36 weeks post menstrual age; (b) 2 weeks postnatal age; (c) age 6 months; (d) age 20 months and (e) age 5 years. (From Johnson *et al.* 1983.)

myelination was beautifully demonstrated: from a virtual absence of white matter in a neonate of 36-weeks postmenstrual age to a level approaching that seen in adults at age 9 yrs. This is shown in Fig. 6.39. A large number of paediatric neurological disorders have also been examined. Hydrocephalus, for example, is readily observed, particularly on SE scans, and it is possible to assess the patency of ventricular shunts used in its treatment. However, it is necessary to be sure that the shunt is of a non-ferromagnetic variety (see § 6.8).

6.6. Whole-body studies

Although, as we have seen, whole-body n.m.r. imaging presents rather greater difficulties than head scanning, publication of the first whole-body cross-section by Damadian *et al.* in 1977 nevertheless preceded the first head-scan by about a year. That first thoracic section took some five hours to produce, using the f.o.n.a.r. point-scanning method (see § 3.2.1). Although crude by present-day standards, the basic anatomical features were recognizable, and it was clear that n.m.r. imaging was about to come of age. Other groups followed rapidly with line (Mansfield *et al.* 1978; Mallard *et al.* 1980) and, finally, planar methods (Edelstein, Hutchison, Johnson, and Redpath 1980; Edelstein, Hutchison, Smith, Mallard, Johnson, and Redpath 1981; Bangert, Mansfield, and Coupland 1981; Luiten 1981; Luiten, Locher, van Uijen, van Dijk, and den Boef 1982; Hawkes, Holland, Moore, Roebuck, and Worthington 1981*a*, *b*; Young *et al.* 1982*a*, *b*) bringing imaging times down to a few minutes with resolutions ~3 mm in slices of about 2 cm thickness. Instrumentational developments have allowed the resolution to be improved to 1–2 mm in slices of 5–10 mm thickness. Additionally, multislicing capabilities enable up to about 15 slices to be simultaneously gathered (see, for example, Crooks *et al.*, 1983).

6.6.1. Thoracic studies

N.m.r. sections through the upper mediastinum, particularly those obtained using SE sequences, clearly demonstrate the principal anatomical features, including the aorta and great arteries arising from the aortic arch, the systemic venous system, and portions of the main airways—notably trachea, main bronchi, lobar bronchi, and occasionally segmental bronchi (see, for example, Alfidi, Haaga, El Yousef, Bryan, Fletcher, Li Puma, Morrison, Kaufman, Richey, Hinshaw, Kramer, Yeung, Cohen, Butler, Ament, and Lieberman 1982; Gamsu, Webb, Sheldon, Kaufman, Crooks, Birnberg, Goodman, Hinchcliffe, and Hedgecock 1983). The presence of rapid flow, as we have seen, generally destroys the signal from blood so the vascular lumina appear void. This is well-illustrated in the SE sagittal section of Fig. 6.40, where the subclavian and carotid arteries can be seen emanating from the aortic arch. The lack of signal from blood vessels is of great value in their distinction from hilar or mediastinal tumours without recourse to the intravenous contrast agents which are generally required for X-ray CT studies (Cohen Creviston Li Puma, Bryan, Haaga, and Alfidi 1983). One possible source of confusion is that of tumours with hilar fat. Fortunately, however, they can be distinguished on the basis of their different relaxation properties: fat is

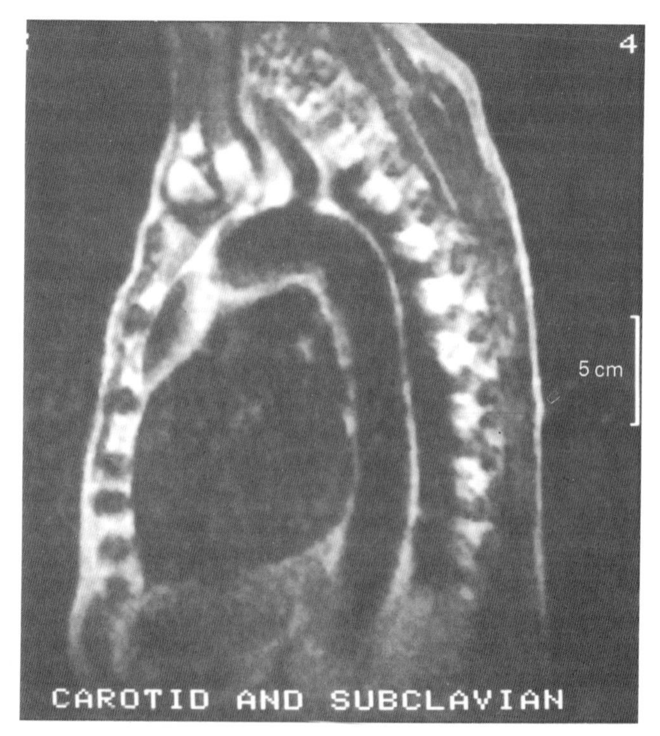

Fig. 6.40. Sagittal thoracic section bisecting the aorta obtained at 21 MHz using a SE sequence (TR = 500 ms). The carotid (proximal to the heart) and subclavian arteries are clearly seen leaving the aorta near the aortic arch. Note lack of signal from vessel lumina. (Courtesy of Siemens Ltd. 1983.)

characterized by a long T_2 and short T_1, whereas tumours virtually always have long T_1 and T_2 values. Thus, a reduction in the repetition period TR will selectively diminish signal from tumour. A second possible source of confusion is with collapsed lung tissue, where oedema can result in similar spin density and relaxation properties. Pulmonary tumours are, however, generally well visualized; in fact, they were the first to be observed *in vivo* by a whole-body n.m.r. imaging technique (Damadian, Goldsmith, and Minkoff 1978).

Other lung disorders—for example, pulmonary oedema—are likely to be visualized (Lauterbur *et al.* 1976), and an accurate quantification of lung water should also be possible (see Hayes, Case, Ailion, Morris, Cutillo, Blackburn, Durney, and Johnson 1982, for details of a small-scale imaging system capable of determining lung water to an accuracy of about 1 per cent). ^{19}F n.m.r. has also been used for lung ventilation studies (see § 6.7). It is interesting to note that, in spite of the relatively-long imaging times required (typically, about 5 min), breathing motions do not degrade the image quality too seriously, and small vessels are depicted with something approaching the static resolution of the system.

Even more interestingly, this is also true to a degree for cardiac studies. Early workers in the field were able to visualize the cardiac wall and interventricular septum for example (Edelstein *et al.* 1980; Hawkes *et al.* 1981*a*; Bydder *et al.* 1982*a*). Presumably, the walls were seen as 'frozen' in the configuration in which they spent most time, namely diastole. Heart conditions which restrict motion or blood flow can allow finer details to be observed in n.m.r. images. For example, in cases of severe coronary artery disease the heart chambers and vascular lumina are often clearly depicted (Herfkens *et al.*, 1983*b*). Similarly, Kaufman, Crooks, Sheldon, Hricak, Herfkens, and Bank (1983) found that the left ventricular wall was clearly visible in one patient because of the dyskinesia resulting from a cardiac aneurysm. Much crisper cardiac images can, however, be obtained on a routine basis if scanning is gated, either from the pulse via a pressure transducer (Steiner, Bydder, Selwyn, Deanfield, Longmore, Klipsten, and Firmin 1983), or perferably from the e.c.g. (see, for example, Alfidi *et al.* 1982; Van Dijk 1984*b*). In the latter case it is usually necessary to equip the e.c.g. leads with r.f. filters to avoid introduction of noise to the receiver coil. However, this is straightforward and does not introduce new problems. Figure 6.41 shows an example of a normal heart study, performed with different types of gating. The improvement in image quality when e.c.g. and/or respiratory gating is employed is evident. Images which have been e.c.g. gated to correspond to diastole and mid-systole are shown in (c) and (d), respectively. Among the many internal details visible are valve leaflets, septae, and papillary muscles. The chambers are clearly distinguished allowing the measurement of ejection fraction from a series of e.c.g.-gated contiguous slices without the need for the artificial contrast enhancement required by X-ray CT and radioisotope studies. The method is well suited to the study of a number of heart conditions. Figure 6.42 shows two examples of cardiomyopathy: (a) and (b) are axial and coronal SE sections through the heart of a patient with a small hypertrophic left ventricle, whereas (c) and (d) are similar sections from a patient with a dilated left ventricle. It has recently been possible to demonstrate cardiac infarcts (Go, MacIntyre, Meaney *et al.* 1985) and much effort has been expended in the development of n.m.r. contrast agents to aid such diagnoses; see § 6.3.5 for examples of progress in this area. As an alternative to gated methods, it is possible to use ultra-high-speed echo-planar imaging techniques to produce images (albeit of lower S/N) in times as a short as a few tens of milliseconds. Cross-sectional movie images showing cardiac motion and blood circulation have already been obtained in an intermediate-sized

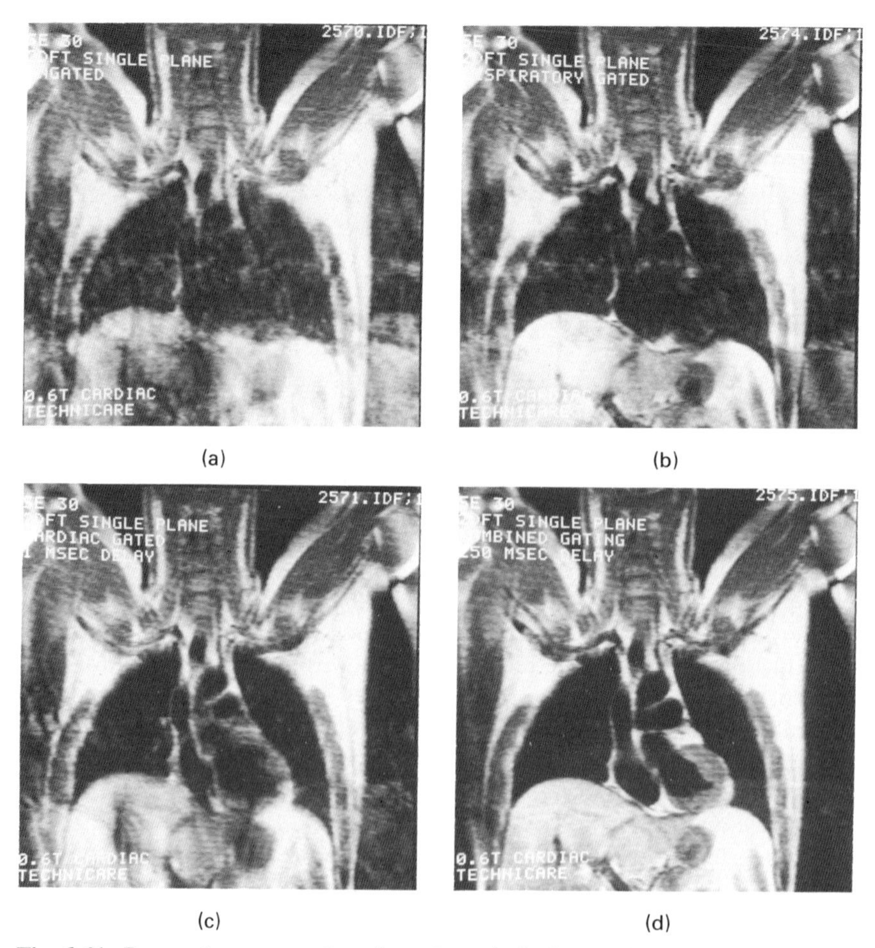

Fig. 6.41. Proton 1 cm coronal sections through the human heart obtained using a two-dimensional Fourier technique at 0.6 T. (a) No gating; (b) respiratory gated; (c) e.c.g. gated to correspond to end diastole; (d) e.c.g. and respiratory gated to correspond to mid systole. (Reproduced with permission from the Technicare Corporation 1984.)

imaging system (see § 6.4.3). Given that ischaemic cardiovascular disease is the single most important medical problem facing the Western world today, the possibilities for a full-sized human system are self-evident.

The absence of signal from rapidly flowing blood enables one to observe obstructions within the vascular system. Clotted blood, for example, gives a high signal due to its relatively short T_1 and long T_2 (see discussion of brain pathology); non-clotted blood has a longer T_2. Herfkens *et al.* (1983*b*) have used this fact to assess the patency of a coronary by-pass. Since vessel walls are often clearly visualized, it is also possible to highlight other vascular defects. In the sagittal section through the normal aorta shown in Fig. 6.40, the anterior and posterior walls appear smooth and uniform. However, it is possible to detect aneurysms as localized dilations and atherosclerotic plaques as regions of non-zero signal intensity within the lumen (Herfkens *et al.* 1983*b*). It is also possible, given the high lipid content of plaques, that they could be identified whilst still fully attached to the vessel wall. Crooks, Sheldon, Kaufman, and Rowan (1982) have demonstrated that this should be possible by imaging anatomical specimens. Quantitative determinations of blood flow velocity and heart wall motion can now be made on a routine basis. See § 6.3.6 and references therein for a discussion of suitable methods.

Breast cancer is another area of major concern. Successful imaging studies have been conducted both with small tissue samples (Bovée *et al.* 1980) and simple mastectomy specimens (Mansfield *et al.* 1979). More recently, Ross, Thompson, Kim, and Bailey (1982) have used n.m.r. imaging and point T_1 determinations for the *in vivo* evaluation of human mammary tissue, and have correlated their findings with those of other diagnostic modalities. They found the shortest T_1 values in cases of extensive fatty replacement, a wide range of T_1 values for mammary dysplasia, elevated T_1 values for neoplasms and markedly elevated T_1 values (beyond the range of neoplasms) for fluid-filled cysts. Planar (10 mm slice; 4 min scan time) n.m.r. imaging methods have also been applied to the study of mammary tissue, using a purpose-built surface coil (El Yousef, Alfidi, Duchesneau, Hubay, Haaga, Bryan, Li Puma, and Ament 1983). The presence of a cyst and an infiltrating ductal carcinoma were demonstrated, and n.m.r. images were found to correlate well with mammograms.

6.6.2. *Abdominal studies*

In spite of the considerable motion in the abdomen and the relatively long imaging times required (~few minutes), the anatomical detail is revealed with remarkable clarity. There are, for example, no linear artefacts caused by stomach movements, which often mar X-ray CT scans at this level. All the principal organs, including the adrenal glands (see, for example, Young *et al.* 1982; Moon, Hricak, Crooks, Gooding, Moss, Engelstad, and Kaufman 1983*b*; Rupp, Reiser, and Stetter 1983) are seen, with their contours often clearly defined by intervening planes of fatty tissue. The main vascular structures: aorta, vena cava, renal vessels, and mesenteric vessels are also well seen. As was the case for head and

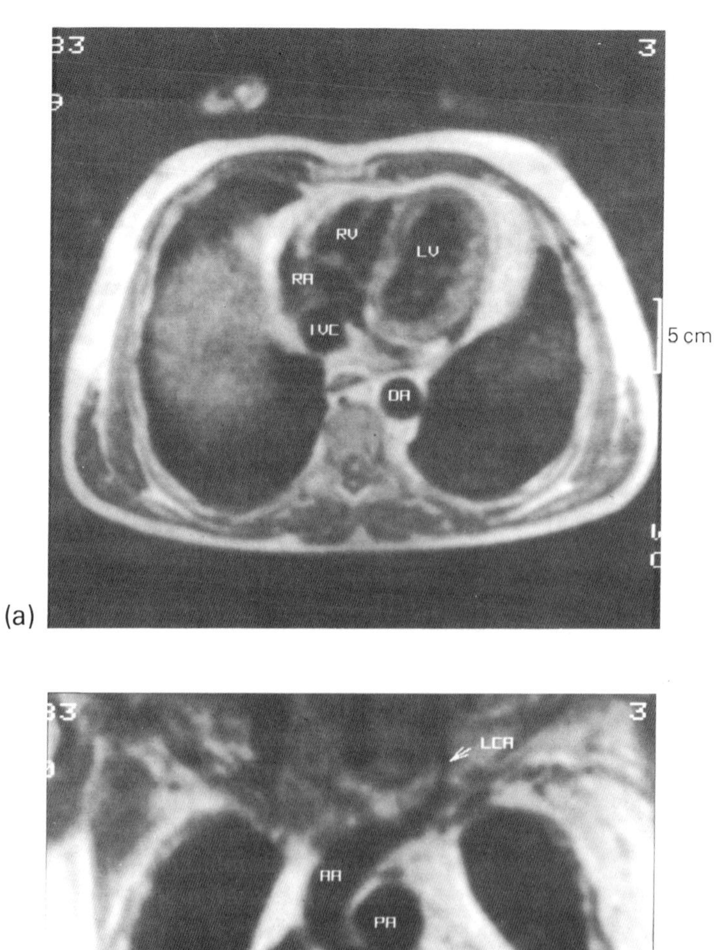

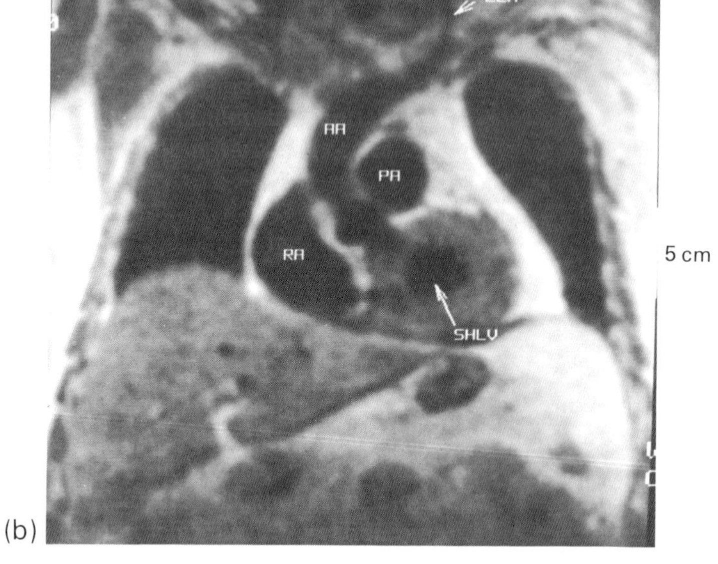

Fig. 6.42. E.c.g.-gated axial and coronal sections through the heart obtained at 21 MHz using a SE sequence (TE = 45 ms, TR = 0.8 s). (a), (b) Patient with a small hypertrophic left ventricle; note the disproportionate thickness particularly evident in (b) of the left ventricular wall relative to the chamber. (c), (d) Patient with a dilated left ventricle, evident in both axial and coronal sections. Key: RA,

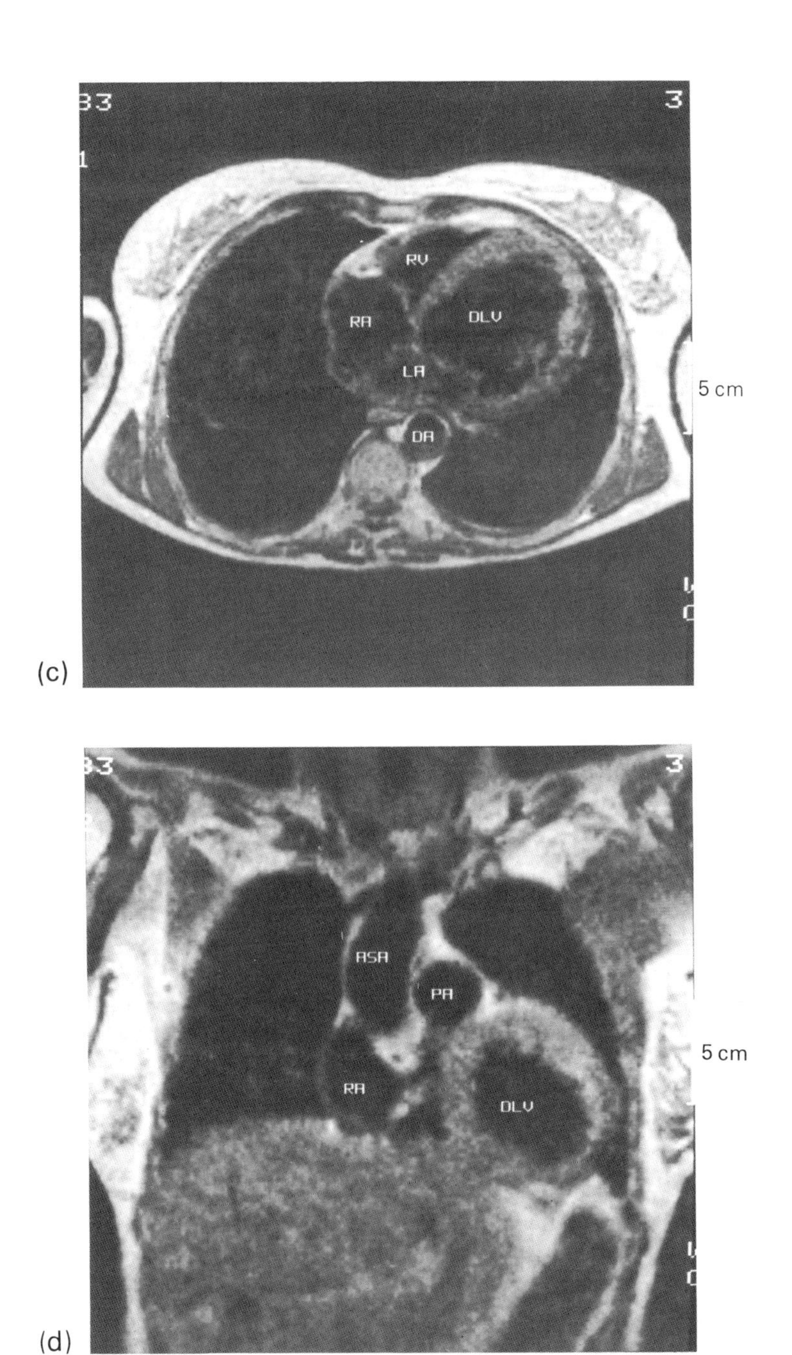

LA right and left atria; RV, LV right and left ventricles; AA(ASA)DA ascending and descending aorta; IVC inferior vena cava; LCA left carotid artery; PA pulmonary artery; SHLV, DLV Small hyperthropic and dilated left ventricles. (Courtesy of Siemens Ltd. 1983.)

thoracic studies T_1- or T_2-related images give much better soft tissue contrast than those based primarily on spin density.

The most intensive n.m.r. clinical trials in the abdominal region have been carried out by the Aberdeen group, who have compiled a table of normal and pathological relaxation times (at 1.7 MHz) which is reproduced here as Table 6.4. It will be seen that there is relatively little overlap between different normal tissues in accordance with the observation that T_1 images afford particularly good tissue contrast. These values increase in the presence of inflammatory or malignant diseases allowing relatively simple diagnosis.

The accurate determination of relaxation times from n.m.r. images is not a straightforward task (see § 6.3.1). Further, as has been previously stated, the parameters are not absolute—T_1, in particular, is strongly frequency-dependent. For this reason, a second table of relaxation times (Table 6.5) is given, based on the study at 8.5 MHz of 109 patients by Rupp *et al.* (1983). Note that in addition to the expected general increase

Table 6.4

Proton T_1 relaxation times at 1.7 MHz for abdominal tissues

Tissue	*T_1 range* (ms)
Kidney	300–340
Urine	600–1000
Acute tubular necrosis	400–420
Simple cyst	600–1000
Carcinoma	400–450
Metastases	400–450
Abscess	390–420
Liver	140–170
Bile	250–300
Blood	340–370
Cirrhosis	180–300
Hepatoma	300–450
Cholangiocarcinoma	300–450
Haemangioma	340–370
Metastases	300–450
Cyst	800–1000
Pancreas	180–200
Pancreatitis	200–275
Pancreatic carcinoma	275–400
Prostate	250–325
Carcinoma	350–450
Spleen	250–290

From Smith 1982*a*.

Table 6.5
Proton T_1 and T_2 relaxation times at 8.5 MHz for abdominal tissues.

Tissue	T_1 (ms)	T_2 (ms)
* Fat	240 ± 20	60 ± 10
Pancreas	290 ± 20	60 ± 40
Pancreatitis	300	150
Pancreatic carcinoma	840	40
Pancreatic pseudocyst	1130	70
* Liver	380 ± 20	40 ± 20
Bile	890 ± 140	80 ± 20
Metastases	570 ± 190	40 ± 10
Abscess	1180	100
* Muscle	400 ± 40	50 ± 10
Spleen	420 ± 50	20
* Kidney	670 ± 60	50 ± 10

Mean values and standard deviations are shown. * denotes mean values determined in more than 10 patients.
Data from Rupp *et al.* 1983.

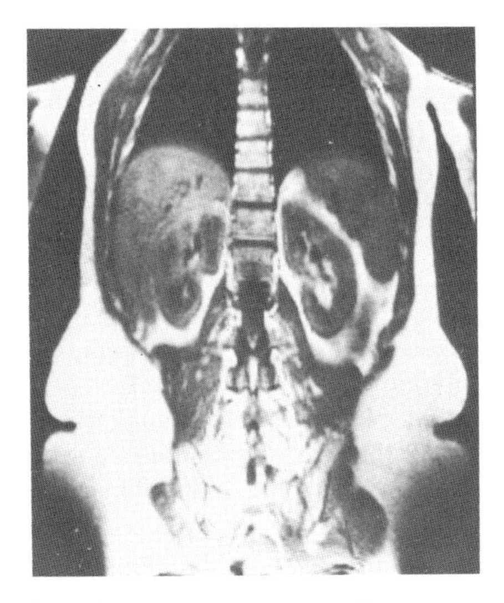

Fig. 6.43. 63 MHz (1.5 T) whole-body n.m.r. 5 mm coronal section obtained using a SR sequence (TR = 400 ms). Note excellent delineation of anatomy and lack of any trace of phase/intensity artefacts due to r.f. attenuation over this very wide field of view image. (Reproduced with permission from the General Electric Co., 1984.)

in T_1s with frequency (see § 6.2) the order of tissues (ranked in ascending order of T_1 value) is different.

In the past many doubts have been expressed concerning the practicality of high-field imaging. In particular, 'skin depth' problems (see § 5.7) were thought to be limiting at frequencies above 10 MHz. Figure 6.43 should do much to dispel such doubts, however. It shows a 5 mm section through the human abdomen obtained at a frequency of 63 MHz (corresponding to a field of 1.5 T). It shows no major phase or intensity artefacts due to incomplete r.f. penetration.

6.6.2.1. The liver. T_1 values for normal liver are short and fall within a fairly narrow range (140–170 ms at 1.7 MHz; 210–270 ms at 6.5 MHz) giving a uniform grey appearance against which blood vessels and bile ducts can be seen (Bydder *et al.* 1982*a*). Hepatic tumours have much longer T_1s (300–450 ms at 1.7 MHz) and are readily distinguished from surrounding tissue both on this basis and also as a result of their characteristically well-defined shapes, see Fig. 6.44. Although hepatoma tends to have a rather shorter T_1 (Young *et al.* 1982), primary tumours cannot always be distinguished from metastatic ones, though both can be differentiated from benign cystic masses (which have much longer T_1s (~600 ms)) and from haemangioma; see Table 6.4. As well as these focal diseases, which are often well-characterized by other imaging modalities, though sometimes with lower spatial resolution (currently ~7 mm for n.m.r., compared with ~25 mm for radioisotope and ultrasound methods), n.m.r. imaging seems to be equally effective in diagnosing diffuse or parenchymal diseases. Hepatitis and cirrhosis both show regions of increased T_1 spreading in a characteristic pattern from the hepatic radicals: an example of cirrhosis is shown in Fig. 6.45. Exceptions to this rule are those types of cirrhosis which involve accumulation of paramagnetic ions; for example, primary biliary cirrhosis, Wilson's disease (copper accumulation), and haemachromatosis (iron accumulation).

The gall bladder is readily distinguished from the liver in n.m.r. images by virtue of the different relaxation properties of bile. Hricak, Filly, Margulis, Moon, Crooks, and Kaufman (1983*b*) have studied the relation between food intake and the intensity of the bile signal. Volunteers who had fasted prior to scanning showed high-intensity signals for the concentrated gall bladder bile, which had a mean T_1 and T_2 at 15 MHz of 594 ms and 104 ms respectively. Following fatty intake (cholecystagogue), a second layer of bile was observed floating on the first. It had a long T_1 of 2646 ms ($T_2 = 126$ ms) and was assumed to be hepatic bile which had replaced the concentrated bile emptied into the colon. A normal gall bladder will concentrate this hepatic bile, removing up to 90 per cent of the water. However, patients with a gall bladder disease—for example,

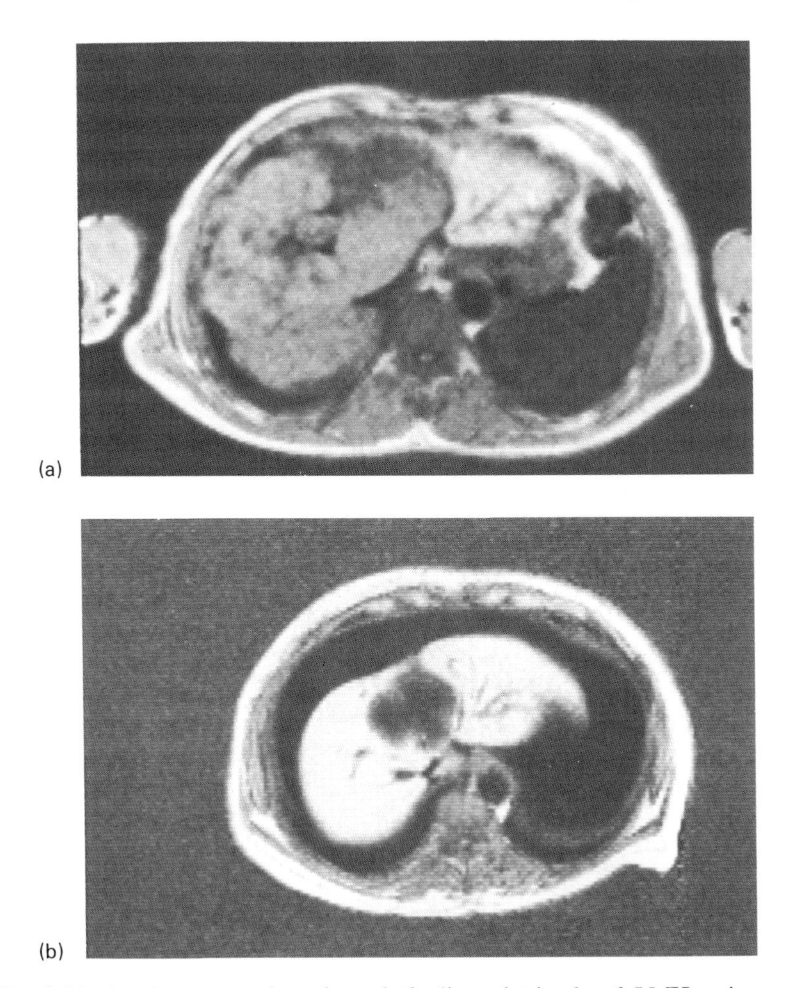

Fig. 6.44. Axial n.m.r. sections through the liver obtained at 6.5 MHz using an IR (TI = 400 ms) method. (a) Patient with hepatoma; (b) patient with liver metastasis from a carcinoma of the breast. Both intrinsic and extrinsic tumours are visible as regions of prolonged T_1, which appear dark in the images. (Courtesy of Picker International Ltd. 1983.)

cholecystitis—lose this ability and can be diagnosed on the basis of such studies. Gallstones are also readily seen as well-defined low-intensity regions (see also the *in vitro* study by Moon, Hricak, Margulis, Bernhoft, Way, Filly, and Crooks 1983*c*).

6.6.2.2. The pancreas. The normal pancreas is not always easily visualized on n.m.r. images, presumably as a consequence of the similarity in

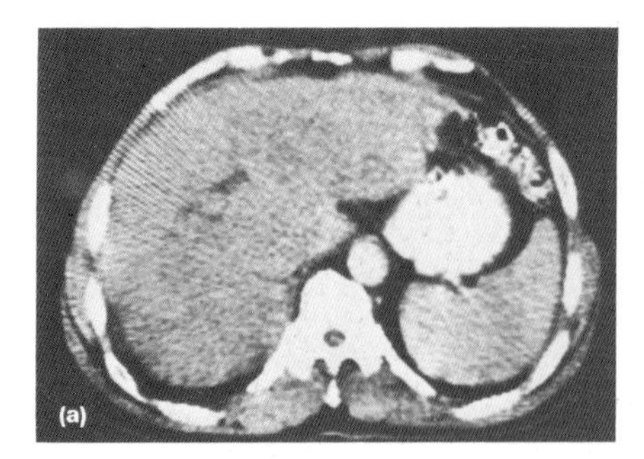

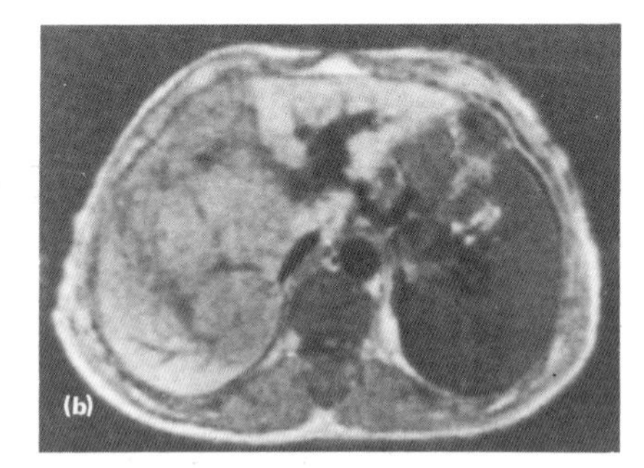

Fig. 6.45. Axial sections comparing X-ray CT (a) and IR n.m.r. (6.5 MHz, IR = 400 ms) (b) scans of a patient with cirrhosis of the liver. In (b) the dark appearance of the liver is due to a generalized increase in relaxation time. (From Bydder *et al.* 1982*a*.)

its n.m.r. parameters with those of the surrounding organs. Thus, at 6.5 MHz the T_1 values of liver and pancreas are similar (Young *et al.* 1982), whilst at 1.7 MHz, Smith *et al.* (1982) have experienced difficulty in distinguishing it from duodenum and bowel. One way around this difficulty involves the use of orally-administered dilute ferric chloride solution to reduce selectively the T_1 of the gut (Gore 1982). Another way of altering the signal intensity is to fill stomach and bowel with mineral oil, a method which has been used in some animal studies (Newhouse *et al.* 1982; see also § 6.4.3). Such contrast agents (see § 6.3.5 for other examples) may also be useful in the diagnosis of pancreatic disease. The success rate of n.m.r. imaging has been so good for this particular application that it may well come to be considered the method of choice.

The inflammation associated with pancreatitis gives a modest uniform increase in T_1 whereas carcinomas in the head, body, or tail of the pancreas are seen as well-defined masses of greatly increased T_1 (Smith *et al.* 1982*a*; Bydder *et al.* 1982*a*; Young *et al.* 1982).

6.6.2.3. The kidneys. The kidneys appear as horse-shoe-shaped organs with a T_1 (1.7 MHz) in the range 300–340 ms. Their most remarkable feature is the visibility of internal structure (Luiten *et al.* 1982; Smith *et al.* 1982*a*; Newhouse *et al.* 1982; Young *et al.* 1982; Alfidi *et al.* 1982; Hricak *et al.* 1983*a*) with medulla and cortex clearly distinguishable. An example is shown in Fig. 6.46. In some animal studies two separate cortical regions can be seen (see § 6.4.3).

In a normal kidney the collecting systems appear empty. However, when the kidney is hydronephrotic, both pelvis and collecting systems are dilated with urine which is easily detected by virtue of its very long T_1

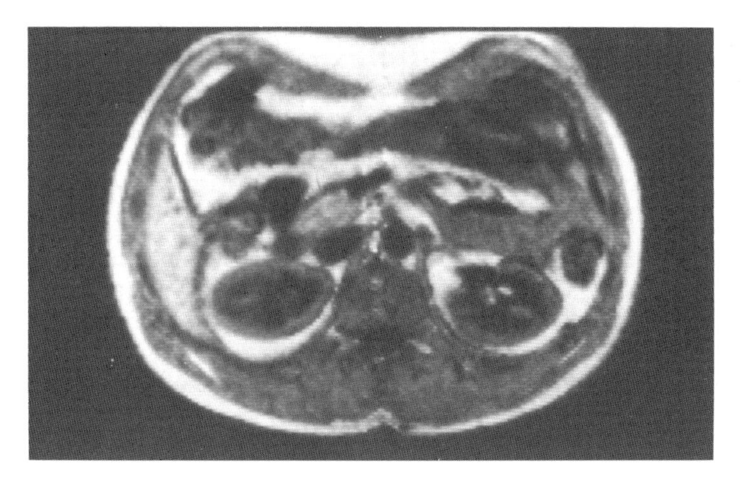

Fig. 6.46. Axial n.m.r. section obtained at 6.5 MHz using an IR sequence (IT = 400 ms) and demonstrating normal kidney anatomy. Note that cortex and medulla are clearly resolved. (Courtesy of Picker International Ltd. 1983.)

(Smith 1982*a*). When the kidney is inflamed, as in the case of tubular necrosis, the collecting system is again empty but the parenchyma is swollen and has a T_1 which is increased to 400–420 ms (1.7 MHz) allowing straightforward diagnosis. It is therefore possible to distinguish renal from post-renal failure.

Solid and cystic renal masses can also readily be distinguished by virtue of their different relaxation properties. For example, Hricak, Williams, Moon, Moss, Alpers, Crooks, and Kaufman (1983*c*) have found that renal cysts generally have T_1 and T_2 values in excess of 2 s and 200 ms respectively at 15 MHz. The smooth outer margin of the cysts is clearly seen, although they are often highlighted by ring artefacts due to excessive contrast with renal parenchyma. (Such artefacts are also seen between the cortex and surrounding perirenal fat. These may be due, in part, to chemical shift differences, however.) Haemorrhagic cysts have shorter T_1s and longer T_2s and can be distinguished on this basis, appearing brighter in SE studies. This ability to identify haemorrhagic cysts is important, since in many cases they are associated with neoplasm and aspiration is therefore indicated. Renal cell carcinoma has very different relaxation properties from normal parenchymas, allowing easy identification. Accurate staging has also proved possible. Kidney function can be studied too and a number of urological contrast agents have been proposed for this purpose (see § 6.3.5).

The adrenal glands are generally well seen in n.m.r. scans as Y-shaped regions delineated by retroperitoneal fat, and it is also possible to distinguish cortex from medulla due to the former's higher lipid content (Moon *et al.* 1983*a*). Lesions are usually well demarcated, metastatic tumours, for example, appearing as low-intensity regions on SE scans. The high spatial resolution over limited fields of view offered by increased gradient strengths has also proved useful in studies of adrenal structure (Young *et al.* 1982).

6.6.2.4. The pelvic region. In the pelvic region it is possible to demonstrate the normal anatomy effectively, although the female reproductive system has proved somewhat difficult. Large blood vessels such as the iliac arteries can be observed, suggesting that there may be some potential for the detection of atheroma in this region. Malignancies of the iliac and para-aortic lymph nodes are recognizable as are cystic and malignant ovaries. It is also possible to detect enlargement of the prostate due to hypertrophy from carcinoma.

Because of the inherent safety of n.m.r. imaging methods (see § 6.8), they may well be of value in foetal scanning, although ultrasound is currently well established in this role. Smith, Adam, and Phillips (1983) have examined six patients in the first trimester of pregnancy and were able to measure biparietal diameters and compare them with ultrasonic measurements. The foetus is visible from ten weeks on and is readily distinguished from the surrounding amniotic fluid, which appears dark due to its long T_1. (See also the foetal studies of goats by Foster, Knight, Rimmington, and Mallard 1983.) For reasons, of safety, orthopaedic studies in the pelvic region, though of lower resolution than that achievable with X-ray CT methods, may also be of importance, in, for example, paediatric hip investigations. Adult studies of the hip joint show little signal from the bony cortex of acetabellum and femur. The latter therefore appears as a low-intensity ring surrounding a region of higher intensity arising from the marrow in the medullary cavity. Muscles, ligaments, and the major blood vessels are well defined. N.m.r. studies of avascular necrosis of the femoral head have been reported (Moon, Genant, Helms, Chafetz, Crooks, and Kaufman 1983*a*; Rupp *et al.* 1983). The signal from the necrotic region is reduced due to a lenthening of T_1. In some cases one can also observe a halo of diminished signal intensity arising from a thickened joint cavity, synovial hypertrophy, or synovial effusion.

6.6.3. *Musculoskeletal applications*

The important anatomical features of the lumbosacral spine are well shown by n.m.r. (Moon *et al.* 1983*a*; Modic *et al.* 1983; Alfidi *et al.* 1982; Pykett 1982). The cortical bone of the vertebrae appears dark, as do the ligaments. Intervertebral discs are well seen, and display a high-intensity centre surrounded by a lower-intensity outer ring on SE scans, suggesting that the annulus fibrosis is being distinguished from the nucleus pulposus. The thecal sac and nerve roots are outlined by high-intensity epidural fat which is symmetrically disposed in healthy subjects. The spinal cord itself appears relatively bright, surrounded by the lower signal of c.s.f.

The flexibility to study different sections without movement of the patient is particularly valuable in cases of spinal injury. Sagittal sections

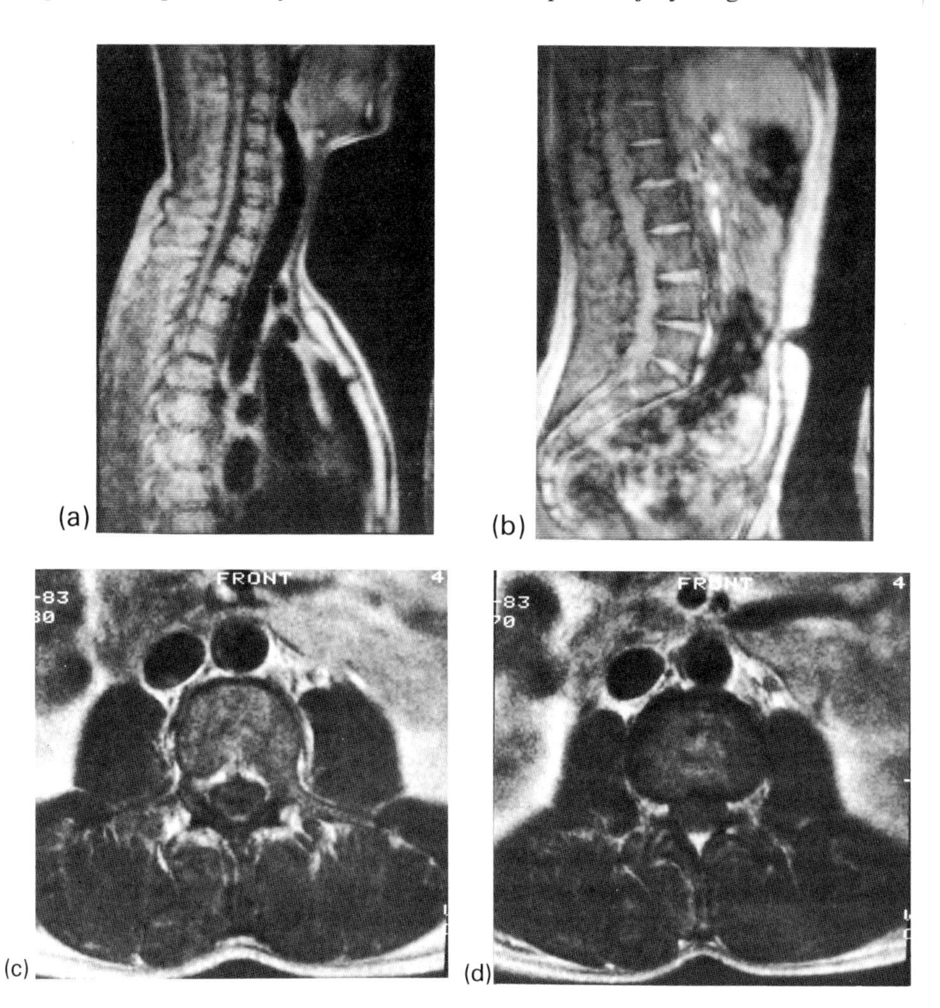

Fig. 6.47. (a), (b) Upper and lower normal spine shown using n.m.r. sagittal sections obtained at 6.5 MHz using an IR (TI = 400 ms) sequence. (Courtesy of Picker International Ltd. 1983.) (c), (d) Transverse axial sections through lumbar vertebra (c) and intervertebral disc (d) obtained using a high-resolution (gradient zoom) SE (TE = 38 ms) n.m.r. scan at 21 MHz. (Courtesy of Siemens Ltd. 1983.)

are especially useful. In some cases, high-resolution, reduced field of view, studies yield additional information allowing, for example, white grey matter contrast to be observed within the spinal cord (Young *et al.* 1982). Sagittal (a, b) and high-resolution axial (c, d) views are illustrated in Fig. 6.47.

Since X-ray CT gives high bone contrast, it is the modality of choice when the cortex is directly involved; for example, in cases of bony overgrowth or fractures with splinters impinging on the spinal cord or nerves. N.m.r. is, however, accurate in cases of cord trauma and spinal blockage. In the latter case, for example, the extent of a metastatic lesion causing the spinal block can be readily assessed. Herniated discs can also be demonstrated by n.m.r.

In rheumatoid and osteoarthritis, the T_1 of the synovial fluid surrounding the joint is seen to increase and there may well be important applications in this area. A number of groups have already demonstrated their ability to produce images of the human knee showing clearly resolved detail (Zeitler and Schittenhelm 1981; Edelstein, Bottomley, Hart, Leue, Schenck, and Redington 1982; Moon *et al.* 1983*a*; Rupp *et al.* 1983). The bony cortex and medullary cavities are well seen as are ligaments, tendons, and blood vessels. Muscle bundles are sharply outlined by surrounding planes of fat and can be distinguished from tendons by virtue of their greater signal intensity on SE images. The cruciate ligament is also well resolved.

A number of bone tumours have been observed, including a giant cell tumour of the femur using s.f.p. methods at 63 MHz (Brady, Gebhardt, Pykett, Buonanno, Newhouse, Burt, Smith, Mankin, Kistler, Goldman, Hinshaw, and Pohost 1982), and of the tibia using SR methods at 6.5 MHz (Brady, Rosen, Pykett, McGuire, Mankin, and Rosenthal 1983). The tumours are clearly distinguished from normal marrow by virtue of their prolonged T_1s. Thinning and destruction of the cortical bone is also clearly demonstrable. Moon *et al.* (1983*a*) have observed a renal cell carcinoma in a vertebral body. It was found to be impinging on the spine causing cord block.

6.7. Multinuclear and chemical shift imaging

Whereas virtually all n.m.r. imaging to date has been of protons, there is in principle no reason why any of the other n.m.r. nuclei should not be imaged. Unfortunately, as discussed in § 2.3.3, those elements which are of greatest biomedical interest have low n.m.r. sensitivities or are present only in small amounts. Even if isotopic enrichment were used, it would therefore not be possible (at least for biological samples) to image them with anything like the resolution which can be achieved with protons.

Nevertheless, we have seen in § 5.7 that the imaging sensitivity depends on the inverse cube of the spatial resolution. (Restated in simpler terms, this says that the n.m.r. signal is directly proportional to the amount of tissue contributing to each pixel). Thus, for example, a tenfold change in resolution allows one to make up a factor of 1000 in imaging sensitivity. Figure 6.48 indicates the dependence of achievable resolution on concentration for a number of the nuclei listed in Table 2.1. Points corresponding to typical physiological concentrations are indicated (see also the article by Kramer (1982)).

Clearly, for multinuclear studies, sensitivity is at a premium. A rational approach therefore would require the selection of the highest practicable operating frequency consistent with sample size and electromagnetic properties (see § 5.7). The magnetic field would then be chosen to bring the nucleus in question into resonance at this optimum frequency. Since, tritium excepted, all nuclei have a magnetogyric ratio γ less than that of the proton, this would require the use of higher field magnets. For example, at an operating frequency of 30 MHz, magnetic fields of 2.66, 1.74, and 2.80 T would be needed for ^{23}Na, ^{31}P, and ^{13}C respectively. Such high fields require the use of superconducting technology. However, superconducting magnets are designed to operate at a particular field strength, and changing fields, though possible, is not a trivial operation (see Table 5.4 for magnet run-up times). A compromise solution is generally adopted with the choice of a single high-field system operating, for whole-body studies, at a fixed field in the range 1.5–2.0 T. In Fig. 6.48, however, results have been given at constant frequency (rather than field) in order to indicate the resolution to be expected for a given nucleus under optimal conditions. It must be emphasized that such optimization is not possible for all nuclei with a single magnet system.

Given that one may be willing to accept longer scan times or lower S/N (since inherent tissue contrast is likely to be much greater than for protons) the prospects for multinuclear imaging start to look rather less gloomy. (See § 5.7 for functional relationship between resolution S/N and imaging time.) Indeed, some *in vivo* ^{23}Na and ^{31}P studies have already been undertaken. Figure 6.49, for example, shows a ^{23}Na image of an isolated Langendorff perfused rat heart (De Layre, Ingwall, Malloy, and Fossel 1981). A projection reconstruction method (12 projections, 230 scans per projection) was used and n.m.r. data acquisition was gated (Fossel, Morgan, and Ingwall 1980) allowing images to be obtained at different times during the cardiac cycle. In Fig. 6.49, (b) corresponds to a midventricular slice with the heart at mid diastole, whereas (c) shows the same slice at mid systole. In these images the cardiac tissue is shown in relief, depicted by the high concentration of ^{23}Na in the Krebs–Henseleit ringer present both in the left ventricular cavity and externally. In a sense,

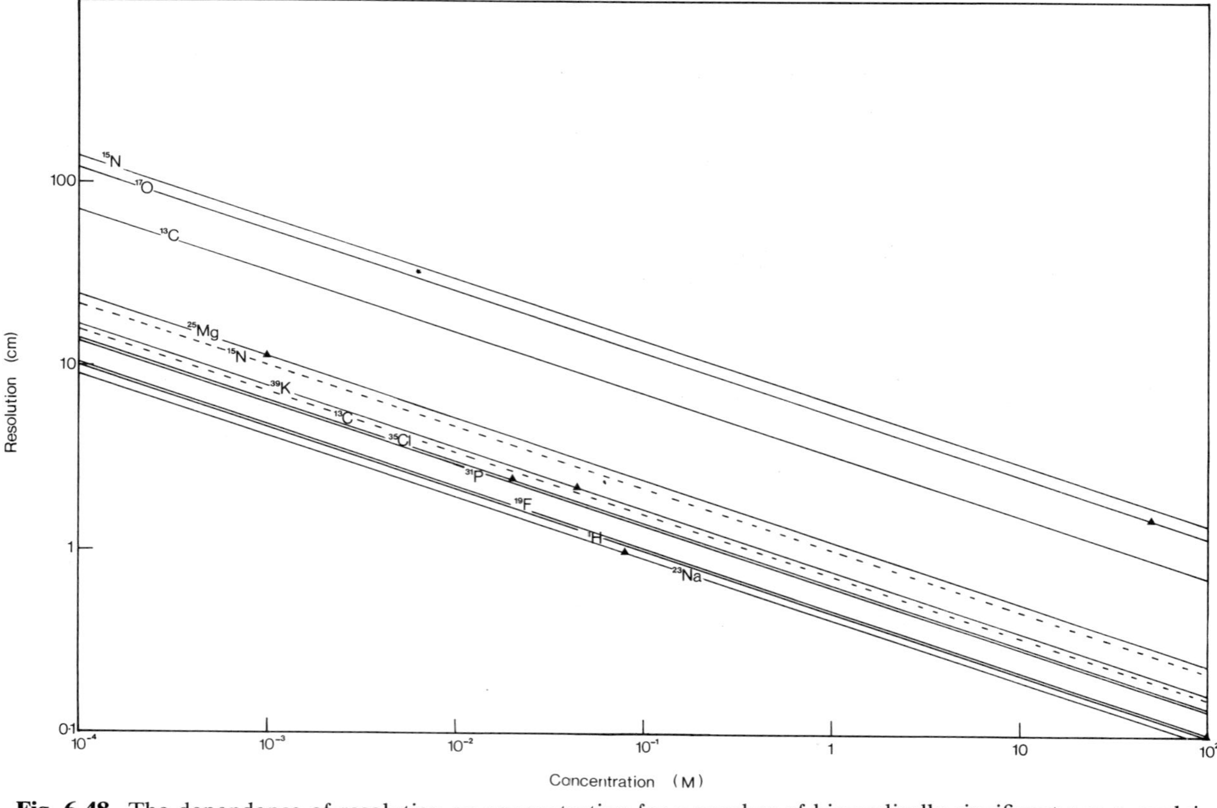

Fig. 6.48. The dependence of resolution on concentration for a number of biomedically significant n.m.r. nuclei. ▲ indicates a typical physiological concentration for the nucleus.

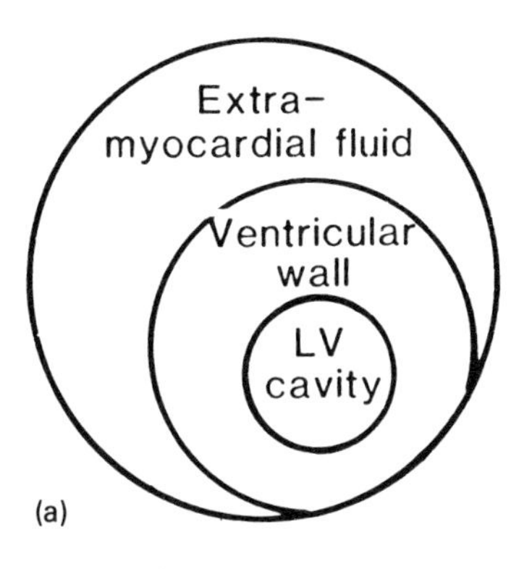

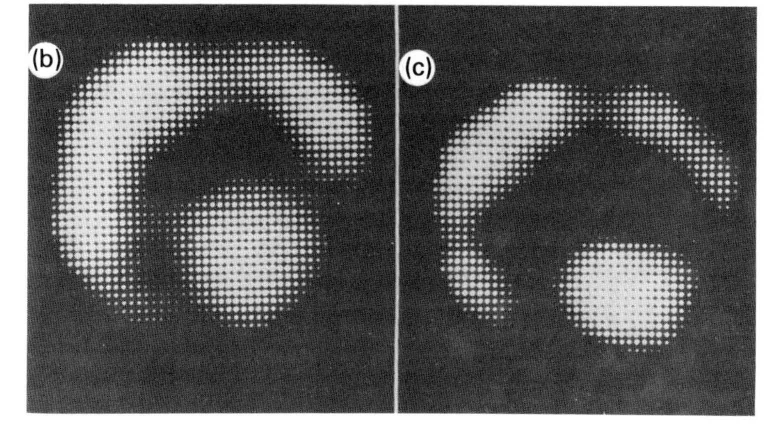

Fig. 6.49. (a) Diagrammatic representation at the midventricular level of an isolated perfused heart in the n.m.r. sample tube. (b) Gated ^{23}Na n.m.r. image of an isolated perfused working rat heart in diastole (note heart is off centre). (c) Equivalent image at systole. (From De Layre *et al.* 1981.)

this is not a true image of the heart at all: no detail is seen within the soft tissue and any other n.m.r. nucleus introduced into the perfusate could have been used to achieve the same effect. However, the important point is that blood has a ^{23}Na content comparable with that of Krebs–Henseleit ringer, indicating that it should be possible to achieve a similar visualization of the cardiac chambers *in vivo*.

More recently, a group at the Neurological Institute of New York have shown ^{23}Na images displaying good soft tissue contrast (Maudsley and Hilal 1984; Maudsley, Hilal, Simon, and Perman 1982; Hilal, Maudsley, Bonn, Simon, Cannon, and Perman 1983*a*; Hilal, Maudsley, Simon, Perman, Bonn, Mawad, Silver, Ganti, Sane, and Chien 1983*b*). Operating at a frequency of 30 MHz, using a spin-echo Fourier method, they

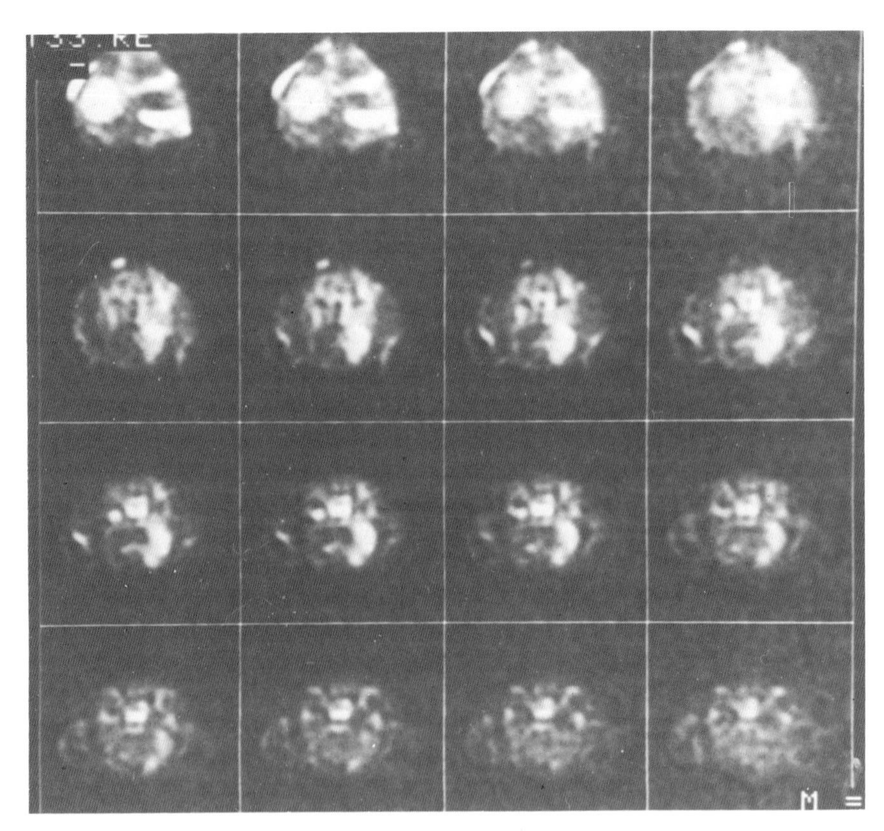

Fig. 6.50. *In vivo* serial ^{23}Na n.m.r. transverse images of a cat's head 12 h following ligation of the middle cerebral artery. Images, commencing upper left, finishing lower right, show sections from the level of the eyes through to the neck. In sections 1–4 the empty orbit on the right side can be seen to contain a pool of serum and c.s.f., and a brighter region of enhanced sodium signal can be seen on the right hand side of the brain corresponding to the region of stroke. Images were obtained at 30 MHz using a Fourier technique with a 100 ms repetition time (TR). The minimum imaging time for a 64^3 array (spatial resolution of about 2.5 mm over a 10 cm object) was 20 min. In this case some signal averaging was performed and the imaging time was $3\frac{1}{2}$ h. (From Maudsley and Hilal 1984.)

investigated a cat-brain stroke model, demonstrating a pronounced increase in tissue sodium as early as six hours following ligation of the middle cerebral artery (Fig. 6.50). A canine model of myocardial ischaemia, in which the left anterior descending coronary artery was ligated for one hour, followed by one hour's reperfusion, was also studied. Hearts were excised before being imaged and showed a focal increase in tissue sodium in the region served by the ligated artery. In each case the contrast resulted mainly from the breakdown of the ATP-driven sodium pump which can lead to an increase in sodium concentration of from 300 to 400 per cent. ^{23}Na images of canine kidneys demonstrated good discrimination between cortex and medulla due to the much higher sodium concentration in the former. In fact, soft-tissue sodium concentrations vary over at least an order of magnitude in the human body (Spector 1956), so one can expect good contrast on this basis alone. T_1 and T_2 differences are likely to be additional sources of tissue contrast. In some conventional ^{23}Na studies it has been possible to use shift reagents, such as dysprosium III tripolyphosphate, which do not penetrate cell membranes and which can therefore distinguish intracellular from extracellular pools (Gupta and Gupta 1982; Balschi, Cirillo, and Springer 1982). With the retention of chemical-shift information (see below) such experiments may be possible in imaging studies, too.

The situation with regard to potassium and chlorine is analogous to that of sodium, though the resolution is likely to be rather lower (see Fig. 6.48). ^{17}O has not received much attention, perhaps because in the body it is mostly found in water, which is more conveniently sampled with conventional proton imaging methods.

One nucleus which does have a sensitivity approaching that of the proton is ^{19}F. It is not present naturally to any significant degree and is therefore a good candidate for tracer studies. For example, the vascular system could be selectively imaged using one of the fluorine-rich blood substitutes such as perfluorobutylamine or perfluorotetrahydrofuran (Holland, Bottomley, and Hinshaw 1977).

^{19}F n.m.r. can also be applied to lung ventilation studies using non-toxic perfluorinated gases such as SF_6, CF_4, C_2F_6 or C_4F_6. In a recent study (Heidelberger and Lauterbur 1982; Rinck, Petersen, Heidelberger, Acuff, Reinders, Bernardo, Hedges, and Lauterbur 1983), a mixture of 80 per cent CF_4, 20 per cent O_2 was used to obtain two-dimensional ^{19}F pulmonary ventilation images of a dog at 3.76 MHz. The imaging time of 25 min and spatial resolution of 1.5 cm were broadly comparable with those achieved in xenon ventilation studies. However, improvements in technique, particularly through the use of higher operating frequencies, should improve substantially on these figures. The n.m.r. relaxation properties of these gases are novel—the spin rotation mechanism predominates, giving for CF_4, $T_1 = T_2 = 20$ ms. The short T_1 allows the use of a very short cycle time but the equally short T_2 virtually precludes the use of spin-echo methods.

Some important drugs contain fluorine and could be imaged at normal dose levels. Stevens, Morris, Iles, Sheldon, and Griffiths (1984) have recently used a surface coil method to monitor the uptake and subsequent metabolism in rat liver and tumours of intravenously administered 5-fluorouracil (see § 2.7.2). It is possible to use ^{19}F-labelled chelators to measure the intracellular concentrations of a number of metal ions such as Ca^{2+} and Mg^{2+}, for example (Smith, Hesketh, Metcalfe, Feeney, and Morris 1983). These studies illustrate again the principle of using a high-sensitivity nucleus (^{19}F) in order to monitor the concentration of a nucleus with lower sensitivity. Although not yet applied in an imaging context, such methods offer a very attractive solution to a seemingly intractable problem.

Another possibility is to label metabolites or drugs with ^{19}F in analogous manner to the use of ^{18}F for positron emission studies (^{18}F fluorodeoxyglucose, for example, is widely used to study glucose transport and metabolism in the mammalian brain). In general, the quantities of ^{19}F material required for n.m.r. imaging (see Fig. 6.48) will be much greater than a typical isotopic dose. Nevertheless, we should stress that ^{19}F is the naturally-occurring, stable isotope, so that fluorine n.m.r. studies are free from the hazards of ionizing radiation. In situations where this, rather than toxicity, has been the limiting factor, larger doses may therefore be acceptable.

Although much information is available from the measurement of spin densities and relaxation times, there is also very considerable interest in the ability of n.m.r. to perform chemical analyses, which arises from the slight shifts in resonant frequency for nuclei in different electronic environments (see § 2.6.1). With suitable modification, most imaging schemes are capable of retaining this chemical-shift information. In particular, proposals for chemical-shift imaging have made for the sensitive point (Bottomley 1982; Scott, Brooker, Fitzsimmons, Bennett, and Mick 1982; see also § 3.2.3), echo-planar (Mansfield 1983), and projection reconstruction methods (Lauterbur, Kramer, House, and Chen, 1975; Bendel, Lai, and Lauterbur 1980; Hall and Sukumar 1982, 1984*a*; see also § 4.1). However, the method which lends itself most naturally to this extension is Fourier imaging, either in its standard form (Brown, Kincaid, and Ugurbil 1982; Maudsley, Hilal, Perman, and Simon 1983; Haselgrove, Subramanian, Leigh, Gyulai, and Chance 1983; Pykett and Rosen 1983; Hall and Sukumar 1984*b*; Hall, Sukumar, and Talagala 1984; see also § 4.2.2.3) or in its rotating frame guise (Cox and Styles 1980; see § 4.2.5).

The impetus for such work derives largely from the success of ^{1}H, ^{13}C, ^{15}N, ^{19}F, and ^{31}P tissue studies using conventional or surface-coil methods (see § 2.7). Indeed, ^{31}P surface-coil techniques have been used clinically to diagnose a number of muscular diseases arising from enzyme deficiencies, and also to monitor ^{31}P metabolite levels in neonates. In a sense, the use of a surface coil may be regarded as a point chemical-shift imaging method: in principle, one could move the subject relative to the coil in order to map out the surface distribution of metabolites. Provided one is interested in surface structures, this method is particularly effective, since, as well as restricting the region from which signal is gathered, it also confines the pick-up of sample noise to the same volume. Note that the term 'surface' can be somewhat misleading—it really means a region located within a radius of the coil centre. In the case of a large coil, such as that used for the neonatal brain studies, this can represent a very substantial fraction of the sample.

With the advent of high-field wide-bore magnets considerable interest has been generated in the probing of internal structures, particularly with cardiovascular applications in mind. Although surface-coil methods can be combined with field profiling (t.m.r.—see § 3.2.2) for this purpose, the inherent S/N advantage is thereby degraded. The precise spatial localization offered by full imaging techniques then starts to look a worthwhile proposition.

Note that the retention of chemical-shift information itself requires the use of high magnetic fields (in order that the chemically-shifted components may be resolved). Additionally, the homogeneity of the main magnetic field needs to be especially good. To resolve the various ^{31}P metabolites, for example, a homogeneity of about 1 p.p.m. is necessary. If, in addition, one wishes to make pH measurements from the shift of the inorganic phosphate resonance (see § 2.7.1) then the requirement increases to 0.1 p.p.m. These figures are easily met in small bore $\leqslant 10$ cm systems, but are at least an order of magnitude better than normally achieved in whole-body (1 m) magnets, even with the use of room temperature shim assemblies (see Table 5.4). Fortunately, it is possible to relax this requirement considerably: what is actually needed is for the homogeneity to be preserved over the volume corresponding to a single pixel. Any slower (spatial) variation in the main magnetic field simply causes a distortion of the image plane in the chemical-shift dimension. For discrete chemically-shifted peaks, the method is self-calibrating: one simply takes a known resonance, for example of phosphocreatine in the case of a ^{31}P image, and moves the frequency axis to bring it into registration with the same resonance in other pixels. Cross-sectional images showing the distribution of each chemically-shifted component—for example, inorganic phosphate, phosphocreatine, and ATP—can be displayed. Alternatively, one can show the spectrum from a particular region of interest corresponding to a single pixel or collection of pixels. In the latter case, one could employ a dual proton–phosphorus imaging system, using the high spatial resolution proton image to guide the ^{31}P pixel selection. In this way one can trade off spatial resolution against S/N in a manner consistent with the local anatomy and pathology.

Figure 6.51 shows two examples of proton chemical-shift images obtained at 61.5 MHz by Pykett and Rosen (1983) using a Fourier chemical-shift imaging method. Figure 6.51(a) is a proton cross-sectional

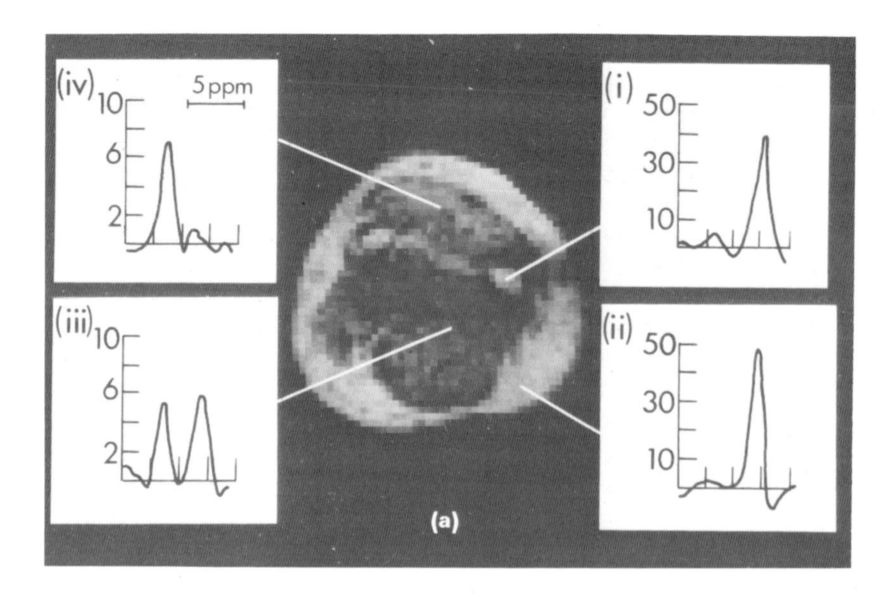

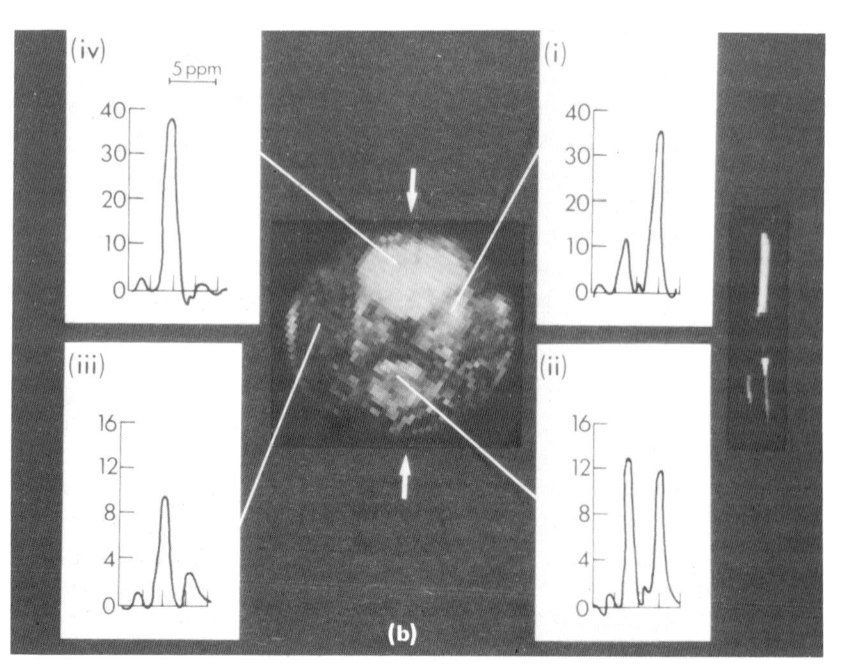

Fig. 6.51. Proton chemical-shift image at 61.5 MHz acquired *in vivo* from the human forearm (a) showing spectra originating from bone marrow (i), subcutaneous fat (ii), extensor (iii) and flexor muscles (iv). (b) Proton chemical shift image corresponding to a coronal section through the head of a normal anaesthetized cat. (i)–(iv) show clearly-resolved peaks from mostly water and lipid resonances. Edge profiles (extreme right) indicate presence of lipid and water in the tongue, but only water in the brain. (From Pykett and Rosen 1983.)

image of a human forearm. It corresponds to the total observable proton spin density—the data has been summed along the frequency axis and there is no chemical-shift resolution. This is displayed in the insets 1–4, which show proton spectra from the regions of interest indicated by the arrows. Peaks due to water (downfield or left-hand side of spectrum) and the —CH_2— backbone of mobile lipids (upfield or right-hand side of spectrum) are readily apparent. Figure 6.51(b) shows a similar image and related series of spectra obtained from the head of a normal anaesthetized cat. Large lipid signals are seen arising from the retro-orbital fat but, very interestingly, no such signals are seen in the brain. The difference presumably arises from the different nature of the lipids—depot fat and membrane respectively. The lack of any lipid signal in brain indicates that, at least for imaging schemes employing spin-echo methods which discriminate against short T_2 components, the contrast observed between white and grey matter cannot be accounted for in terms of differences in myelin content (see § 6.5.1). Maudsley, Hilal, Simon, and Wittekoek (1983) have also produced proton chemical-shift images of cat heads using Fourier chemical-shift imaging methods. Separate cross-sectional images corresponding to the distribution of water —CH_2— and CH_3 protons were generated.

High-resolution proton n.m.r. spectra can contain a tremendous wealth of metabolic information. However, in tissues this is normally masked by the water and mobile lipid protons present, typically in concentrations 10^4–10^5 times higher than the metabolites of interest. Nevertheless, with the aid of solvent-suppression techniques this problem can be overcome satisfactorily and *in vivo* surface coil studies of proton metabolites such as lactic acid have been reported (see also Aue, Müller, Cross, and Seelig 1984). In principle, the same kind of studies could be carried out using chemical-shift imaging methods, though the field homogeneity requirements are severe—the entire proton chemical-shift spectrum spans a range of only 10 p.p.m.

A great deal of interest has been shown in ^{31}P chemical-shift imaging. Bottomley (1982), for example, has used the sensitive point method to record ^{31}P spectra from phantoms containing ATP. The chemical-shift resolution obtained was, however, rather poor. Scott *et al.* (1982) have described a similar system based on the sensitive point method. ^{31}P n.m.r. imaging studies have also been performed on test-tube phantoms using projection reconstruction (Bendel *et al.* 1980; see also § 4.1) and Fourier methods (Brown *et al.* 1982; Cox and Styles 1980; Hoult 1977, 1979; Maudsley *et al.* 1983; see also § 4.2). More recently, Haselgrove *et al.* (1983) have demonstrated the *in vivo* application of the latter method, showing ^{31}P n.m.r. spectra obtained in a 7-in magnet at 24.3 MHz from rat legs and gerbil heads. Spatial resolution was achieved in one dimension only, the spectra thus corresponding to the total concentration of metabolites in successive planes normal to the applied gradient direction. The use of a spin-echo sequence with a relatively long TE value (50 ms) resulted in the loss of the ATP signal, due to its short T_2. Nevertheless, as shown in Fig. 6.52, the authors were able to demonstrate the loss of phosphocreatine and concomitant increase in inorganic phosphate in ischaemic muscle and brain.

^{13}C is another nucleus which has been widely studied, particularly in cells and perfused organs (see § 2.7.2). As indicated in Fig. 6.48, natural-abundance imaging studies of most metabolites which occur at concentrations of 10 mM or less are not practicable. However, it is possible to study the distribution and nature (whether saturated or unsaturated, for example) of mobile fats. This may be of importance in the investigation of

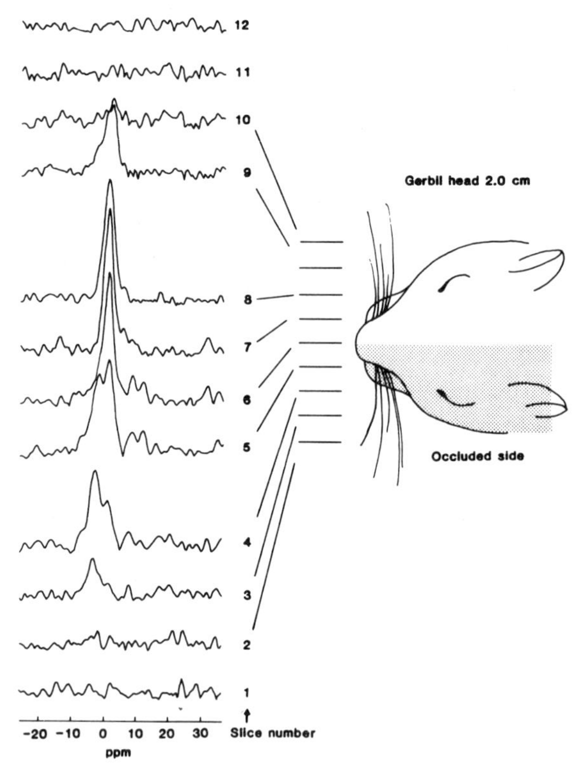

Fig. 6.52. ^{31}P n.m.r. one-dimensional chemical shift images at 24.3 MHz of a gerbil head, in which the right carotid artery was occluded to form unilateral ischaemia. Numbers refer to slice positions as illustrated. The peak at 0 p.p.m. arises from phosphocreatine and the downfield peak (to its left) from inorganic phosphate. (From Haselgrove *et al.* 1983 Copyright 1983 by the AAAS.)

diseases such as muscular dystrophy, where muscle tissue is replaced by fat. Equally, glycogen can be present in large quantities and could be studied by this method (see Stevens, Iles, Morris, and Griffiths 1982). For the majority of metabolites, however, ^{13}C-enriched material would be required. This is expensive, though not prohibitively so (D-1-^{13}C glucose, for example, costs £580/g, 1984 prices) and the spectra are potentially highly-informative. No *in vivo* ^{13}C imaging studies have been reported to date, though Hall and Sukumar (1982) have used a modified projection-reconstruction method to obtain chemical-shift images of ^{13}C phantoms. They improved their signal with the INEPT polarization transfer method (Morris and Freeman 1979). Note that it is also possible to improve the sensitivity by using an indirect method of observation. Thus, provided ^{13}C–^{1}H couplings can be identified, one can observe the ^{13}C nucleus through its effect on the proton spectrum.

6.8. Hazards of n.m.r.

Although n.m.r. imaging does not involve the use of any form of ionizing radiation, and is considered to be an extremely safe technique which could be used during pregnancy, for paediatric examinations, and for adult pelvic studies, it is nevertheless of great importance that an attempt should be made to assess the level of risk involved. The various academic and industrial concerns involved in the development of the subject approached the National Radiation Protection Board (NRPB) for advice at an early stage, supplying details of their various techniques and agreeing to keep exposure records for volunteers and patients. With the benefit of this information, the NRPB were able in 1980 to issue guidelines for exposure to n.m.r. clinical imaging (NRPB 1981). A revised document was issued in 1983 (NRPB 1983).

A number of recent reviews detailing possible mechanisms of interaction and assessing their relative importance are also available (Budinger 1979, 1981; Saunders 1982; Mansfield and Morris 1982) and we refer the interested reader to these for detailed discussion. In what follows we attempt to give a feel for likely hazards associated with the principal features of n.m.r. imaging systems—namely: large static magnetic fields, time varying magnetic fields, and radiofrequency irradiation.

6.8.1. Static magnetic fields

The literature on the biological effects of magnetic fields is extensive (Barnothy 1964; Tenforde 1979) and frequently contradictory. Often, insufficient data was collected to allow statistically-meaningful conclusions to be drawn, and in many cases inadequate or inappropriate controls were used. Following an extensive literature survey of detrimental health effects from static magnetic fields, Budinger (1981) concluded that 'no experimental protocol has been found that when repeated by other investigators gives similar positive results'.

One proposed interaction mechanism involves the reorientation of biological structures arising from their anisotropic magnetic susceptibilities. (For cellular and sub-cellular units the energy of interaction lies in the range 10^{-14}–10^{-12} ergs, which compares with the thermal quanta (at 37 °C) of 4.28×10^{-14} ergs). Such behaviour is indeed observed, for example, in preparations of retinal rods, DNA, sickled cells, and liquid crystals, of which the lyotropic variety provide models for cell membranes. However, the situation is likely to be rather different in living tissues, where the cohesive forces holding the sub-units together, from the molecular level up, are (hopefully) very much greater than the disruptive thermal ones. Although gross structural disturbance is extremely unlikely, it has been postulated that membranes may be distorted in such a way that their permeability is affected; no such effects have thus far been observed. In similar vein, it has been suggested that enzyme conformations may be changed with dire consequences on the kinetics of the reactions they catalyse. We have undertaken a careful study of one particular enzyme, β-galactosidase, which cleaves lactose into dextrose and galactose (Thomas and Morris 1981). *O*-nitrophenyl-β-D galactopyranoside was used as a substrate and the reaction was followed spectrophotometrically both under laboratory conditions and in a field of 1 T. The reaction rates were found to be identical. A considerable body of corroborative evidence for the lack of effect stems from the use of high-resolution proton n.m.r. for the study of enzyme systems, an important area of research of many years standing. Conformational changes and kinetic parameters are precisely the quantities which one aims to measure in these experiments. No magnetic field variation is observed, and results agree well with those obtained using other techniques (see Saunders and Cass (1983), however, for examples of a class of enzymes which do demonstrate magnetic effects).

Of course, if ferromagnetic structures are present, then the polarizing forces are many orders of magnitude stronger, and we should expect to see alignment effects. Haematite crystals have been found in bacteria (Blakemore, Frankel, and Wolfe 1979) which use the earth's magnetic field to orient themselves. Such 'magnetic guidance systems' have also been postulated in bees and birds though, as yet, no such role has emerged in human studies, though the presence of magnetic deposits in the sphenoid sinus has been reported (Baker, Mather, and Kennaugh 1983).

One magnetic interaction which is well documented is the magnetohydrodynamic effect, in which an e.m.f. is generated when electrically-conductive fluids flow in a magnetic field. This manifests itself most notably as an additional 'flow potential' in the T-wave region of the e.c.g. Figure 6.53 shows an example of this: the flow potential has an amplitude of about 0.75 mV and was recorded externally from squirrel monkeys exposed to a static field of about 10 T (Beischer 1969). It was estimated that the peak flow potential present across the aortic arch was ~50 mV but no effect on the heart-rate was observed over a fifteen minute period. Gaffey, Tenforde, and Dean (1980) have studied T-wave abnormalities in rats, dogs, and baboons as a function of field strength, finding a threshold near 0.3 T followed by a proportional increase in e.m.f. with field consistent with the hydrodynamic origins of the effect. In humans, if we assune a peak aortic blood velocity $\sim 0.63\,\mathrm{m\,s^{-1}}$ and an aortic diameter of 0.025 m, the flow potential generated across the aorta can be estimated at about $16\,\mathrm{mV\,T^{-1}}$ (Saunders 1982). Note that the potential across an individual cell would be very much less, and certainly well below the cell depolarization threshold of 40 mV. No e.c.g. abnormalities have been observed in humans with the magnetic fields used for n.m.r. imaging (~0.05–1.5 T) or for t.m.r. (~1.8 T).

Another possible biological effect of magnetic fields is the force exerted on the moving charges associated with the action potential in nervous tissue. It can be shown, however, that these effects should be negligible at fields below 10 T (Budinger 1981).

Based on these and other findings, the NRPB recommended that field

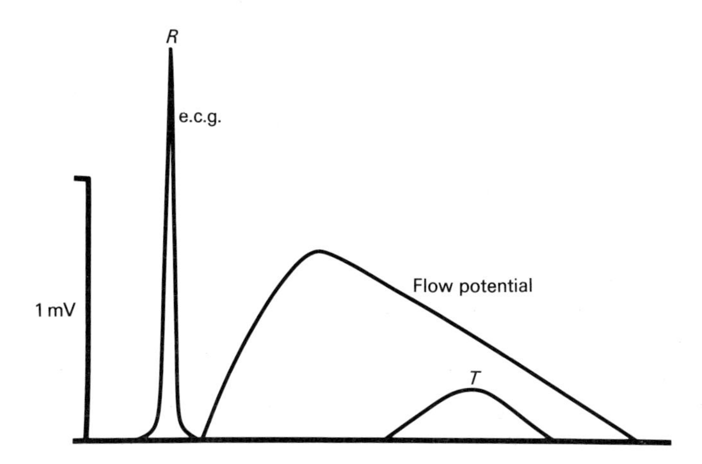

Fig. 6.53. Diagrammatic illustration of the relative positions of externally-recorded e.c.g. and flow potential in squirrel monkey exposed to a static field of 10 T. (From Saunders 1982.)

strengths be restricted to less than 2.5 T for the present. However, it is anticipated that this limit could safely be revised upwards if the need arose.

The above discussion was concerned with possible biological effects. From this standpoint, static magnetic fields at the strengths currently employed for n.m.r. imaging would appear to be intrinsically safe. Nevertheless, there are important associated hazards which arise from the phenomenon well-known to every schoolboy—that of magnetic attraction. Any ferromagnetic object, such as a pair of scissors, a bunch of car keys, a spanner, etc., will, on entering the non-homogeneous fringe field of a magnet, experience a force directed toward the magnet centre, whence it may arrive with considerable velocity. This 'missile effect' constitutes an extremely serious hazard, not only to the patient being scanned, but also to the magnet itself. In the latter case, the worry is not simply over impact-damage caused by the collision of a large ferromagnetic object, such as an oxygen cylinder, but also the equal and opposite force exerted on the magnet windings. Removal of the attached item may not be straightforward either, possibly requiring partial or total running down of the field. The solution to the problem is to keep ferromagnetic objects remote from the magnet by restricting access to entrances protected by archway metal detectors. This strategy also excludes cardiac arrest, life support, or patient monitoring equipment from the magnet vicinity. Until non-magnetic counterparts are developed, this represents a severe limitation which may preclude scanning of critically-ill patients at risk from cardiac seizure or reliant on life-support systems. Even for relatively low-risk patients, rapid removal from the magnet and transport to the emergency unit should be an important design consideration. Some progress is being made in this area: for example, the team at University College Hospital, London, in collaboration with Oxford Research Systems have developed and marketed a non-magnetic neonatal incubation unit suitable for neonatal brain studies.

Whilst the use of a metal detector offers some degree of protection it is by no means a perfect solution. Many research workers have had the frustrating experience of discovering that the magnetic strips on cash or telephone credit cards have been wiped clean by an injudiciously-close approach to a magnet. Conventional wrist watches are similarly prone to such lapses of concentration. Far more serious is the risk of scanning a patient with a small ferromagnetic surgical implant. New, Rosen, Brady, Buonanno, Kistler, Burt, Hinshaw, Newhouse, Pohost, and Taveras (1983) investigated twenty-one neurosurgical aneurysm and other haemostatic clips in current or common recent use. They found fringe field forces on sixteen of the clips which, in the case of five clips, were judged to be of sufficient strength for there to be serious risk of dislodgement and hence of haemorrhage or damage to sensitive areas. A variety of other surgical and dental devices and materials were also investigated, including stainless-steel suture wire, dental amalgam, titanium forceps, and a ventricular shunt connector. Only in the latter case was a force demonstrated. In addition to the danger of forces in the inhomogeneous fringe field and torques in the uniform main field, magnetic implants give rise to another reason for concern—they cause image artefacts, the scale of which depends on the mass of material and degree of magnetization. Thus, for example, gold, steel suture wire, and dental amalgam produced no effects, whereas dental stainless steel (type 316) as used in dentures and orthodontic braces yielded large artefacts which obscured a major proportion of the facial region. New *et al.* (1983) concluded that future risks could be minimized or eliminated through the use of high nickel (10–14 per cent) stainless steels, titanium, or tantalum. Needless to say, prospective candidates for n.m.r. imaging studies should be carefully screened for magnetic implants, and in cases where doubt exists X-rays should be taken.

Concern has also rightly been expressed with regard to the safety of personnel with implanted cardiac pacemakers: Pavlicek, Geisinger, Castle, Borkowski, Meaney, Bream, and Gallagher (1983) have made a detailed study of six popular models. They found that all the pacemakers had ferromagnetic components, and in two cases torques were judged to be sufficiently strong to cause movement of the devices within the chest wall unless restrained by formation of fibrotic tissue. This is reason enough for excluding patients with pacemakers from n.m.r. study. However, there are other factors to be taken into consideration.

Previously devices were operated at a fixed (synchronous) rate. Modern or 'demand' pacemakers, however, operate asynchronously, cutting in only when the heart ceases to beat autonomously. In order that they can be remotely reprogrammed (i.e. without necessitating surgery) it is possible to change them to operate in synchronous mode. This is effected with a reed switch closed by a 0.1 T hand-held magnet. Any demand pacemaker placed in an imaging magnet will thus automatically revert to its synchronous mode. This in itself is not hazardous; nor is there much chance of inadvertently reprogramming the pacemaker's pulse parameters, since the communication channel which operates in the 100–200 KHz range is protected with a digital security code. If, however, the reed switch is faulty and fails to close, it is possible that electromagnetic interference, generated by gradient switching or r.f. pulse trains, could mimic cardiac activity and hence inhibit the demand pacemaker in the presence of a heart which requires pacing. Although manufacturers provide screening to guard against this possibility, some n.m.r. pulse sequences approach the maximum safe r.f. level. Another possibility is the direct stimulation of the heart by voltages induced in the pacemaker leads—currently-proposed imaging schemes fall well below this threshold, however.

Although many pacemakers may well function in strong magnetic fields, and some patients fitted with them have been imaged, it would seem prudent to exclude them from future studies. Further, in order to protect other personnel, it is necessary to contain within the n.m.r. unit, and restrict access to, those regions in which the field strength is sufficient to activate the reed switch of a cardiac pacemaker. In their study, Pavlicek *et al.* (1983) found the minimum field required was 17 G. A suitable limit might therefore be 5 or perhaps 10 G.

A final point concerns the safety of patients in the eventuality of a sudden field collapse caused by power failure or magnet quench. One has to be sure that no e.m.f. sufficient to cause cardiac arrest can be generated. Given the high inductance of imaging magnets and proper safety design there should be no worry in this regard (typically, a superconducting field will collapse over a period of about 10 s during a quench). In the case of a superconducting system one has the additional responsibility of dealing safely with the large volume of helium gas generated during a quench (800 l of liquid helium yields 560 000 l of gas at S.T.P.). Fortunately, this is produced over a period of about 30 s rather than instantaneously so that, provided the helium boil-off/quench vent is of sufficient bore (typically, 100 mm) and the room well-ventilated (in case the safety pressure plates (bursting discs) in the magnet turret blow) there should be no problem.

6.8.2. *Time-varying magnetic fields*

The effects associated with time-varying fields are easily understood in principle; they arise from the currents produced by Faraday induction. Unfortunately, the heterogeneity of the human body, at both the macroscopic and microscopic levels, makes it difficult to assess conductive paths and their associated hazards. Nevertheless, order of magnitude calculations can easily be made and these can be supplemented with observations of effects using externally-applied electrodes. The crucial quantity is tissue current density, and Table 6.6 gives threshold values for a number of important effects including ventricular fibrillation. Budinger (1979) estimates that a magnetic field changing at a rate of 1 $\mathrm{T s^{-1}}$ will induce a current density of 1 $\mu\mathrm{A\,cm^{-2}}$ (= 0.01 $\mathrm{A m^{-2}}$). Although this figure will vary in proportion to the loop radius, being greatest for tissue at the periphery of the trunk, it nevertheless provides a useful guideline. Much of the work using external electrodes has been concerned with evaluating the effects of a.c. mains electric shock and has therefore been carried out at frequencies of either 50 or 60 Hz. In fact, thresholds are strongly

Table 6.6
Threshold tissue current density at 60 Hz.

Electric phosphene	0.1–0.5 A m^{-2}
Skin sensation	1–10 A m^{-2}
Muscle contraction	1–10 A m^{-2}
Ventricular fibrillation	1–10 A m^{-2}

From Saunders 1982.

frequency-dependent as shown in Fig. 6.54 for electric phosphene (visual light flash) production. Thresholds tend to be at a minimum in the frequency range up to a few tens of Hz, increasing rapidly above 100 Hz. The duration of the current pulse is also of importance; the minimum threshold occurring for continuously applied a.c. currents. As the pulse duration is reduced, so the threshold tends to increase, often in an approximately exponential fashion with a time constant (K) dependent on the type of tissue being excited. For voluntary muscle fibres, values of K in the range 1–10 ms have been reported. Thus, under the worst possible conditions of low-frequency, continuously-applied a.c. fields, the minimum threshold current for ventricular fibrillation would appear to be ~ 1 Am^{-2}, corresponding to a rate of change of field $(\partial B/\partial t) \sim 100$ Ts^{-1}.

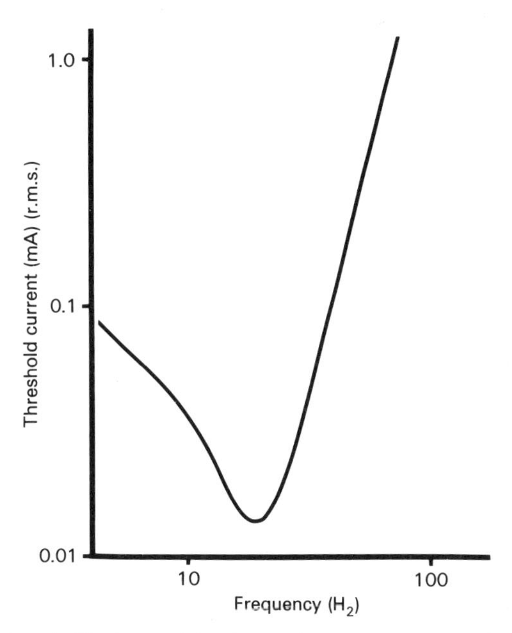

Fig. 6.54. The threshold alternating electrode current density (r.m.s.) for electric phosphene production in man. (From Saunders 1982.)

Taking into account the time constant K, the NRPB recommended in 1980 that $\partial B/\partial t$ be limited to $\sim 20\ \mathrm{Ts}^{-1}$ for pulses of duration >10 ms. It should be mentioned that a number of volunteers have been exposed to switched gradients giving a $\partial B/\partial t \sim 12\ \mathrm{Ts}^{-1}$ for $\sim 100\ \mu$s durations repeated at the rate of 64 Hz (Thomas and Morris 1981). No effects, adverse or otherwise, have been detected.

6.8.3. R.f. fields

Another potential hazard associated with n.m.r. imaging arises from the use of resonant r.f. pulses for excitation of the nuclear spin system. For ^{1}H studies, frequencies in the range 0.5–85 MHz have been used. At these frequencies nervous tissue is not excited and the principal effects are confined to tissue heating through resistive loss. Indeed, this very principle is used in shortwave (typically 27 MHz) diathermy treatment to stimulate blood circulation and promote healing of damaged tissue.

The biologically important quantity is the increase in tissue temperature but, since this is difficult to measure, the United States and European safety levels are based on the principle that, for long term exposures, the r.f. heating should not exceed the basal metabolic rate. This is roughly $1\ \mathrm{Wkg}^{-1}$, corresponding to a total power of about 70 W for a typical adult. The NRPB's recommendations were that mean r.f. power absorption be kept below this limit. Budinger (1981) also comes to similar conclusions, recommending that r.f. levels be limited such that mean power absorption is $<1.5\ \mathrm{Wkg}^{-1}$ for prolonged exposure and $<4\ \mathrm{Wkg}^{-1}$ for exposures of <10 min at frequencies <10 MHz. By way of comparison, moderate and heavy exercise generate $\sim 5\ \mathrm{Wkg}^{-1}$ and $15\ \mathrm{Wkg}^{-1}$ excess heat respectively.

The electromagnetic characteristics (relative permittivity and conductivity) of tissues have been measured at various frequencies (see, for example, Bottomley 1978; Bottomley and Andrew 1978; Durney, Johnson, Barber, Massoudi, Iskander, Lords, Ryser, Allen, and Mitchell 1978), and it is possible to calculate r.f. heating effects based on a number of tissue models. There is an extensive literature on the subject, stimulated by the concern for the safety of personnel in the vicinity of radio and radar stations. In these situations it is possible to use the 'far field' approximation in which one considers plane electromagnetic waves impinging on the subject. Such calculations indicate the existence of a resonance effect when the object dimensions are $\sim 0.4\ \lambda$, i.e. at about 70 MHz for adult humans. Power absorption is most marked at the tissue surface and localized 'hot spots' can occur. Unfortunately, these calculations are inappropriate for n.m.r. imaging exposures, which are essentially 'near field'. A number of attempts have been made to model the n.m.r.

situation (Bottomley and Andrew 1978; Budinger 1979; Bottomley and Edelstein 1981; Mansfield and Morris 1982; see also § 5.7). One important point which emerges is that the r.f. power absorbed is proportional to the square of the frequency and inversely proportional to both the interval between pulses and to the 90° pulse width (the pulse width is proportional to B_1, whereas power is proportional to B_1^2). Thus, a long 90° pulse, for example a 1 ms selective pulse, results in a tenfold lower power absorption than a 100 μs non-selective pulse. At low frequency, all standard n.m.r. imaging sequences, with the possible exception of the steady-state free precession method, fall well within the safety standard. However, at high frequency, sequences with heavy duty cycles, for example multiple spin-echo methods in which non-selective 180° pulses are used for each refocusing, absorbed power levels can become important, and at frequencies $\geq$100 MHz (equivalent to a static field of 2.35 T) this may well be a limiting factor.

The theoretical models also suggest that whereas r.f. penetration of the body is virtually complete at frequencies <10 MHz, at high frequency the conductivity of the body shields the interior ('skin depth' effect) confining the energy to surface layers. Thus, at frequencies >10 MHz it may be necessary to place stricter limitations on r.f. field levels. However, although the effect is readily demonstrated in phantoms, whole-body ^{1}H images at up to 64 MHz have so far failed to demonstrate any significant 'skin depth' artefacts. Presumably the many tissue interfaces present in the body restrict conduction paths and reduce the shielding effect. This is also of importance in regard to the setting of safety levels in switched magnetic field gradients and only goes to emphasize our basic lack of understanding of conduction in the human body.

An alternative empirical approach to the study of power deposition in n.m.r. imaging involves the measurement of the change in quality factor Q of the receiver coil when the subject is introduced (Mansfield and Morris 1982; Saunders 1982; Budinger 1981).

An r.f. heating equivalent to the basal metabolic rate will, if continuously applied, result in a mean temperature rise of about 1 °C. Of course, this will vary according to the ability of the individual tissues to dissipate heat. Those with poor vascularization (e.g. the testes), or none at all (e.g., the eye lens), will be most at risk. However, it is calculated that, even in the latter case, the temperature rise will be less than 2 °C, giving a final temperature well short of that necessary to cause cataracts (43 °C).

Those experiments based on the use of a single pulse to excite the spins normally require very little r.f. power. The steady-state free precession method (see § 3.3.6), on the other hand, involves a much heavier duty cycle and powers can approach the basal metabolic rate (Bottomley and Edelstein 1981). Nevertheless, there should be no real problems in

Table 6.7
Recommended guidelines for n.m.r. imaging

Supervision of exposed persons

Volunteers participating in experimental trials of NMR imaging techniques should be medically assessed and pronounced suitable candidates before exposure. There is no reason to exclude those volunteers with a history of epilepsy or cardiac disease but this should be judged at the time of examination. Volunteers should be freely consenting and fully informed. The guidance in paragraph 3.2.3 of the WHO Technical Report 611 (1977) regarding exposure to ionizing radiation is equally applicable to NMR exposures. Volunteers who are exposed frequently should be checked at regular intervals for evidence of abnormal changes in electrocardiogram recordings.

Patients should be exposed only with the approval of a registered medical practitioner who should be satisfied that the exposure is likely to contribute to the treatment of the patient or is part of a research project approved by a local research ethical committee.

Although there is no evidence to suggest that the embryo is sensitive to magnetic and radiofrequency at the process, the static magnetic field should not exceed 2.5 tesla to the whole or a substantial portion of the body.

Staff operating equipment should not be exposed for prolonged periods to more than 0.02 tesla to the whole body or 0.2 tesla to the arms or hands. For short periods totalling less than 15 minutes, these limits may be increased to 0.2 tesla for the whole body and 2 tesla to the arms and hands. A number of such short exposures in any day is permissible, provided reasonable intervals occur between examinations.

Time-varying magnetic fields (excluding radiofrequency fields). For periods of magnetic flux density change exceeding 10 ms, exposures should be restricted to less than 20 tesla s^{-1} rms for all persons. For periods of change less than 10 ms the relationship $(dB/dt)^2\ t < 4$ should be observed where dB/dt is the rms value of the rate of change of magnetic flux density in any part of the body in tesla sec^{-1} and t is the duration of the change of the magnetic field in seconds. For sinusoidally varying magnetic fields or other continuously varying periodic fields the duration of intensities encountered in NMR imaging, it might be prudent to exclude pregnant women during the first trimester. There is no need to exclude women who will subsequently undergo abortion.

Persons fitted with cardiac pacemakers should not be exposed (Pavlicek et al, 1983) unless corrective procedures are available and they have been made aware of the potential hazard before being exposed.

The exposure of individuals with large metallic implants, such as hip prostheses, should be stopped immediately if discomfort is experienced around the site of the implant. Intracranial metallic clips, particularly those used to treat aneurysms, may present a hazard if the clips are made of magnetic materials (New *et al.* 1983). Patients fitted with such clips should be excluded.

Appropriate resuscitation equipment should be available during imaging and a registered medical practitioner trained in the techniques of resuscitation should be available at short notice, although there is no need for him to be present during the exposure.

Exposure conditions
Exposures should be kept within the following limits:

Static magnetic fields. For those exposed to the imaging the change can be considered to be the half period of the waveform.

Radiofrequency fields. Exposure should be such as to avoid any significant rise in the temperature of the sensitive tissues of the body. Acceptable exposures should not result in a rise in body temperature of more than 1 °C as shown by skin and rectal temperature or more than 1 °C in any mass of tissue not exceeding 1 gram in the body. This may be ensured by limiting the mean specific absorption rate in the whole body to 0.4 $W\,kg^{-1}$ and the specific absorption rate in any mass of tissue not exceeding 1 gram to 4 $W\,kg^{-1}$. These limits apply to volunteers, patients and staff exposures. NRPB have made more general proposals for the limitation of exposures to radiofrequency and microwave radiations (NRPB, 1982).

Other precautions. Notices should be posed where appropriate to warn passers-by of the presence of magnetic and radiofrequency fields and of the possibility that these may affect the functioning of cardiac pacemakers and other electronic equipment. Loose metallic objects should not be permitted in the vicinity of strong magnetic fields.

(From N.R.P.B. 1983.)

choosing pulse widths and repetition rates to comply with the proposed safety standard.

A more severe problem arises if one wishes to record ^{1}H decoupled ^{13}C spectra in a t.m.r.-type experiment (see § 2.7.2). Because the coupling constants are large (~150 Hz), substantial power is required and great care needs to be exercised to avoid heating tissue to dangerous levels. Some of the more efficient n.m.r. decoupling sequences such as WALTZ 8 may prove useful in reducing the required power level (see Levitt, Freeman, and Frenkiel 1983).

Metallic objects are good absorbers of r.f. and patients with metallic implants should initially be excluded from n.m.r. imaging studies because of the risk of localized heating effects.

6.8.4. Conclusions

As well as studies of the hazards associated with individual aspects of n.m.r. imaging systems, some data is available for exposure to the full n.m.r. imaging process. Thomas and Morris (1981) have used the Ames test to examine the viability of normal and mutant strains of *E. coli*; Wolff, Crooks, Brown, Howard, and Painter (1980) have looked for evidence of increased sister chromatid exchange in Chinese hamster ovary cells, and Cooke and Morris (1981) have looked at gross lesions, sister chromatid exchange, and amodal cells in human lymphocytes exposed *in vitro*. None of these investigations has indicated any deleterious effect with any of the wide range of imaging systems studied. It is therefore most unlikely that there is any genetic effect.

Some groups have performed animal behavioural studies with similar negative results. However, the most compelling evidence for the safety of the method lies in the large number of patients and volunteers who have received n.m.r. scans. Some individuals, usually those scientists responsible for the early development of the subject, have accumulated many tens, perhaps hundreds, of hours of n.m.r. imaging exposure with no immediate or subsequent deleterious effect. High-resolution ^{31}P and ^{13}C experiments in which one is primarily concerned with monitoring the metabolic status of the tissues, also provide good evidence for the safety of the basic n.m.r. method at high field (1.8 T = 32 MHz for ^{31}P).

The revised guidelines for n.m.r. imaging experiments, recommended by the NRPB (1983) are summarized in Table 6.7. The main difference between Table 6.7 and the guidelines of the US Bureau of Radiological Health is the more stringent (3 T s^{-1} as against 20 T s^{-1}) switched field requirement of the American body. The West German official recommendations are similar (Kommission für Nichtionisierende Strahlen 1984). A paper explaining the reasons behind the NRPB guidelines has recently been published by two members of the NRPB's staff (Saunders and Smith 1984).

It should be clear that n.m.r. promises to be an extremely safe imaging modality and, whilst it is always prudent to proceed cautiously, there is every reason to expect that it will be of great value in those applications (paediatric and antenatal, for example) where methods based on ionizing radiation should be avoided. Some concern has been expressed over patients with cardiac histories in view of the induced potential caused by switching gradients and blood flow. However, it should be stressed that no deleterious effects have so far been observed and, with proper precautions, cardiac studies may be an important area of application. Similar anxiety has been expressed over epileptic patients, but the Aberdeen team have investigated a number of such patients with no record of seizures either during or immediately following n.m.r. scanning (Smith 1982).

References

Alfidi, R. J., Haaga, J. R., El Yousef, S. J., Bryan, P. J., Fletcher, B. D., Li Puma, J. P., Morrison, S. C., Kaufman, B., Richey, J. B., Hinshaw, W. S., Kramer, D. M., Yeung, H. N., Cohen, A. M., Butler, H. E., Ament, A. E., and Lieberman, J. M. (1982). *Radiology* **143,** 175.

Andrew, E. R., Bottomley, P. A., Hinshaw, W. S., Holland, G. N., Moore, W. S., and Simaroj, C. (1977). *Phys. med. Biol.* **22,** 971.

Asato, R., Handa, H., Hashi, T., Hatta, J., Komocke, M., and Yazaki, T. (1983). *Stroke* **14,** 191.

Aue, W. P., Müller, S., Cross, T. A., and Seelig, J. (1984). *J. magn. Reson.* **56,** 164.

Bailes, D. R., Young, I. R., Thomas, D. J., Straughan, K., Bydder, G. M., and Steiner, R. E. (1982). *Clin. Radiol.* **33,** 395.

Baker, R. R., Mather, J. G., and Kennaugh, J. H. (1983). *Nature, Lond.* **301,** 78.

Balschi, J. A., Cirillo, V. P., and Springer, C. S. (1982). *Biophys. J.* **38,** 344.

Bangert, V., Mansfield, P., and Coupland, R. E. (1981). *Br. J. Radiol.* **54,** 152.

Barnothy, M. F. (ed.) (1964). *Biological effects of magnetic fields*, vols. 1 and 2. Plenum Press, New York.

Beall, P. T. (1982). *Physiol. Chem. Phys.* **14**(4), 399.

———, and Hazlewood, C. F. (1976). *Science* **192,** 904.

Beischer, D. E. (1969). In *Biological effects of magnetic fields*, Vol. 2, p. 259 (ed. M. F. Barnothy). Plenum Press, New York.

Belton, P. S., Packer, K. J., and Sellwood, T. C. (1973). *Biochim. biophys. Acta* **304,** 56.

Bendel, P., Lai, C-N., and Lauterbur, P. C. (1980). *J. magn. Reson.* **38,** 343.

Blakemore, R., Frankel, R., and Wolfe, F. (1979). *Science* **203,** 1355.

Block, R. E., and Maxwell, G. P. (1974). *J. magn. Reson.* **14,** 329.

Bloembergen, N., and Morgan, L. O. (1961). *J. chem. Phys.* **34,** 842.

Bo, W. J., Meschan, I., and Krueger, W. A. (1980). *Basic atlas of cross-sectional anatomy*. W. B. Saunders Co., Philadelphia, Pa.
Bottomley, P. A. (1978). *J. Phys. E: Scient. Instrum.* **11,** 413.
——— (1979). *Cancer Res.* **39,** 468.
——— (1981). *Experientia* **37,** 768.
——— (1982). *J. magn. Reson.* **50,** 335.
———, and Andrew, E. R. (1978). *Phys. med Biol.* **23,** 630.
———, and Edelstein, W. A. (1981). *Med. Phys.* **8,** 510.
Bovée, W., Huisman, P., and Smidt, J. (1974). *J. natn. Cancer Inst.* **52,** 595.
———, Creyghton, J. H. N., Getreuer, K. W., Korbee, D., Lubregt, S., Smidt, J., Wind, R. A., Lindeman, J., Smid, L., and Posthuma, H. (1980). *Phil. Trans. R. Soc.* B **289,** 535.
Brady, T. J., Rosen, B. R., Pykett, I. L., McGuire, M. H., Mankin, H. J., and Rosenthal, D. I. (1983). *Radiology* **149,** 181.
———, Goldman, M. R., Pykett, I. L., Buonanno, F. S., Kistler, J. P., Newhouse, J. H., Burt, C. T., Hinshaw, W. S., and Pohost, G. M. (1982). *Radiology* **144**(2) 343.
———, Rosen, B. R., Gold, H. K., Khaw, B. A., Fallon, J. T., Goldman, H. R., Ter Penning, B. J., Yasuda, T., and Haber, E. 'Work in progress', Annual meeting of Society of Magnetic Resonance in Medicine, San Francisco, Aug. 1983. Abstr. 10.
———, Gebhardt, M. C., Pykett, I. L., Buonanno, F. S., Newhouse, J. H., Burt, C. T., Smith, R. J., Mankin, J. H., Kistler, J. P., Goldman, M. R., Hinshaw, W. S., and Pohost, G. M. (1982). *Radiology* **144**(3) 549.
Brasch, R. C. (1983). *Radiology* **147,** 781.
———, London, D. A., Wesbey, G. E., Tozer, T. N., Nitecki, D. E., Williams, R. D., Doemeny, J., Tuck, L. D., and Lallemand, D. P. (1983*a*). *Am. J. Roentgenol.* **141,** 773.
———, Nitecki, D. E., Brant-Zawadzki, M., Enzmann, D. R., Wesbey, G. E., Tozer, T. N., Tuck, L. D., Cann, C. E., Fike, J. R., and Sheldon, P. (1983*b*). *Am. J. Roentgenol.* **141,** 1019.
Brown, T. R., Kincaid, B. M., and Ugurbil, K. (1982). *Proc. natn. Acad. Sci. U.S.A.* **79,** 3523.
Bryan, R. N., Willcott, M. R., Schneiders, N. J., Ford, J. J., and Derman, H. S. (1983). *Radiology* **149,** 189.
Budinger, T. F. (1979). *IEEE Trans. Nucl. Sci.* **26,** 2821.
——— (1981). *J. comput. assist. Tomogr.* **5,** 800.
Buonanno, F. S., Pykett, I. L., Brady, T. J., Vielma, J., Burt, C. T., Goldman, M. R., Hinshaw, W. S., Pohost, G. M., and Kistler, J. P. (1983). *Stroke* **14,** 178.
———, ———, Vielma, J., Brady, T. J., Burt, C. T., Goldman, M. R., Newhouse, J. H., New, P. F. J., Hinshaw, W. S., Pohost, G. M., and Kistler, J. P. (1982). *Proc. int. symp. NMR imaging*, p. 147. Bowman Gray School Medicine, Winston-Salem.
Bydder, G. M., Doyle, F. H., Young, I. R., and Hall, A. S. (1982*a*). *Proc. int. symp. NMR imaging*, p. 107, Bowman Gray School Medicine, Winston-Salem.
———, Goatcher, A., Hughes, J. M. B., Orr, J. S., Pennock, J. M., Steiner, R. E., and Tripathi, A. (1982*b*). *J. Physiol.* **332,** 46P.

———, Steiner, R. E., Young, I. R., Hall, A. S., Thomas, D. J., Marshall, J., Pallis, C. A., and Legg, N. J. (1982*c*). *Am. J. Roentgenol.* **139,** 215.

———, ———, Thomas, D. J., Marshall, J., Gilderdale, D. J., and Young, I. R. (1983). *Clin. Radiol.* **34,** 173.

Carr, D. H., Brown, J., Bydder, G. M., Weinmann, H-J., Speck, U., Thomas, D. J., and Young, I. R. (1984). *Lancet* **i,** 484.

de Certaines, J., Herry, J. Y., Lancien, G., Benoist, L., Bernard, A. M., and Le Clech, G. (1982). *J. nucl. Med.* **23,** 48.

———, Moulinoux, J. P., Benoist, L., Bernard, A-M., and Rivet, P. (1982). *Life Sci.* **31,** 505.

Chang, D. C., Hazlewood, C. F., Nichols, B. L., and Rorscharch, H. E. (1972). *Nature, Lond.* **235,** 170.

Clow, H., and Young, I. (1978). *Br. J. Radiol.* **52,** 680.

Cohen, A. M., Creviston, S., Li Puma, J. P., Bryan, P. J., Haaga, J. R., and Alfidi, R. J. (1983). *Radiology* **148,** 739.

Cooke, P., and Morris, P. G. (1981). *Br. J. Radiol.* **54,** 622.

Cottam, G. L., Vasek, A., and Lusted, D. (1972). *Res. Commun. Chem. Pathol. Pharmacol.* **4,** 495.

Cox, S. F. J., and Styles, P. (1980). *J. magn. Reson.* **40,** 209.

Crooks, L. E., Sheldon, P., Kaufman, L., and Rowan, W. (1982). *IEEE Trans.* N.S. **29,** 1181.

———, Hoenninger, J., Arakawa, M., Kaufman, L., McRee, R., Watts, J., and Singer, J. H. (1980). *Radiology* **136,** 701.

———, Mills, C. M., Davis, P. L., Brant-Zawadzki, M., Hoenninger, J., Arakawa, M., Watts, J., and Kaufman, L. (1982). *Radiology* **144,** 843.

———, Arakawa, M., Hoenninger, J., Watts, J., McRee, R., Kaufman, L., Davis, P. L., Margulis, A. R., and De Groot, J. (1982). *Radiology* **143,** 169.

———, Ortendahl, D. A., Kaufman, L., Hoenninger, J., Arakawa, M., Watts, J., Cannon, C. R., Brant-Zawadzki, M., Davis, P. L., and Margulis, A. R. (1983). *Radiology* **146,** 123.

———, and Kaufman, L. (1984). *Br. med. Bull.* **40**(2), 167.

Damadian, R. (1971). *Science* **171,** 1151.

———, Goldsmith, M., and Minkoff, L. (1977). *Physiol. Chem. Phys.* **9,** 97.

———, ———, and ——— (1978). *Physiol. Chem. Phys.* **10,** 285.

———, Zaner, K., Hor, D., and DiMaio, R. (1974). *Proc. natn. Acad. Sci. U.S.A.* **71,** 1471.

Davis, P. L., Kaufman, L., and Crooks, L. E. (1982). *Proc. int. symp. NMR imaging,* p. 101. Bowman Gray School Medicine, Winston-Salem.

———, Kaufman, L., Crooks, L. E., and Miller, T. R. (1981). *Invest. Radiol.* **16**(5), 354.

De La Paz, R. L., Brady, T. J., Buonanno, F. S., New, P. F. J., Kistler, J. P., McGinnis, B. D., Pykett, I. L., and Taveras, J. M. (1983). *J. comput. assist. Tomogr.* **7**(1), 126.

De Layre, J. L., Ingwall, J. S., Malloy, C., and Fossel, E. T. (1981). *Science* **212,** 935.

Dick, D. A. T. (1971). *Membranes Ion Transport* **3,** 211.

Doyle, M., Rzedzian, R., Mansfield, P., and Coupland, R. E. (1983). *Br. J. Radiol.*, **56,** 925.
Doyle, F. H., Gore, J. C., Pennock, J. M., Bydder, G. M., Orr, J. S., Steiner, R. E., Young, I. R. Burl, M., Clow, H., Gilderdale, D. J., Bailes, D. R., and Walters, P. E. (1981). *Lancet* **ii,** 53.
Droege, R. T., Wiener, S. N., Rzeszotarski, M. S., Holland, G. N., and Young, I. R. (1983). *Radiology* **148,** 763.
Durney, C. H., Johnson, C. C., Barber, P. W., Massoudi, H., Iskander, M. F., Lords, J. L., Ryser, D. K., Allen, S. J., and Mitchell, J. C. (1978). *Radiofrequency radiation dosimetry handbook* (2nd edn). USAF School of Aerospace Medicine (RZP), Aerospace Medical Division, Brooks Air Force Base, Texas, 78235.
Edelstein, W. A., Bottomley, P. A., Hart, H. R., and Smith, L. S. (1983). *J. comput. assist. Tomogr.* **1**(3), 391.
———, Hutchison, J. M. S., Johnson, G., and Redpath, T. (1980). *Phys. med. Biol.* **25,** 748.
———, ———, Smith, F. W., Mallard, J. R., Johnson, G., and Redpath, T. W. (1981). *Br. J. Radiol.* **54,** 149.
———, Bottomley, P. A., Hart, H. R., Leue, W. M., Schenck, J. F., and Redington, R. W. (1982). *Proc. int. symp. NMR imaging,* p. 139. Bowman Gray School Medicine, Winston-Salem.
Feinberg, D. (1983). *Sci. prog. Soc. Mag. Res. Med.* pp. 128–9.
Fossel, E. T., Morgan, H. E., and Ingwall, J. S. (1980). *Proc. natn. Acad. Sci. U.S.A.* **77,** 3654.
Foster, K. R., Resing, H. A., and Garroway, A. N. (1976). *Science* **194,** 324.
Foster, M. A., Knight, C. H., Rimmington, J. E., and Mallard, J. R. (1983). *Radiology,* **149,** 193.
Frey, H. E., Knispel, R. R., Kruuv, J., Sharp, A. R., Thompson, R. T., and Pintar, M. M. (1972). *J. natn. Cancer Inst.* **49,** 903.
Fulton, A. B. (1982). *Cell* **30,** 345.
Gaffey, C. T., Tenforde, T. S., and Dean, E. E. (1980). *Bioelectromagnetics* **1,** 209.
Gamsu, G., Webb, W. R., Sheldon, P., Kaufman, L., Crooks, L. E., Birnberg, F. A., Goodman, P., Hinchcliffe, W. A., and Hedgecock, M. (1983). *Radiology* **147,** 473.
Go, R. T., MacIntyre, W. J., Meaney, T. F. *et al.* (1985). *Magn. res. Med.*, in press.
Goldman, M. R., Brady, T. J., Pykett, I. L., Burt, C. T., Buonanno, F. S., Kistler, J. P., Newhouse, J. H., Hinshaw, W. S., and Pohost, G. M. (1982*a*). *Circulation* **66**(5), 1012.
———, ———, ———, ———, Newhouse, J. H., Buonanno, F., Kistler, J. P., Hinshaw, W. S., and Pohost, G. M. (1982*b*). *Proc. int. symp. NMR imaging,* p. 171. Bowman Gray School Medicine, Winston-Salem.
Gordon, R. E., Hanley, P. E., and Shaw, D. (1983). *Prog. NMR spectrosc.* **15,** 1.
Gore, J. C. (1982). *Proc. int. symp. NMR imaging,* p. 15. Bowman Gray School Medicine, Winston-Salem.

———, Doyle, F. H., and Pennock, J. M. (1982). In *NMR imaging* (eds. Partain, C. L., Price, R. R., Rollo, F. D., James, A. E.) W. B. Saunders, Philadelphia.

Griffiths, L. K., Rosen, G. M., Rauckman, E. J., and Drayer, B. P. (1983). Annual meeting of Society of Magnetic Resonance in Medicine, San Francisco, August, p. 144.

Gupta, R. K., and Gupta, P. (1982). *J. magn. Reson.* **47,** 344.

Haase, A., Malloy, C., and Radda, G. K. (1983). *J. magn. Reson.* **55,** 164.

Hall, L. D., and Sukumar, S. (1982). *J. magn. Reson.* **50,** 161.

———, and ——— (1984*a*). *J. magn. Reson.* **56,** 326.

———, and ——— (1984*b*). *J. magn. Reson.* **56,** 314.

———, ———, and Talagala, S. L. (1984). *J. magn. Reson.* **56**, 275.

Hansen, G., Crooks, L. E., Davis, P., De Groot, J., Herfkens, R., Margulis, A. R., Gooding, C., Kaufman, L., Hoenninger, J., Arakawa, M., McRee, R., and Watts, J. (1980). *Radiology* **136,** 695.

Haselgrove, J. C., Subramanian, V. H., Leigh, J. S., Gyulai, L., and Chance, B. (1983). *Science* **220** (4602), 1170.

Hawkes, R. C., Holland, G. N., Moore, W. S., and Worthington, B. S. (1980). *J. comput. assist. Tomogr.* **4,** 577.

———, ———, ———, and ——— (1982). *Proc. int. symp. NMR imaging*, p. 115. Bowman Gray School Medicine, Winston–Salem.

———, ———, ———, Roebuck, E. J., and Worthington, B. S. (1981*a*). *J. comput. assist. Tomogr.* **5,** 605.

———, ———, ———, ———, and ——. (1981*b*). *J. comput. assist. Tomogr.* **5,** 613.

Hayes, C. E., Case, T. A., Ailion, D. C., Morris, A. H., Cutillo, A., Blackburn, C. W., Durney, C. H., and Johnson, S. A. (1982). *Science* **216,** 1313.

Hazlewood, C. F. (1976) In *Cell-associated water* (eds. Drost-Hansen, N., and Clegg, J. S.) Academic Press, New York.

———, Chang, D. C., Medina, D., Cleveland, G., and Nichols, B. L. (1972). *Proc. natl Acad. Sci.* U.S.A. **69,** 1478.

Hedvig, H., Crooks, L., Sheldon, P., and Kaufman, L. (1983). *Radiology* **146,** 425.

Heidelberger, E., and Lauterbur, P. C. (1982). Society of Magnetic Resonance in Medicine, August meeting.

Held, G., Noack, F., Pollak, V., and Melton, B. (1973). *Z. Naturf.* **28C,** 59.

Herfkens, R., Sievers, R., Kaufman, L., Sheldon, P. E., Ortendahl, D. A., Lipton, M. H., Crooks, L. E., and Higgins C. B. (1983*a*). *Radiology* **147,** 761.

———, Higgins, C. B., Hricak, H., Lipton, M. J., Crooks, L. E., Lanzer, P., Botvinick, E., Brundage, B., Sheldon, P. E., and Kaufman, L. (1983*b*). *Radiology* **147,** 749.

———, Davis, P. L., Crooks, L. E., Kaufman, L., Price, D., Miller, T., Margulis, A. R., Watts, J., Hoenninger, J., Arakawa, M., and McRee, R. (1981). *Radiology* **141,** 211.

Hilal, S. K., Maudsley, A. A., Bonn, J., Simon, H. E., Cannon, P., and Perman, W. H. (1983*a*). *Proc. Soc. magn. reson. Med.* (Aug.).

———, ———, Simon, H. E., Perman, W. H., Bonn, J., Mawad, M. E., Silver, A. J., Ganti, S. R., Sane, P., and Chien, I. C. (1983*b*). *Am. J. Neuroradiol.* **4**(3), 245.

Hinshaw, W. S., Bottomley, P. A., and Holland, G. N. (1977). *Nature, Lond.* **270,** 722.

———, Andrew, E. R., Bottomley, P. A., Holland, G. N., Moore, W. S., and Worthington, B. S. (1978). *Br. J. Radiol.* **51,** 273.

———, ———, ———, ———, ———, and Worthington, B. S. (1979). *Br. J. Radiol.* **52,** 36.

Holland, G. N., Bottomley, P. A., and Hinshaw, W. S. (1977). *J. magn. Reson.* **28,** 133.

———, Moore, W. S., and Hawkes, R. C. (1980). *J. comput. assist. Tomogr.* **4,** 1.

Hollis, D. P., Saryan, L. A., Eggleston, J. C., and Morris, H. P. (1975). *J. natl. Cancer Inst.* **54,** 1469.

———, Bulkey, B. H., Nunnally, R. L., Jacobus, W. E., and Weisfeldt, M. L. (1978). *Clin. Res.* **26,** 240A.

Hoult, D. I. (1977). *J. magn. Reson.* **26,** 165.

——— (1979). *J. magn. Reson.* **33,** 183.

Hounsfield, G. N. (1973). *Br. J. Radiol.* **46,** 1016.

Hricak, H., Crooks, L., Sheldon, P., and Kaufman, L. (1983*a*). *Radiology* **146,** 425.

———, Filly, R. A., Margulis, A. R., Moon, K. L., Crooks, L. E., and Kaufman, L. (1983*b*). *Radiology* **147,** 481.

———, Williams, R. D., Moon, K. L., Moss, A. A., Alpers, C., Crooks, L. E., and Kaufman, L. (1983*c*). *Radiology* **147,** 765.

Huberty, J., Engelstad, B., Wesbey, G., Moseley, M., Young, G., Tuck, D., Brito, A., Hattner, R., and Brasch, R. (1983). Abstracts for Society of Magnetic Resonance in Medicine meeting in San Francisco, Aug. 1983, p. 175.

Inch, W. R., McCredie, J. A., Knispel, R. R., Thompson, R. T., and Pintar, M. M. (1974). *J. natn. Cancer Inst.* **52,** 353.

Johnson, M. A., Pennock, J. M., Bydder, G. M., Steiner, R. E., Thomas, D. J., Hayward, R., Bryant, D. R. T., Payne, J. A., Levene, M. I., Whitelaw, A., Dubowitz, L. M. S., and Dubowitz, V. (1983). *Am. J. Roentgenol.* **141,** 1005.

Kaufman, L., Crooks, L., Sheldon, P., Hricak, H., Herfkens, R., and Bank, W. (1983). *Circulation* **67**(2) 251.

Kiricuta, Jr. I. C., and Simplaceanu, V. (1975). *Cancer Res.* **35,** 1164.

———, Demco, D., and Simplaceanu, V. (1973). *Arch. Geschwülstf.* **42,** 226.

Koenig, S. H., and Schillinger, W. E. (1969). *J. biol. Chem.* **244,** 3283.

Kolata, G. B. (1976). *Science* **192,** 1220.

Kommission für Nichtionisierende Strahlen (1984). *Bundesgesondheitsblatt* **27,** 92.

Koutcher, J. A., Goldsmith, M., and Damadian, R. (1978). *Cancer* **41,** 174.

Kramer, D. M. (1982). 'Imaging of elements other than hydrogen', in *NMR imaging in medicine* (eds. Kaufman, L., Crooks, L. E., and Margulis, A. R.) Igaku-Shoin, New York.

Kuntz, I. D. (1971). *J. Am. chem. Soc.* **93,** 514.

———, Brassfield, T. S., Law, G. D., and Purcell, G. V. (1969). *Science* **163,** 1329.

Lauterbur, P. C. (1973). *Nature, Lond.* **242,** 190.

———, Dias, M. H. M., and Rudin, A. M. (1978). In *Electrons to tissue: frontiers of biological energetics.* (eds. Dutton, P. L., Leigh, J. S., and Scarpa, A.). Academic Press, New York.

———, Frank, J. A., and Jacobson, M. J. (1976). *Dig. int. conf. Med. Phys. 4th, Physics in Canada* **32,** Abstr. 33.9.

———, Kramer, D. M., House, Jr. W. V., and Chen, C-N. (1975). *J. Am. chem. Soc.* **97,** 6866.

Levene, M. I., Whitelaw, A., Bubowitz, V., Bydder, G. M., Steiner, R. E., Randell, C. P., and Young, I. R. (1982). *Br. med. J.* **285,** 774.

Levine, S., and Payan, H. (1966). *Exper. Neurol.* **16,** 255.

Levitt, M. H., Freeman, R., and Frenkiel, T. (1983). In *Advances in Magnetic Resonance*, Vol. 11, pp. 47–110 (ed. J. S. Waugh). Academic Press, New York.

Ling, G. N. (1962). *A physical theory of the living state. The association–induction hypothesis.* Blaisdell, New York.

Ling, C. R., Foster, M. A., and Hutchison, J. M. S. (1980). *Phys. med. Biol.* **25,** 748.

———, Foster, M. A., and Mallard, J. R. (1979). *Br. J. Cancer* **40,** 898.

Loeffler, W., and Oppelt, A. (1981). *Eur. J. Radiol.* **1,** 338.

Luiten, A. L. (1981). *Medicamundi* **26,** 98.

———, Locher, P. R., van Uijen, C. M. J., van Dijk, P., and den Boef, J. H. (1982). *Proc. int. symp. NMR imaging*, p. 133. Bowman Gray School Medicine, Winston-Salem.

McGinnis, B. D., Brady, T. J., New, P. F. J., Buonanno, F. S., Pykett, I. L., De La Paz, R. L., Kistler, J. P., and Taveras, J. M. (1983). *J. comput. assist. Tomogr.* **7,** (4) 575.

Mallard, J. R., Hutchison, J. M. S., Edelstein, W. A., Ling, C. R., Foster, M. A., and Johnson, G. (1980). *Phil. Trans. R. Soc.* B **289,** 519.

Mansfield, P. (1983). *J. Phys. D: Appl. Phys.* **16,** L235.

———, and Maudsley, A. A. (1976). *Proc. Ampere Congr., 19th*, Heidelberg, p. 247. Groupment Ampere, Heidelberg.

———, and ——— (1977). *Br. J. Radiol.* **50,** 188.

———, and Morris, P. G. (1982). 'NMR imaging in biomedicine', in *Advances in magnetic resonance*, suppl. 2 (ed. J. S. Waugh). Academic Press, New York.

———, and Pykett, I. L. (1978). *J. magn. Reson.* **29,** 355.

———, ———, Morris, P. G., and Coupland, R. E. (1978), *Br. J. Radiol.* **51,** 921.

———, Morris, P. G., Ordidge, R., Coupland, R. E., Bishop, H. M., and Blamey, R. W. (1979). *Br. J. Radiol.* **52,** 242.

Mathur-de Vré, R., Grimée, R., and Rosa, M. (1983). *Biosci. Rep.* **3,** 599.

Maudsley, A. A. and Hilal, S. K. (1984). *Br. med. Bull.* **40**(2), 165.

———, ———, Perman, W. H., and Simon, H. E. (1983). *J. magn. Reson.* **51,** 147.

———, ———, Simon, H. E., and Perman, W. H. (1982). *Proc. Soc. magn. reson. Med.* (August).

———, ———, ———, and Wittekoek, S. (1983). *Proc. Soc. magn. reson. Med.* (August).

Modic, M. T., Weinstein, M. A., Pavlicek, W., Starnes, D. L., Duchesneau, P. M., Boumphrey, F., and Hardy, R. J. (1983). *Radiology* **148,** 757.

Moon, K. L., Genant, H. K., Helms, C. A., Chafetz, N. I., Crooks, L. E., and Kaufman, L. (1983*a*). *Radiology* **147,** 161.

———, Hricak, H., Crooks, L. E., Gooding, C. A., Moss, A. A., Engelstad, B. L., and Kaufman, L. (1983*b*). *Radiology* **147,** 155.

———, ———, Margulis, A. R., Bernhoft, R., Way, L. W., Filly, R. A., and Crooks, L. E. (1983*c*). *Radiology* **148,** 753.

Moore, W. S., and Holland, G. N. (1980). *Br. med. Bull.* **36,** 297.

———, ———, and Kreel, L. (1980). *J. comput. assist. Tomogr.* **4,** 1.

Morris, G. A., and Freeman, R. (1979). *J. Am. chem. Soc.* **101,** 760.

National Radiological Protection Board (1981). *Radiography* **47,** 258.

——— (1983). *Br. J. Radiol.* **56,** 974.

New, P. F. J., Rosen, B. R., Brady, T. J., Buonanno, F. S., Kistler, J. P., Burt, C. T., Hinshaw, W. S., Newhouse, J. H., Pohost, G. M., and Taveras, J. M. (1983). *Radiology* **147,** 139.

Newhouse, J. H., Pykett, I. L., Brady, T. J., Burt, C. T., Goldman, M. R., Buonanno, F. S., Kistler, J. P., and Pohost, G. M. (1982). *Proc. int. symp. NMR imaging.* p. 121. Bowman Gray School Medicine, Winston-Salem.

Ordidge, R. J., Mansfield, P., and Coupland, R. E. (1981). *Br. J. Radiol.* **54,** 850.

———, ———, Doyle, M., and Coupland, R. E. (1982*a*). *Proc. int. symp. NMR imaging,* p. 89. Bowman Gray School Medicine, Winston-Salem.

———, ———, ———, and ——— (1982*b*). *Br. J. Radiol.* **55,** 729.

Packer, K. J., (1977). *Phil. Trans. R. Soc.* B **278,** 59.

Parker, D. L., Smith, V., Sheldon, P., Crooks, L. E., and Fussell, L. (1983). *Med. Phys.* **10**(3), 321.

Pavlicek, W., Geisinger, M., Castle, L., Borkowski, G. P., Meaney, T. F., Bream, B. L., and Gallagher, J. H. (1983). *Radiology* **147,** 149.

Pykett, I. L. (1982). *Scient. Am.* **246,** 54.

———, and Rosen, B. R. (1983). *Radiology* **149,** 197.

———, Buonanno, F. S., Brady, T. J., and Kistler, J. P. (1983). *Stroke* **14** (2), 173.

———, Rosen, B. R., Buonanno, F. S., and Brady, T. J. (1983). *Phys. med. Biol.* **28**(6) 723.

Rinck, P. A., Petersen, S. B., Heidelberger, E., Acuff, V., Reinders, J., Bernardo, M. L., Hedges, L. K., and Lauterbur, P. C. (1983). Annual meeting of Society of Magnetic Resonance in Medicine, August.

Ross, R. J., Thompson, J. S., Kim, K., and Bailey, R. A. (1982). *Radiology* **143,** 195.

Runge, V. M., Clanton, J. A., Smith, F. W., Hutchison, J., Mallard, J., Partain, C. L., and James, A. E. (1983*a*). *Roentgenology* **141,** 943.

———, Stewart, R. G., Clanton, J. A., Jones, M. M., Lukehart, C. M., Partain, C. L., and James, A. E. (1983*b*). *Radiology* **147,** 789.

Rupp, N., Reiser, M., and Stetter, E. (1983). *Eur. J. Radiol.* **3,** 68.

Rutar, V. (1982). *Appl. Spectrosc.* **36,** 259.

Rzedzian, R., Mansfield, P., Doyle, M., Guilfoyle, D., Chapman, B., Coupland, R. E., Chrispin, A., and Small, P. (1983). *Lancet* **ii,** 1281.

Saryan, L. A., Hollis, D. P., Economou, J. S., and Eggleston, J. C. (1974). *J. natn. Cancer Inst.* **52,** 599.

Saunders, R. (1982). *Proc. int. symp. NMR imaging*, p. 65. Bowman Gray School Medicine, Winston-Salem.

Saunders, R. D., and Cass, A. (1983). N.R.P.B. Internal Report No M96.

———, and Smith, H. (1984). *Br. Med. Bull.* **40**(2), 148.

Schara, M., Šentjurc, M., Auersperg, M., and Golouh, R. (1974). *Br. J. Cancer* **29,** 483.

Scott, K. N., Brooker, H. R., Fitzsimmons, J. R., Bennett, H. F., and Mick, R. C. (1982). *J. magn. Reson.* **50,** 339.

Singer, J. R. (1982). *Proc. int. symp. NMR imaging*, p. 185. Bowman Gray School Medicine, Winston-Salem.

———, and Crooks, L. E. (1983). *Science* **221,** 654.

Sipponen, J. T., Kaste, M., Ketonen, L., Sepponen, R. E., Katevuo, K., and Sivula, A. (1983). *J. comput. assist. Tomogr.* **7**(4), 585.

Smith, F. W. (1982*a*). *Proc. int. symp. NMR imaging*, p. 125. Bowman Gray School Medicine, Winston-Salem.

——— (1982*b*). *Lancet* **i,** 974.

———, Adam, A. H., and Phillips, W. D. P. (1983). *Lancet* **i,** 60.

Smith, G. A., Hesketh, R. T., Metcalfe, J. C., Feeney, J., and Morris, P. G. (1983). *Proc. natn. Acad. Sci. U.S.A.* **80,** 7178.

Solomon, I. (1955). *Phys. Rev.* **99,** 559.

Spector, W. S. (1956). *Handbook of biological data.* W. B. Saunders, Philadelphia.

Spetzler, R. F., Zabramski, J. M., Kaufman, B., and Yeung, H. N. (1983). *Stroke* **14,** 185.

Stark, D. D., Bass, N. M., Moss, A. A., Bacon, B. R., McKerrow, J. H., Cann, C. E., Brito, A., and Goldberg, H. I. (1983). *Radiology* **148,** 743.

Steiner, R. E., Bydder, G. M., Selwyn, A., Deanfield, J. Longmore, D. B., Klipsten, R. H., and Firmin, D. (1983). *Br. Heart J.* **50,** 202.

Stevens, A. M., Iles, R. A., Morris, P. G., and Griffiths, J. R. (1982). *FEBS Lett.* **150,** 489.

———, Morris, P. G., Iles, R. A., Sheldon, P. W., and Griffiths, J. R. (1984). *Br. J. Cancer* **50,** 113.

Taylor, D. G., and Bore, C. F. (1981). *J. comput. assist. Tomogr.* **5,** 122.

Tenforde, T. S. (ed.) (1979). *Magnetic field effects on biological systems.* Plenum Press, New York.

Thomas, A., and Morris, P. G. (1981). *Br. J. Radiol.* **54,** 615.

Van Dijk, P. (1984*a*). *J. comput. assist. Tomogr.* **8**(3), 429.

——— (1984*b*). *Diagn. Imag. clin. Med.* **53,** 29.

Wesbey, G. E., Brasch, R. C., Engelstad, B. L., Moss, A. A., Crooks, L. E., and Brito, A. C. (1983). *Radiology* **149**(1) 175.

Wind, R. A., and Yannoni, C. S. (1979). *J. magn. Reson.* **36,** 269.

Wolff, S., Crooks, L. E., Brown, P., Howard, R., and Painter, R. B. (1980). *Radiology* **136,** 707.

Woodhouse, D. R., Derbyshire, W., and Lillford, P. (1975). *J. magn. Reson.* **19,** 267.
Wüthrich, K. (1976). *NMR in biological research: peptides and proteins.* Elsevier, Amsterdam.
Young, I. R., Hall, A. S., Pallis, C. A., Bydder, G. M., Legg, N. J., and Steiner, R. E. (1981). *Lancet* **ii,** 1063.
———, Randell, C. P., Kaplan, P. W., James, A., Bydder, G. M., and Steiner, R. E. (1983). *J. comput. assist. Tomogr.* **7**(2), 290.
———, Bailes, D. R., Collins, A. G., and Gilderdale, D. J. (1982*a*). *Proc. int. Symp. NMR imaging*, p. 93. Bowman Gray School Medicine, Winston-Salem.
———, ———, Burl, M., Collins, A. G., Smith, D. T., McDonnell, M. J., Orr, J. S., Banks, L. M., Bydder, G. M., Greenspan, R. H., and Steiner, R. E. (1982*b*). *J. comput. assist. Tomogr.* **6**(1), 1.
El Yousef, S. J., Alfidi, R. J., Duchesneau, R. H., Hubay, C. A., Haaga, J. R., Bryan, P. J., Li Puma, J. P., and Ament, A. E. (1983). *J. comput. assist. Tomogr.* **1**(2), 215.
Zeitler, E., and Schittenhelm, R. (1981). *Electromedica* **3,** 134.
Zimmerman, J. R., and Brittin, W. E. (1957). *J. phys. Chem.* **61,** 1328.

Index

Coronary Circulation in Nonsmokers and Smokers